Advanced Cultivation Technologies for Horticultural Crops Production

Advanced Cultivation Technologies for Horticultural Crops Production

Guest Editors

Nay Myo Win
Inkyu Kang
Jingi Yoo

MDPI

Basel • Beijing • Wuhan • Barcelona • Belgrade • Novi Sad • Cluj • Manchester

Guest Editors

Nay Myo Win
Department of Horticulture & Life Science
Yeungnam University
Gyeongsan
Republic of Korea

Inkyu Kang
Department of Horticultural Science
Kyungpook National University
Daegu
Republic of Korea

Jingi Yoo
Department of Horticulture & Life Science
Yeungnam University
Gyeongsan
Republic of Korea

Editorial Office
MDPI AG
Grosspeteranlage 5
4052 Basel, Switzerland

This is a reprint of the Special Issue, published open access by the journal *Agriculture* (ISSN 2077-0472), freely accessible at: https://www.mdpi.com/journal/agriculture/special_issues/O20I0E134N.

For citation purposes, cite each article independently as indicated on the article page online and as indicated below:

Lastname, A.A.; Lastname, B.B. Article Title. *Journal Name* **Year**, *Volume Number*, Page Range.

ISBN 978-3-7258-7711-9 (Hbk)
ISBN 978-3-7258-7712-6 (PDF)
https://doi.org/10.3390/books978-3-7258-7712-6

Cover image courtesy of Nay Myo Win

Contents

Nay Myo Win, Jingi Yoo and In-Kyu Kang
Advanced Cultivation Technologies for Horticultural Crops Production
Reprinted from: *Agriculture* **2026**, *16*, 787, https://doi.org/10.3390/agriculture16070787 1

Xin-Qiang Zheng, Xue-Han Zhang, Jian-Gao Zhang, Rong-Jin Zheng, Jian-Liang Lu, Jian-Hui Ye and Yue-Rong Liang
Effects of Photovoltaic-Integrated Tea Plantation on Tea Field Productivity and Tea Leaf Quality
Reprinted from: *Agriculture* **2026**, *16*, 125, https://doi.org/10.3390/agriculture16010125 5

Gleydson de Freitas Silva, Tayd Dayvison Custódio Peixoto, Miguel Ferreira Neto, Antônio Gustavo de Luna Souto, Ricardo André Rodrigues Filho, Kariolania Fortunato de Paiva Araújo, et al.
Effect of Nitrogen Sources on the Phenological Phases of Italian Zucchini Under Salt Stress
Reprinted from: *Agriculture* **2026**, *16*, 62, https://doi.org/10.3390/agriculture16010062 23

Shengyi Zhao, Chen Fang, Tianzheng Hua and Yong Jiang
Detecting the Maturity of Red Strawberries Using Improved YOLOv8s Model
Reprinted from: *Agriculture* **2025**, *15*, 2263, https://doi.org/10.3390/agriculture15212263 41

Mira Radunić, Maja Jukić Špika, Jelena Gadže, Smiljana Goreta Ban, Juan Carlos Díaz-Pérez and Dan MacLean
Microclimate Condition Influence on the Physicochemical Properties and Antioxidant Activity of Pomegranate (*Punica granatum* L.): A Case Study of the East Adriatic Coast
Reprinted from: *Agriculture* **2025**, *15*, 2210, https://doi.org/10.3390/agriculture15212210 57

Agda Malany Forte de Oliveira, Vander Mendonça, Patrycia Elen Costa Amorim, Raires Irlenizia da Silva Freire, Lucas Rodrigues Bezerra da Silva, David Emanoel Gomes da Silva, et al.
Organic and Mineral Fertilization on the Photosynthetic, Nutritional, and Productive Efficiency of (*Ficus carica* L.) Subjected to Conduction Systems in a Semi-Arid Region of Brazil
Reprinted from: *Agriculture* **2025**, *15*, 2128, https://doi.org/10.3390/agriculture15202128 74

Nay Myo Win, Juhyeon Park, Seonae Kim, Youngsuk Lee, Van Giap Do, Jung-Geun Kwon, et al.
Effects of Mechanical Pruning on Tree Growth, Yield, and Fruit Quality of 'Arisoo' Apple Trees
Reprinted from: *Agriculture* **2025**, *15*, 2118, https://doi.org/10.3390/agriculture15202118 97

Jinze Pei, Hongyu Yang, Menghan Huang, Xiaojun Yan, Xinran Zeng, Lijin Guo and Liangquan Wu
Multi-Objective Nitrogen Optimization in Tea Cultivation: A Pathway to Achieve Sustainability in Cash Crop Systems
Reprinted from: *Agriculture* **2025**, *15*, 1949, https://doi.org/10.3390/agriculture15181949 110

Chengyao Jiang, Kexin Zhang, Yue Ma, Yu Song, Mengyao Li, Yangxia Zheng, et al.
Comparison of Ray Tracing Software Performance Based on Light Intensity for Spinach Growth
Reprinted from: *Agriculture* **2025**, *15*, 1852, https://doi.org/10.3390/agriculture15171852 126

Álvaro Iglesias-Ganado, Jorge Poveda, Oscar Santamaría, Sara Rodrigo, María I. Pozo and Jorge Martín-García
Effects of Seed Priming with *Talaromyces ruber* Extracts on Tomato (*Solanum lycopersicum*) Growth
Reprinted from: *Agriculture* **2025**, *15*, 1868, https://doi.org/10.3390/agriculture15171868 143

Ilias Karampatzakis, Fotis Bilias, Chrysanthi Polychroniadou, Georgia Tanou, Panagiotis Kekelis, Aphrodite Theofilidou, et al.
Assessing Nutrient Losses and Recycling in Sweet Cherry Orchards: A Yield-Based Approach
Reprinted from: *Agriculture* **2025**, *15*, 1312, https://doi.org/10.3390/agriculture15121312 . . . . **156**

Carolina Ragoni Maniero, Marco Antonio Tecchio, Harleson Sidney Almeida Monteiro, Camilo André Pereira Contreras Sánchez, Giuliano Elias Pereira, Juliane Barreto de Oliveira, et al.
Phenological Performance, Thermal Demand, and Qualitative Potential of Wine Grape Cultivars Under Double Pruning
Reprinted from: *Agriculture* **2025**, *15*, 1241, https://doi.org/10.3390/agriculture15121241 . . . . **172**

agriculture

Editorial

Advanced Cultivation Technologies for Horticultural Crops Production

Nay Myo Win [1], Jingi Yoo [1,*] and In-Kyu Kang [2,*]

1 Department of Horticulture & Life Science, Yeungnam University, Gyeongsan 38541, Republic of Korea; naymyowin@yu.ac.kr
2 Department of Horticultural Science, Kyungpook National University, Daegu 41566, Republic of Korea
* Correspondence: jingiyoo@yu.ac.kr (J.Y.); kangik@knu.ac.kr (I.-K.K.)

The agricultural sector faces the challenge of meeting global food security demands while maintaining environmental stability and sustainability [1]. Climate change, including rising daily temperatures and unpredictable seasonal precipitation, significantly affects the productivity and conventional farming practices of horticultural crops [2,3]. Therefore, the introduction and adoption of advanced cultivation technologies and farming strategies are essential for enhancing productivity, maintaining biodiversity, and ensuring food security and the long-term sustainability of agricultural systems [4]. The use of advanced farming technologies, such as climate-smart agriculture, precision agriculture, controlled-environment agriculture, mechanization and automation, artificial intelligence, and digital farming tools, enables farmers to optimize crop yield and quality while reducing excessive input use, minimizing greenhouse gas emissions, improving profitability, and promoting efficient and sustainable agricultural production [3,4].

This Special Issue of *Agriculture* presents recent developments, practical applications, and innovative research, providing a platform for improving productivity, sustainability, and ecosystem resilience in modern horticultural crop farming systems. A total of eleven articles are included in this Special Issue.

In the first article, Maniero et al. [5] investigated the phenological and productive performance of wine grape (*Vitis vinifera* L.) cultivars under a double-pruning management system across two production cycles in a subtropical climate in Southeastern Brazil. The results showed that the duration of phenological stages under double pruning varied among cultivars and between production cycles. In addition, several cultivars responded positively to double pruning by exhibiting increased antioxidant activity, supporting their suitability for premium wine production. This study provides valuable insights into productivity and cultivar adaptability under double-pruning management and supports the selection of suitable wine grape cultivars for cultivation in subtropical climate regions.

In the second article, Karampatzakis et al. [6] developed a predictive framework to evaluate nutrient uptake, distribution, and losses in sweet cherry trees in relation to fruit yield in two commercial orchards. The study showed that the relationship between fresh fruit yield and total nutrient uptake is nonlinear and that differences in tree age are a major factor influencing yield variability. Additionally, the results indicated that the non-removal of plant residues, such as fallen leaves and other plant parts, can support their reuse as part of an integrated fertilization strategy. This work highlights the potential of nutrient recycling as a sustainable orchard management approach and reduces reliance on external inputs.

In the third article, Jiang et al. [7] evaluated light intensity performance for spinach growth under different planting arrangements and lamp-source distances using three

 https://doi.org/10.3390/agriculture16070787

light-tracing software programs: LightTools 8.6, TracePro 7.4.3, and Ansys Lumerical FDTD Solution 2024. The results showed that the simulation outputs differed significantly among the software programs under the same planting arrangement. However, spinach leaves received higher light intensity and plants exhibited faster growth under a rectangular planting arrangement across different lighting distances. The study further identified TracePro as the most suitable software for analyzing changes in the light environment and monitoring light availability to spinach leaves during the growth process in plant factory systems.

In the fourth article, Iglesias-Ganado et al. [8] investigated the application of fungal (*Talaromyces ruber*) extracts to enhance the growth of tomato (*Solanum lycopersicum*) plants. The study demonstrated that seed priming with fungal extracts significantly stimulated early-stage root development and improved seedling vigor and overall agronomic performance in tomato. These findings indicate that the application of fungal bioactive extracts to seeds represents a promising approach for improving tomato seedling quality in commercial nursery production.

In the fifth article, Pei et al. [9] examined nitrogen management strategies in relation to economic and environmental outcomes in tea plantations in China. The study aimed to reduce excessive nitrogen application, lower greenhouse gas emissions, and optimize ecological and economic benefits. The results indicated that a nitrogen application rate of 120 kg/ha/yr represents an optimal level for achieving both environmental and economic advantages, reducing nitrogen input by more than 41% compared with conventional application guidelines while maintaining yield. This finding provides a practical approach for improving nitrogen management, supporting green agricultural practices, and promoting the sustainable production of tea.

In the sixth article, Win et al. [10] reported that mechanical pruning reduced the time required for pruning in apple trees but also decreased leaf nitrogen content and light penetration within the canopy, resulting in smaller fruit size with poorer coloration. The study suggested that supplementary hand pruning following mechanical pruning can enhance light penetration within the canopy, improve fruit size, and produce higher-quality apples. These findings indicate that combining mechanical and hand pruning can reduce field workloads in apple orchards and help address labor shortages in modern agriculture. However, the long-term effects of mechanical pruning on plant hormone regulation and nutrient metabolism require further investigation.

In the seventh article, Oliveira et al. [11] investigated the effects of mineral fertilizers (nitrogen, phosphorus, and potassium) and organic fertilizers (cattle manure, organic compost, chicken litter, and sheep manure) on the productive efficiency of fig trees under different training systems in semi-arid regions of Brazil. The results demonstrated that the application of mineral fertilizers and chicken litter increased yield and enhanced the photosynthetic performance of fig trees. The effects were more pronounced in plants trained under a three-branch training system. The study concluded that the combined application of mineral and organic fertilizers, together with an appropriate training system, represents an effective strategy for improving physiological and developmental processes and optimizing fig tree productivity.

In the eighth paper, Radunić et al. [12] investigated the effects of different microclimate conditions on the physiochemical and antioxidant properties of pomegranate trees across four orchards located along the eastern Adriatic coast of Croatia. The study indicated that variations in orchard location and environmental conditions significantly influenced yield and fruit size. Higher rainfall during the flowering stage negatively affected pollination, aril number, and fruit set. In contrast, orchard sites with higher spring rainfall after the fruit set stage were more favorable for enhancing phenolic content and fruit growth. Therefore,

 https://doi.org/10.3390/agriculture16070787

this study provides important insights for selecting suitable orchard locations and cultivars to improve pomegranate productivity under variable environmental conditions.

In the ninth article, Zhao et al. [13] developed an improved GAM-YOLOv8s model to detect the maturity of red strawberry. The results demonstrated that the GAM-YOLOv8s model achieved superior maturity detection performance and higher accuracy than other mainstream YOLO models under varying lighting conditions. This development is important for enabling precise detection of strawberry ripeness, which can improve marketable yield and reduce pre- and postharvest losses, thereby contributing to the advancement of modern agricultural practices.

In the tenth article, Silva et al. [14] investigated nitrogen fertilization strategies using nitrate and ammoniacal nitrogen at different phenological stages of Italian zucchini cv. Caserta under salt stress conditions. The study showed that the application of ammoniacal nitrogen during the vegetative phase resulted in higher yield, whereas its application during the reproductive phase caused flower abortion. Importantly, maintaining a balanced ratio of ammonium and nitrate throughout the entire plant growth cycle produced higher yield than the application of nitrate fertilizer alone. Therefore, this study provides valuable insights into fertilizer management strategies for improving zucchini production under salt stress conditions.

In the eleventh article, Zheng et al. [15] evaluated the installation of photovoltaic (PV) power generation systems in tea plantations (PVtea) in China. The results showed that this approach significantly enhanced the resistance of tea plants to frost damage in spring and heat stress in summer. In addition, the PVtea system promoted the accumulation of secondary metabolites in tea plants and increased amino acid content in tea leaves. Therefore, the integration of PV systems with tea plantations not only improves tea productivity but also supports clean energy generation, providing a potential solution to land-use conflicts through the combined use of agricultural land for crop production and renewable energy development.

Taken together, the articles included in this Special Issue examine recent and advanced farming technologies aimed at improving the productivity and profitability of horticultural crops while maintaining the sustainability of agricultural systems. These collective advances in cultivation strategies enhance understanding of how to achieve productive, economically viable, and ecologically sustainable farming systems in the future. Moreover, these contributions are expected to support and encourage further research on this important topic.

Author Contributions: Conceptualization, N.M.W., J.Y. and I.-K.K.; writing—original draft preparation, N.M.W., J.Y. and I.-K.K.; writing—review and editing, N.M.W., J.Y. and I.-K.K. All authors have read and agreed to the published version of the manuscript.

Funding: This research received no external funding.

Conflicts of Interest: The authors declare no conflicts of interest.

References

1. Pawlak, K.; Kołodziejczak, M. The role of agriculture in ensuring food security in developing countries: Considerations in the context of the problems of sustainable food production. *Sustainability* **2020**, *12*, 5488. [CrossRef]
2. Roy, A.; Moradkhani, H.; Mekonnen, M.; Moftakhari, H.; Magliocca, N. Towards strategic interventions for global food security in 2050. *Sci. Total Environ.* **2024**, *954*, 176811. [CrossRef] [PubMed]
3. Bhatnagar, S.; Chaudhary, R.; Sharma, S.; Janjhua, Y.; Thakur, P.; Sharma, P.; Keprate, A. Exploring the dynamics of climate-smart agricultural practices for sustainable resilience in a changing climate. *Environ. Sustain. Indic.* **2024**, *24*, 100535. [CrossRef]
4. Chen, X. The role of modern agricultural technologies in improving agricultural productivity and land use efficiency. *Front. Plant Sci.* **2025**, *16*, 1675657. [CrossRef] [PubMed]

https://doi.org/10.3390/agriculture16070787

5. Maniero, C.R.; Tecchio, M.A.; Monteiro, H.S.A.; Sanchez, C.A.P.C.; Pereira, G.E.; de Oliveira, J.B.; Brito, S.d.N.S.; Domingues Neto, F.J.; Leonel, S.; Silva, M.d.S.; et al. Phenological performance, thermal demand, and qualitative potential of wine grape cultivars under double pruning. *Agriculture* **2025**, *15*, 1241. [CrossRef]
6. Karampatzakis, I.; Bilias, F.; Polychroniadou, C.; Tanou, G.; Kekelis, P.; Theofilidou, A.; Giannopoulos, G.; Pavlatou-Ve, A.; Aschonitis, V. Assessing nutrient losses and recycling in sweet cherry orchards: A yield-based approach. *Agriculture* **2025**, *15*, 1312. [CrossRef]
7. Jiang, C.; Zhang, K.; Ma, Y.; Song, Y.; Li, M.; Zheng, Y.; Pan, T.; Lu, W. Comparison of ray tracing software performance based on light intensity for spinach growth. *Agriculture* **2025**, *15*, 1852. [CrossRef]
8. Iglesias-Ganado, Á.; Poveda, J.; Santamaría, O.; Rodrigo, S.; Pozo, M.I.; Martín-García, J. Effects of seed priming with *Talaromyces ruber* extracts on tomato (*Solanum lycopersicum*) growth. *Agriculture* **2025**, *15*, 1868. [CrossRef]
9. Pei, J.; Yang, H.; Huang, M.; Yan, X.; Zeng, X.; Guo, L.; Wu, L. Multi-objective nitrogen optimization in tea cultivation: A pathway to achieve sustainability in cash crop systems. *Agriculture* **2025**, *15*, 1949. [CrossRef]
10. Win, N.M.; Park, J.; Kim, S.; Lee, Y.; Do, V.G.; Kwon, J.-G.; Kwon, S.-I.; Yoo, J.; Kang, I.-K.; Kweon, H.-J. Effects of mechanical pruning on tree growth, yield, and fruit quality of 'Arisoo' apple trees. *Agriculture* **2025**, *15*, 2118. [CrossRef]
11. Oliveira, A.M.F.d.; Mendonça, V.; Amorim, P.E.C.; Freire, R.I.d.S.; Silva, L.R.B.d.; Silva, D.E.G.d.; Ferreira, F.N.; Bonou, S.I.; Silva, L.d.A.; Fernandes, P.D.; et al. Organic and mineral fertilization on the photosynthetic, nutritional, and productive efficiency of (*Ficus carica* L.) subjected to conduction systems in a semi-arid region of Brazil. *Agriculture* **2025**, *15*, 2128. [CrossRef]
12. Radunić, M.; Jukić Špika, M.; Gadže, J.; Goreta Ban, S.; Díaz-Pérez, J.C.; MacLean, D. Microclimate condition influence on the physicochemical properties and antioxidant activity of pomegranate (*Punica granatum* L.): A case study of the East Adriatic Coast. *Agriculture* **2025**, *15*, 2210. [CrossRef]
13. Zhao, S.; Fang, C.; Hua, T.; Jiang, Y. Detecting the maturity of red strawberries using improved YOLOv8s model. *Agriculture* **2025**, *15*, 2263. [CrossRef]
14. Silva, G.d.F.; Peixoto, T.D.C.; Ferreira Neto, M.; Souto, A.G.d.L.; Rodrigues Filho, R.A.; Araújo, K.F.d.P.; Cavalcante, J.S.J.; Pereira, K.T.O.; Moreira, R.C.L.; Fernandes, P.D.; et al. Effect of nitrogen sources on the phenological phases of Italian zucchini under salt stress. *Agriculture* **2026**, *16*, 62. [CrossRef]
15. Zheng, X.-Q.; Zhang, X.-H.; Zhang, J.-G.; Zheng, R.-J.; Lu, J.-L.; Ye, J.-H.; Liang, Y.-R. Effects of photovoltaic-integrated tea plantation on tea field productivity and tea leaf quality. *Agriculture* **2026**, *16*, 125. [CrossRef]

https://doi.org/10.3390/agriculture16070787

agriculture

MDPI

Article

Effects of Photovoltaic-Integrated Tea Plantation on Tea Field Productivity and Tea Leaf Quality

Xin-Qiang Zheng [1], Xue-Han Zhang [1], Jian-Gao Zhang [2], Rong-Jin Zheng [3], Jian-Liang Lu [1], Jian-Hui Ye [1] and Yue-Rong Liang [1,*]

1 Tea Research Institute, Zhejiang University, Hangzhou 310058, China; xqzheng@zju.edu.cn (X.-Q.Z.); 12216079@zju.edu.cn (X.-H.Z.); jllu@zju.edu.cn (J.-L.L.); jianhuiye@zju.edu.cn (J.-H.Y.)

2 Zhejiang Zhongxin Energy Development Co., Ltd., Hangzhou 310002, China; 13906505017@139.com

3 Institute of Agricultural Bioenvironmental Engineering, Zhejiang University, Hangzhou 310058, China; rongjinzheng@zju.edu.cn

* Correspondence: yrliang@zju.edu.cn

Abstract

Agrivoltaics integrates photovoltaic (PV) power generation with agricultural practices, enabling dual land-use and mitigating land-use competition between agriculture and energy production. China has 3.43 million hectares of tea fields, offering significant potential for PV-integrated tea plantations (PVtea) to address land scarcity in clean energy development. This study aimed to investigate the impact of PV modules above tea bushes in PVtea on the yield and quality of tea, as well as tea plant resistance to environmental stresses. The PV system uses a single-axis tracking system with a horizontal north–south axis and ±45° tilt. It includes 70 UL-270P-60 polycrystalline solar panels (270 Wp each), arranged in 5 columns of 14 panels, spaced 4500 mm apart, covering 280 m^2. The panels are mounted 2400 mm above the ground, with a total capacity of 18.90 kWp (656 kWp/ha). Tea yield, quality-related components, leaf photosystem II (PSII) activity, and plant resistance to environmental stresses were investigated in comparison to an adjacent open-field tea plantation (control). The mean photosynthetic active radiation (PAR) reaching the plucking table of PVtea was 52.9% of the control, with 32.0% of the control on a sunny day and 49.0% on a cloudy day, accompanied by an increase in ambient relative humidity. These changes alleviated the midday depression of leaf PSII activity caused by high light, resulting in a 9.3–15.3% increase in leaf yield. Moreover, PVtea summer tea exhibited higher levels of amino acids and total catechins, resulting in tea quality improvement. Additionally, PVtea enhanced the resistance of tea plants to frost damage in spring and heat stress in summer. PVtea integrates photovoltaic power generation with tea cultivation practices, which not only facilitates clean energy production—an average annual generation of 697,878.5 kWh per hectare—but also increases tea productivity by 9.3–15.3% and the land-use equivalence ratio (LER) by 70%.

Keywords: *Camellia sinensis*; agrovoltaics; tea quality; tea yield; frost damage; heat damage

1. Introduction

In the face of escalating global climate change and energy crises, the development of clean energy and the pursuit of sustainable development have emerged as a global imperative [1,2]. Photovoltaic (PV) power generation, a critical component of clean energy, has witnessed rapid advancement in recent years. Nevertheless, traditional PV power plants require substantial land areas, leading to increasing conflicts with agricultural land-use [3]. As an innovative industrial model, agrovoltaics (PV agriculture) integrates PV power generation with agricultural practices, which not only facilitates clean energy production but also ensures food security and enhances land efficiency, thereby offering an effective solution to the aforementioned challenges.

The concept of agrovoltaics originated in Europe and has since experienced rapid development in recent years, particularly in countries such as China, Japan, and the United States [4]. At its core, agrovoltaics involves utilizing agricultural land for PV power generation to achieve "dual land use," thereby ensuring that agricultural productivity is maintained while simultaneously generating clean energy. Currently, the primary models of agrovoltaics include PV greenhouses [5], agrovoltaic systems over open fields [6], and fishery-photovoltaics [7]. These models have demonstrated significant success in various sectors, including vegetable cultivation, floriculture, and aquaculture.

The tea plant (*Camellia sinensis* (L.) O. Kuntze) is an important cash crop that belongs to the family Theaceae in botanical classification. It prefers to grow in partial shade under the forest canopy, making it well-suited for growing under diffused light conditions. Scientists have conducted extensive research on the impact of artificial shading on the growth, yield, and quality of tea plants. Shading modifies the microenvironment and can provide plants with some protection from frequent heat, drought, frost, and hail induced by climate change, and has the potential to improve plant growth, yield, and quality. When compared to no-shade control, black, blue, and red shade nets increased plant growth index (PGI), net photosynthetic rate, and stomatal conductance, and decreased air and leaf surface temperatures in summer, but they reduced cold damage in winter [8]. Tea plants grown under shade trees at higher elevations attained better quality and higher productivity of the leaf [9]. The best method to achieve improvements in the quality and commercial value of Gyokuro-styled green tea grown in Australia is to shade all visible light intensities to one per cent of ambient levels in as little as two weeks [10]. These suggest that tea plants exhibit characteristics of weak light tolerance.

Tea plants are widely planted in China, with 3.43 million hectares of tea fields in 2023 [11]. Building PV modules in the extensive tea fields for PV power generation can effectively address the land scarcity challenge faced by clean energy production. However, the impact of PV-integrated tea plantation (PVtea) on tea productivity and quality has not been available. There have been no comprehensive studies to date that simultaneously evaluate microclimate, physiology, yield, quality, and climate resilience in agrivoltaic systems for perennial crops, limiting the development of agrivoltaic technologies in this field. This study was set to install PV modules in existing tea gardens to examine their effects on the growth of tea plants, as well as the yield and quality-related phytochemicals, including catechins, amino acids, caffeine, and chlorophylls. The findings will provide a scientific basis for the large-scale promotion and application of this technology in tea fields. In the present paper, a PVtea is defined as a model that integrates PV modules above tea plants, enabling simultaneous production of tea and clean energy on the same land, which will significantly contribute to alleviating the land shortage issue for photovoltaic power stations. It is estimated that if one-third of China's tea plantations were converted into PVtea, the annual electricity generation would reach 675 billion kilowatt-hours, equivalent to six times the power output of the Three Gorges Hydropower Station.

2. Materials and Methods

2.1. Construction of PVtea Plantation

The experimental tea field, situated in the Fuyang District of Hangzhou City, China (30°04′89″ N, 119°86′22″ E), was grown with 18-year-old tea plants (*C. sinensis* cv. Xiangshanzao). The construction of a single-axis tracking system of PV modules with a

https://doi.org/10.3390/agriculture16010125

north–south-oriented horizontal axis allowed for an inclination adjustment range of ±45°. Model UL-270P-60 polycrystalline silicon solar panels with a rated power of 270 Wp and dimensions of 1650 × 992 × 40 mm (Ningbo Eureka Solar Technology Development Co., Ltd., Nignbo, China) were used in the PV power generation system. In the tea field with 280 m^2 (20 m × 14 m), 70 PV panels were installed in 5 columns of PV modules, with 14 panels for each module and a center-to-center distance of 4500 mm between adjacent PV modules. The distance between the lowest point of the panels and the ground was 2400 mm. The total installed power capacity in the 288 m^2 tea field was 18.90 kWp, which fits 656'kWp per ha. There were two independent plots in the experimental design (Figure 1), and the installation was finished in November 2017. Based on 1000 effective sunlight hours per year for photovoltaic power generation at the test site, the annual electricity generation for each hectare of tea field is approximately 656,000 kWh. Observations on tea plant resistance to environmental stresses, as well as leaf yield and leaf quality, were conducted from 2018 to 2023.

Figure 1. PV modules over the tea field.

To further validate the impact of PVtea on tea productivity and quality, this study also examined the PVteas located in Jingshan Town (30°39′59″ N, 119°79′41″ E), Yuhang District, Zhejiang Province, where 35-year-old *C. sinensis* cv. Fudingdabai was grown, and in Mumen Town (32°10′28″ N, 106°30′08″ E), Wanguan, Sichuan Province, where a 12-year-old *C. sinensis* cv. Mingshan 213 was grown, whose installations were completed in September 2021, and leaf yield and leaf quality were assessed in 2024 and 2025, respectively.

The agronomic management of the tea fields was conducted according to local farming practices. Organic fertilizer (fermented rapeseed cake, 3000 kg/ha) was applied in November. Compound fertilizer ($N:P_2O_5:K_2O$ = 10:4:10) was applied at a rate of 300 kg/ha in early March, 150 kg/ha in mid-May, and 150 kg/ha in late July. Weeding was performed prior to fertilizer application. There were no obvious differences in pest and disease incidence between PVtea and the control (without PV shade); therefore, no additional pest and disease control measures were implemented in the PVtea fields.

2.2. Investigation on Ambient Temperature and Humidity

TH10R-EX temperature and humidity monitors (Pingyang Miaoguan Sci-Tech Ltd., Wenzhou, China) were installed 20 cm above the plucking tables in both the PVtea fields and the open-field tea plantations (control) to continuously monitor and record ambient temperature and relative humidity.

2.3. Observation on Tea Sprouting in Spring

During the early growing stage of tea plant sprouting in spring, a metal frame measuring 30 cm × 30 cm was randomly placed on the plucking table of the tested tea bushes.

 https://doi.org/10.3390/agriculture16010125

The number of shoots with fish leaves, shoots with one leaf and a bud, shoots with two leaves and a bud, and shoots with three leaves and a bud within the frame were counted separately. This counting process was repeated three times, and the sprouting index was calculated using the following formula:

$$\text{Sprouting index} = (A \times 1 + B \times 2 + C \times 3 + D \times 4)/\text{total number of shoots}$$

Herein, A is the number of shoots with fish leaf sprouting; B is the number of shoots with one leaf and a bud; C is the number of shoots with two leaves and a bud; and D is the number of shoots with three leaves and a bud [12].

2.4. Investigation on the Yield of Fresh Tea Leaf

A metal wire frame (1 m × 1 m) was randomly placed on the picking table of the tested tea bushes, and the tea shoots within the frame were plucked and weighed. Each harvest was replicated three times, and the mean value (g/m^2) was used to assess the fresh tea leaf yield. To eliminate marginal effects, a 1.5 m buffer zone was set aside around the PVtea field, and random sampling was carried out inside the buffer zone. The adjacent open-field tea plants attached to the same tea garden were used as a control.

2.5. Measurement of Tea Leaf Length, Width, and Leaf Area

The fourth leaf beneath the apical bud on tea shoots was collected for testing. An LI-3000C leaf area meter (Ecotek Co., Ltd., Beijing, China) was used to measure the maximum length, maximum width, average width, and leaf area according to the manufacturer's instructions. Ten leaves were analyzed for each treatment.

2.6. Test of Activity of Leaf Photosystem II (PSII)

Tea plant growth depends on the photosynthesis of tea leaves, which is closely related to the activity of PSII. The efficiency of PSII photochemical reactions can be assessed by the ratio of variable fluorescence (Fv) to maximum fluorescence (Fm), denoted as Fv/Fm [13]. This ratio was measured using a fluorimeter (Handy PEA, Hansatech Instruments Ltd., Norfolk, UK) following the manufacturer's protocol after 30 min of dark adaptation. For in vivo measurements of Fv/Fm, we selected the fourth leaf below the apical bud on tea shoots. Six leaves from six different plants were sampled and tested for each treatment, as outlined in previous studies [14].

2.7. Determination of Quality-Related Chemical Components in Tea

2.7.1. Leaf Extraction

Tea shoots with two leaves and a bud (200 g) were plucked from the tested tea bushes and were fixed by heating in an M1-231E microwave oven (Midea Group Co., Ltd., Foshan City, China) at 800 W (2450 MHz) for 1 min. Three leaf samples from each treatment were collected for subsequent analysis. The samples were dried at 80 °C, ground into powder, and sieved through a 12-mesh sifter. A total of 0.15 g of the ground tea powder was transferred to a 50 mL centrifuge tube with 25 mL of 50% ethanol, mixed, and incubated in a water bath at 70 °C for 20 min, with gentle shaking every 5 min. After extraction, the extract was cooled to room temperature and centrifuged at 12,000 rpm for 10 min. The supernatant, designated as the tea extract, was collected and stored at 4 °C for further analysis.

2.7.2. Determination of Amino Acids

The amino acid concentration was assessed using the ninhydrin assay method [15]. A total of 2 mL of the tea extract solution was mixed with 1 mL of a reagent mixture (comprising 20 g/L ninhydrin and 0.8 g/L $SnCl_2 \cdot 2H_2O$) and 1 mL of buffer solution (0.067 M Na_2HPO_4 and 0.067 M KH_2PO_4, adjusted to pH 8.0) in a 50 mL volumetric flask. This mixture was then subjected to a 15 min reaction period in a boiling water bath. For the control, a similar setup was used, but with distilled water instead of the tea extract. Following the reaction, the solution was transferred to a quartz cuvette with a black aperture and a 1 cm light path for spectrophotometric analysis. The absorbance was measured at 570 nm using an HP8453E UV-VIS spectrophotometer (Hewlett Packard Company, Palo Alto, CA, USA). To establish a calibration curve, theanine (Sigma-Aldrich, St. Louis, MO, USA) was used as the reference standard. The amino acid content in the tea samples was quantified as theanine equivalents based on their absorbance values at 570 nm.

2.7.3. Determination of Catechins and Caffeine

The concentrations of catechins and caffeine were analyzed using high-performance liquid chromatography (HPLC-20AD System, Shimadzu, Kyoto, Japan) under conditions detailed in an earlier study [16]. The analysis was conducted with the following parameters: a sample injection volume of 10 μL; a TC-C18 column (5 μm particle size, 4.6 × 250 mm, Agilent Technologies Inc., Santa Clara, CA, USA); oven temperature at 35 °C; a linear gradient elution program that increased mobile phase B from 18% to 80% over 35 min; a flow rate maintained at 1 mL/min; mobile phase A consisted of acetonitrile/acetic acid/water in a ratio of 6/1/193 (*v*/*v*/*v*), while mobile phase B comprised acetonitrile/acetic acid/water in a ratio of 60/1/139 (*v*/*v*/*v*); and a detection wavelength at 280 nm. Identification and quantification of individual catechins and caffeine were achieved by comparing retention times and peak areas with those of certified reference standards, and their contents were determined based on the dry weight of the samples.

2.7.4. Analysis of Photosynthetic Pigments

The third leaf beneath the apical bud on tea shoots was sampled from tested tea bushes (3 samples each treatment). The primary veins were severed and discarded. A total of 200 mg of the leaves without primary veins was grounded in liquid nitrogen, mixed with 10 mL pre-cooled acetone and 50 mg Polyvinylpolypyrrolidone (PVPP), and then ultrasonically extracted for 30 min in a dark ice bath (JP-060S, Skymen Cleaning Equipment Co., Ltd., Shanghai, China). A total of 1.2 mL of the resultant mixture was placed into a 1.5 mL centrifuge tube and subjected to centrifugation at 12,000 rpm and 4 °C for 15 min. The supernatant was placed into a brown HPLC vial. The analysis of various photosynthetic pigments, including chlorophyll a, chlorophyll b, neoxanthin, violaxanthin, lutein, and β-carotene, was conducted using an LC-20AT HPLC System equipped with a UV-visible detector (Shimadzu, Kyoto, Japan). An Agilent Technologies Inc. (Santa Clara, CA, USA) TC-C18 column (5 μm particle size, 4.6 × 150 mm dimensions) was used to separate the photosynthetic pigments. The HPLC operating parameters were set as follows: the column was maintained at 35 °C, with an injection volume of 10 μL. For the mobile phase composition, solvent A consisted of 3% acetonitrile, 0.5% acetic acid, and 96.5% water (*v*/*v*), while solvent B comprised 75% acetonitrile, 20% methanol, and 5% chloroform (*v*/*v*). A linear gradient elution profile was employed, where the proportion of solvent B increased from 90% to 100% over the first 15 min, remained at 100% for an additional 15 min, and then returned to 90% within the subsequent 5 min. The flow rate was kept constant at 1 mL/min, and detection was performed at a wavelength of 440 nm. Identification and quantification of the detected pigments were achieved by comparing their retention times and peak areas with those of authentic reference standards.

https://doi.org/10.3390/agriculture16010125

2.8. Statistical Analysis

The data were expressed as mean ± standard deviation (DS), which was calculated by Microsoft Excel (Version 2019). The statistics were carried out on IBM SPSS Statistics 24.0 (2016).

3. Results

3.1. Effects of PV Modules on Photosynthetically Active Radiation (PAR)

Based on the year-round PAR test results (October 2018–September 2019), the mean values of PAR on tested PVtea and open tea field (control) were 202.0 μmol/m^2·s and 382.9 μmol/m^2·s, indicating the PVtea was 52.8% of the control. The lowest PAR appeared in December, and the highest was in August (Figure 2).

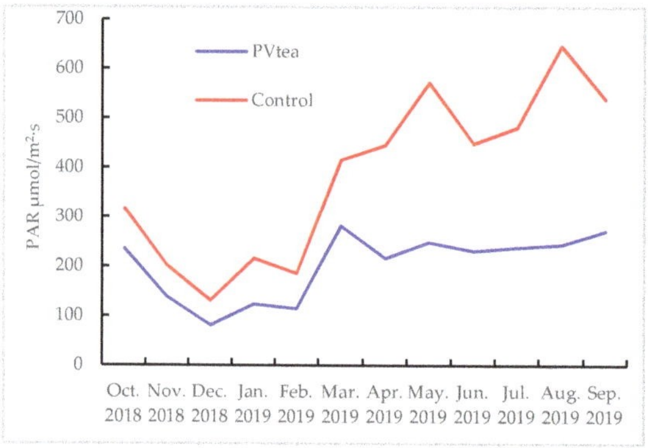

Figure 2. The year-round changes in PAR measured on Fuyang Tea Farm.

The diurnal variation in PAR is primarily driven by changes in the solar altitude angle. The difference in PAR between PVtea and the control increased progressively over time, reaching its peak at noon, followed by a gradual decline in the afternoon (Figure 3). The difference in PAR between PVTea and control was also influenced by the weather. On a sunny day, the mean daily PAR for PVtea was 421.2 μmol/m^2·s, being 32.0% of the control (1314.3 μmol/m^2·s). On a cloudy day, however, the mean daily PAR for PVtea was 326.1 μmol/m^2·s, being 49.0% of the control (665.5 μmol/m^2·s). At noon, the PAR for PVTea was about 600 μmol/m^2·s on both sunny days and cloudy days, though the PAR for control on sunny days was greater than that on cloudy days (Figures 3 and 4).

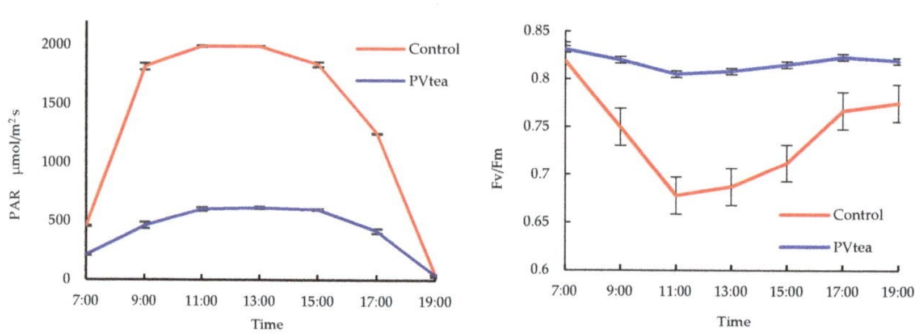

Figure 3. Diurnal changes in PAR and Fv/Fm for a sunny day measured on Fuyang Tea Farm on 23 May 2023.

https://doi.org/10.3390/agriculture16010125

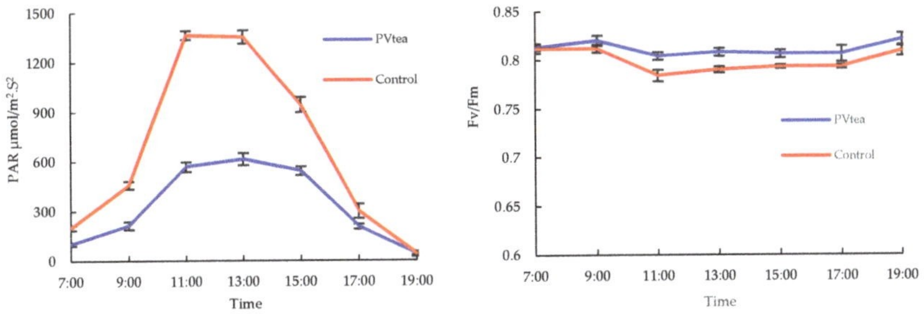

Figure 4. Diurnal changes in PAR and Fv/Fm for a cloudy day measured on Fuyang Tea Farm on 23 September 2023.

Minimal variations in Fv/Fm, a key indicator of photosystem II activity, were observed in PVTea, with values remaining consistently around 0.8 on both sunny and cloudy days. Although a significant suppression of Fv/Fm occurred at noon under sunny conditions, this effect was markedly alleviated during cloudy days (Figures 3 and 4).

3.2. *Effects of PV Modules on Ambient Temperature and Humidity*

In the spring, the temperature on the tea-plucking table in the PVtea was 0.5–2 °C higher than that in the control field at night, but lower during the day (Figure 5). However, during the periods when the sun is directly shining on the plucking table (such as from 9:00 to 10:00 in the morning and from 1:30 to 2:30 in the afternoon), there is little difference in temperature between the PVtea and control.

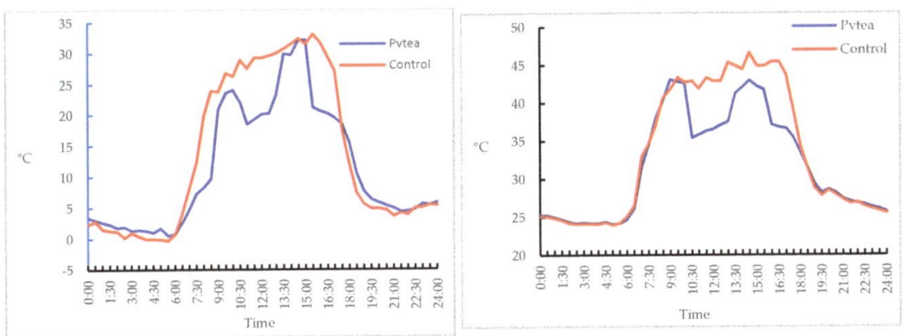

Figure 5. Diurnal temperature changes for spring (**left**) and summer (**right**) measured on Fuyang Tea Farm on 1 April 2019 and 27 July 2018, respectively.

During the summer, there was no significant difference in temperature between the PVtea and control before 10:00 a.m. and after 6:00 p.m. However, from 10:30 a.m. to 5:30 p.m., the temperature in the PVtea was markedly lower than that of the control. The largest temperature difference was 8.3 °C, which occurred at 4:00 p.m. (PVtea: 37.1 °C, control: 45.4 °C), followed by that at 1:00 p.m., with a difference of 7.8 °C (PVtea: 37.5 °C, control: 45.3 °C) (Figure 5). During extreme weather conditions, such as at 13:00 on 16 August 2025, a temperature difference of 13.1 °C was observed between the PVtea and the control group, with the PVtea temperature recorded at 37.4 °C and the control temperature at 51.7 °C.

The above results demonstrated that the temperature in the PVtea was significantly lower than that of the control during midday when sunlight intensity peaked in both spring and summer seasons. In spring, when ambient temperatures are relatively low, the PV modules mitigate nighttime convective heat radiation and reduce air flow, which

https://doi.org/10.3390/agriculture16010125

minimizes heat loss from the PVtea fields and results in higher temperatures in the PVtea compared to the control.

For diurnal changes in relative humidity, no significant differences were observed between PVtea and the control before 7:30 and after 19:00 in spring, as well as before 9:30 and after 18:00 in summer. However, during the daytime periods from 8:00 to 17:00 in spring and from 10:30 to 17:30 in summer, the relative humidity in PVtea was significantly higher than that in the control (Figure 6).

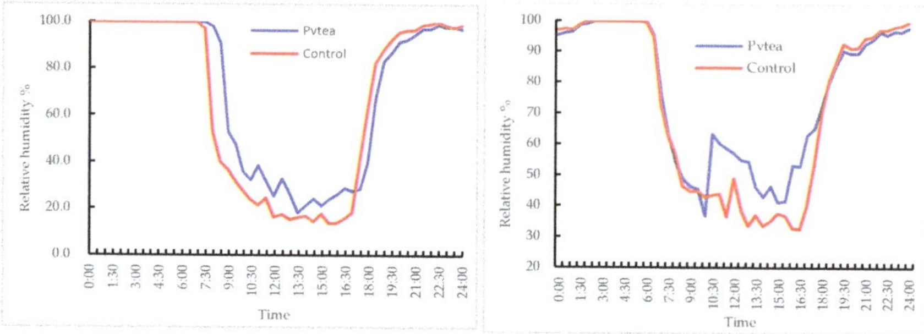

Figure 6. Diurnal relative humidity changes for spring (**left**) and summer (**right**) measured on Fuyang Tea Farm on 1 April 2019 and 27 July 2018, respectively.

3.3. Effects of PV Modules on the Sprouting of Spring Shoots

The observation results indicated that although there was no significant difference in the number of shoots with one leaf and a bud or two leaves and a bud on 3 April 2018, the number of shoots with three leaves and a bud in PVtea was significantly higher than that in the control group, resulting in a higher sprouting index for PVtea (Table 1). On 26 March 2019, observations revealed that the number of shoots with two leaves and a bud, as well as those with three leaves and a bud, were greater in PVtea than in the control, while the number of shoots with one leaf and a bud was lower. The first plucking of Longjing tea began on 28 March 2018 and 23 March 2019 for PVtea, which was two days earlier than in the control group in both years. These findings suggest that the implementation of PVtea effectively promoted the sprouting of tea shoots during the spring season.

Table 1. Comparison of sprouting indexes measured on Fuyang Tea Farm (mean ± SD, n = 3) ①.

Date	Treatment	Number of Shoots with One Leaf and a Bud	Number of Shoots with Two Leaves and a Bud	Number of Shoots with Three Leaves and a Bud	Sprouting Index
3 April 2018	PVtea	18.00 ± 7.57	123.00 ± 14.22	60.00 ± 13.58 *	3.21 ± 0.11 *
	Control	17.00 ± 7.00	120.00 ± 20.81	42.00 ± 10.97	3.14 ± 0.10
26 March 2019	PVtea	49.33 ± 15.63	18.67 ± 14.22 *	1.00 ± 1.73 *	2.22 ± 0.13 *
	Control	56.00 ± 15.87	12.33 ± 8.08	0	2.01 ± 0.22

①: SD, standard deviation. *: significantly different at $p < 0.05$ compared with the control.

3.4. Effects of PV Modules on Fresh Leaf Yield

Fresh leaf yield tests demonstrated that PVtea exhibited significantly higher tea yields compared to the control group (Figure 7). In 2023, the total fresh leaf yield was 1199.2 ± 21.0 g/m^2 for PVtea and 1040.3 ± 15.2 g/m^2 for the control, indicating a 15.3% increase in yield for PVtea. The Qingming tea, which was harvested before the Qingming Festival (BQF, 5 April 2023; 4 April 2024), was of particular significance for the production of Longjing tea, a premium green tea from Hangzhou, China, due to its superior quality

https://doi.org/10.3390/agriculture16010125

and high market value. The BQF leaf yield for PVtea was 82.0 ± 2.3 g/m^2, which was significantly higher than that of the control. The fresh leaf yield in 2024 showed the same trend as in 2023. The total fresh leaf yield of PVtea in 2024 was 866.4 ± 30.1 g/m^2, being 9.3% higher than the control (792.4 ± 24.2 g/m^2). These results showed that PVtea was 9.3–15.3% higher than the control, suggesting that the installation of photovoltaic modules not only does not diminish tea yield, but also enhances tea yield by mitigating direct sunlight radiation through moderate shading.

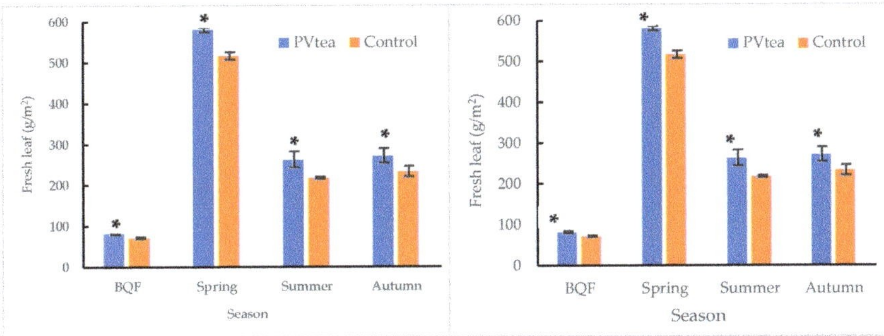

Figure 7. Comparison of fresh leaf yield for different tea seasons in 2023 (**left**) and 2024 (**right**). BQF indicates before the Qingming Festival. Values are means ± SD. *: being significantly different at $p = 0.05$ compared with the control.

Tests on the tea leaf yield of PVteas at Jingshan Town, Zhejiang Province, and at Mumen Town, Wangcang, Sichuan Province, showed the same trends as the above results (Supplementary Tables S1 and S2, and Supplementary Figures S1 and S2).

3.5. Effect of PV Modules on Weight of Shoots and Leaf Size

Tests showed that the weight of 100 shoots with two leaves and a bud in PVtea was 271.7 ± 4.9 g, compared to 261.6 ± 3.94 g for the control, with no statistically significant difference between them (Figure 8). However, the leaf size of the third and fourth leaves beneath the apical bud in PVtea was 22.5 ± 1.4 cm^2 and 31.0 ± 1.3 cm^2, respectively, which were significantly bigger than those of the control group (Figure 9). These results suggest that though the growth rate of PVtea shoots was not significantly different from that of the control group at the initial stage, it markedly exceeded that of the control group at the latter stage, leading to significantly larger size in the third and fourth leaves compared to the control group.

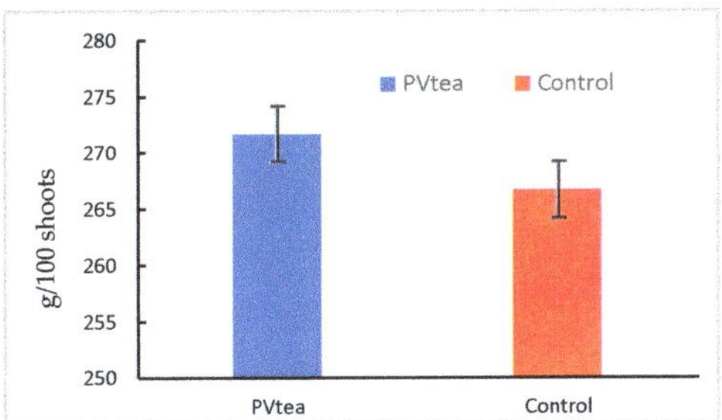

Figure 8. Effect of PV module on shoot fresh weight with two leaves and a bud in 2024. Bar indicates SD.

https://doi.org/10.3390/agriculture16010125

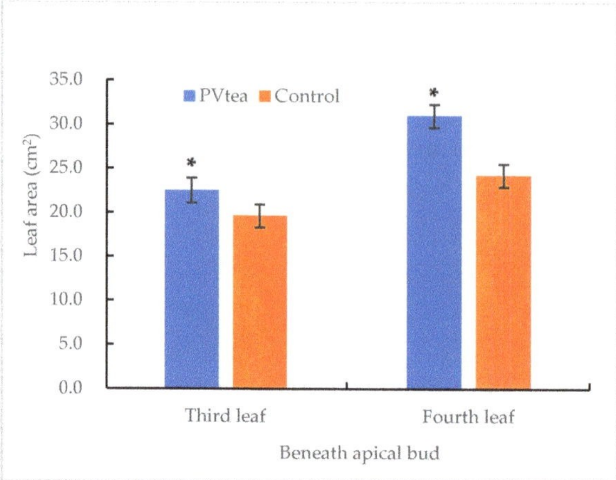

Figure 9. Effect of PV module on the leaf size in 2024. Bar indicates SD. *: being significantly different at $p = 0.05$ compared with the control.

3.6. *Effect of PV Modules on Tea Quality-Related Phytochemicals*

The test of tea quality-related chemical components showed that the contents of caffeine, amino acid, and catechins in PVtea spring tea did not significantly differ from those in the control group (Figure 10). In contrast, the amino acids and total catechin levels in PVtea summer tea were significantly higher than those in the control group (Figure 11). Contents of chlorophylls a and b, as well as lutein in PVtea, were significantly higher than in the control (Figure 12). These findings suggest that PV modules can enhance the quality-related key chemical components in summer tea, resulting in an improvement of tea quality.

Sensory quality of tea samples was assessed by the National Tea Quality Inspection and Testing Center of China, and the test reports for partial samples were added as Supplementary Table S3. They show that the sensory quality of the made teas was improved by the shade of PV modules.

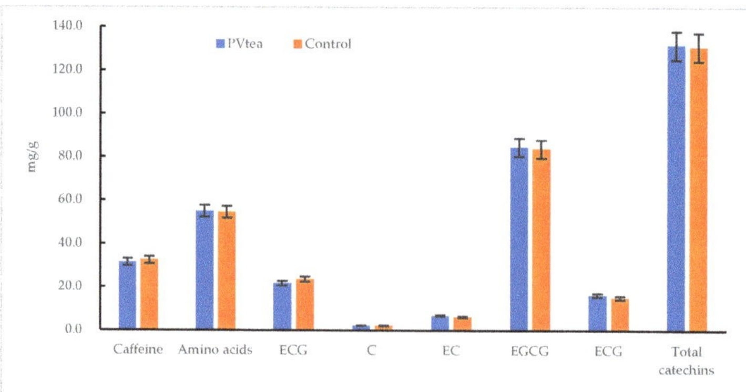

Figure 10. Effect of PV modules on phytochemicals of spring tea leaves in 2023.

https://doi.org/10.3390/agriculture16010125

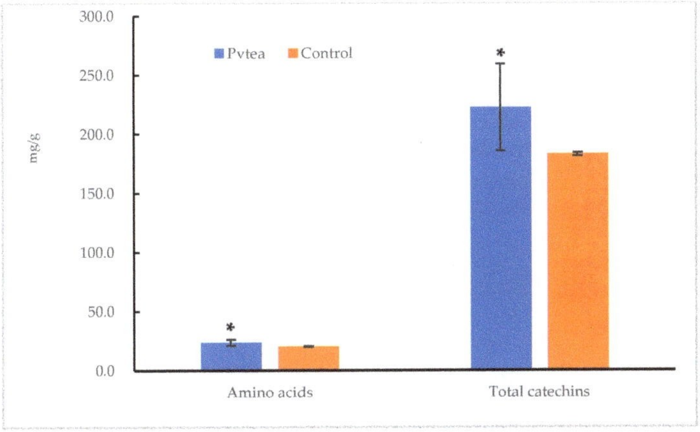

Figure 11. Effect of PV modules on phytochemicals of summer tea leaves in 2023. Bar indicates SD. *: being significantly different at $p = 0.05$ compared with the control.

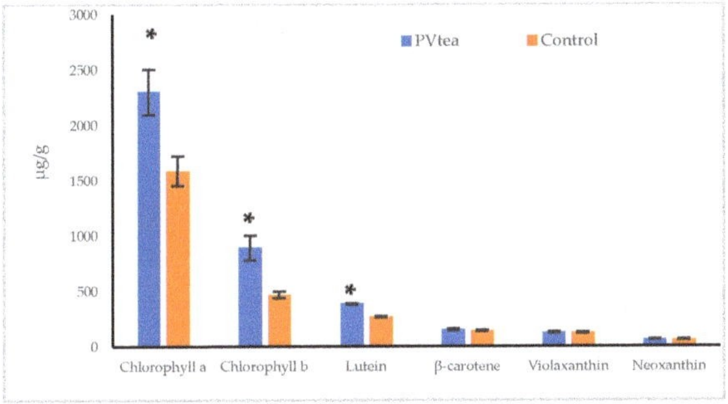

Figure 12. Concentration of pigments in summer tea leaves in 2023. Bar indicates SD. *: being significantly different at $p = 0.05$ compared with the control.

3.7. *Effect of PV Modules on Resistance of Tea Plants to Environmental Stress*

3.7.1. Resistance to Frost Damage in the Early Spring

Frost damage is a common natural disaster in tea-growing regions of the middle and lower reaches of the Yangtze River, occurring approximately every 3 to 5 years. Two spring cold-weather events occurred during the experimental period. One occurred in early April 2018, in which tea plants in the control group were seriously injured by frost, while the tea plants in the PVtea were not subjected to any form of damage or adverse effects (Figure 13). Another cold spell occurred in early April 2020, in which the frost damage of tea plants showed the same trend as that in 2018. These suggest that installing PV modules in tea fields can effectively prevent frost damage to tea plants.

3.7.2. Resistance to Heat Damage in Summer

Heat waves have frequently occurred in test regions. In the summers of 2022 and 2024, the experimental area experienced prolonged periods of high temperatures exceeding 42 °C. During these periods, the surface temperature of the tea-plucking tables reached over 55 °C, causing significant heat damage to the tea plants. In contrast, the tea plants under PV modules grew well, with no adverse effects (Figure 14). These findings indicate that PV modules installed in tea fields can effectively mitigate heat damage to tea plants during extreme summer conditions.

https://doi.org/10.3390/agriculture16010125

Figure 13. PV modules prevent tea plants from frost damage under cold weather conditions.

Figure 14. PV modules prevent tea plants from heat damage under extreme summer conditions.

 https://doi.org/10.3390/agriculture16010125

4. Discussions

4.1. PV Modules Increase Tea Productivity by Mitigating Photodamage to Tea Leaves from Intense Sunlight

Photosynthesis in tea plants is a dynamic process intricately regulated by sunlight intensity, with the light saturation point (LSP) and light compensation point (LCP) serving as critical physiological thresholds. These parameters define the optimal and minimal light conditions required for efficient energy conversion and carbon assimilation, directly impacting tea plant growth and leaf yield, as well as secondary metabolite synthesis, such as catechins and chlorophylls. The LSP denotes the irradiance beyond which increased light no longer enhances the net photosynthetic rate (Pn). Tea plants generally exhibit LSP values of 775–973 $\mu mol/m^2 \cdot s$ under optimal growing conditions [17]. Excessively high light intensities beyond LSP induce photoinhibition, characterized by reduced chlorophyll levels and impaired photosystem II (PSII) functionality, as evidenced by declines in the Fv/Fm ratio under intense illumination [17,18]. For example, light intensities exceeding 550 $\mu mol/m^2 \cdot s$ suppress catechins biosynthesis and downregulate genes associated with photosynthetic electron transport (e.g., *Cs*psbA, *Cs*psbD) [18]. Tea plants exhibit distinct diurnal variations in photosynthesis, particularly under summer sunny conditions. Photosynthetic activity declines at midday due to excessive light intensity and high temperatures, i.e., photosynthetic noon break [19], which arises from multiple stressors: intense solar radiation induces photoinhibition by damaging PSII and reducing D1 protein repair efficiency [14,19]. The LCP represents the irradiance level at which photosynthetic carbon fixation equals respiratory carbon loss, marking the threshold below which Pn becomes negative. The LCP values for tea plants range between 5 and 7 $\mu mol/m^2 \cdot s$ under normal soil moisture conditions [17]. However, environmental stressors like drought elevate LCP, reflecting reduced photosynthetic efficiency under environmental stress [20].

Moderate light intensities (250–350 $\mu mol/m^2 \cdot s$) optimize photosynthetic performance by balancing energy capture and metabolic demand. At this range of light intensities, tea plants exhibit enhanced electron transport rates (ETR), transpiration rates (Tr), and upregulated expression of photosynthetic genes (e.g., *Cs*GLK1), which synergistically boost the synthesis of non-esterified catechins like epigallocatechin (EGC) [18]. Furthermore, shading practices by reducing light to 250 $\mu mol/m^2 \cdot s$ have been shown to mitigate photoinhibition under high-temperature stress by regulating canopy microclimates and preserving chlorophyll integrity [1]. This aligns with findings that shaded tea leaves accumulate higher concentrations of aroma precursors and antioxidants such as gallic acid (GA), through coordinated activation of shikimate pathway genes (*Cs*aroB, *Cs*aroDE) [18,20].

Solar radiation intensity profoundly shapes the photosynthetic dynamics of tea plants through its effects on LSP and LCP. Cultivation strategies must balance light exposure to avoid photoinhibition while ensuring sufficient irradiance to sustain metabolic activity and secondary metabolite production. In agricultural practices, managing light intensity through canopy shading or intercropping is pivotal for optimizing tea yield and quality. For example, black net shading not only prevents photodamage but also promotes the accumulation of theanine and volatile organic compounds (VOCs), which are critical for quality teas. Artificial shading can ameliorate cold-induced depression of photosynthesis in overwintering tea leaves, and the photosynthetic rates of overwintered leaves were higher during and after shading using lawn cloth, resulting in an increased yield in the first flush [21]. Low light conditions further increase the contents of amino acids and caffeine in shoots, but decrease the contents of tea polyphenols and the ratio of tea polyphenols to amino acids, which improves the taste of green tea [22]. The photosynthesis rate of mature leaves on shaded tea bushes exposed to 30–65% PAR was significantly higher ($p < 0.05$) than that of unshaded plants during midday on bright, clear days [19].

https://doi.org/10.3390/agriculture16010125

Figures 3 and 4 in the present study show that the Fv/Fm value in the control decreased substantially at noon on sunny days when PAR was high, but remained relatively stable on cloudy days. The highest levels of PAR in PVtea, however, were less than 600 μmol/m^2·s on both sunny and cloudy days, resulting in a continuously high Fv/Fm value (>0.8). The findings of this study demonstrate that appropriately installing photovoltaic modules above tea plants can effectively alleviate the photoinhibition effect induced by intense sunlight and high temperatures at midday during summer. This intervention also enhances the humidity of the tea field's microclimate (Figures 5 and 6), thereby promoting tea plant photosynthesis and increasing the concentration of key quality-related chemical components such as amino acids, catechins, and chlorophylls (Figures 11 and 12), achieving dual benefits of boosting yield and improving product quality.

4.2. *PVtea Is a Sustainable Model to Combat the Negative Impacts of Extreme Weather on Tea*

4.2.1. Protecting Tea Plants Against Frost Damage

Tea plants are very vulnerable to spring frost damage (SFD), which often occurs in early spring and seriously affects the output and quality of tea in China and Japan [23,24]. Great efforts were made to minimize the SFD-induced loss in tea production, including application of remote sensing, geographic information systems [25], and satellite-based minimum temperature estimation [26,27] for predicting the occurrence of SFD and assessing the SFD-induced economic loss, and to protect tea plants from frost damage by cultivating tea cultivars with sprouting periods that avoid the occurrence of SFD [28], using oscillating frost protective fans or sprinkler irrigation in tea fields [29–32].

The critical temperature threshold (CTT) for SFD to tea plants is 3.0–3.7 °C. Frost damage occurs when the initial budburst of tea shoots coincides with periods in which the daily minimum temperature falls below the CTT [33,34]. SFD impairs the photosynthetic capacity of tea plants by reducing the chlorophyll fluorescence parameter Fv/Fm, which serves as an indicator of PSII activity [32]. Furthermore, Fv/Fm can be used to evaluate the effectiveness of frost protection measures against SFD. Tea leaves protected by sprinkler irrigation exhibited Fv/Fm values ranging from 0.6 to 0.7, which is significantly higher than those observed in leaves subjected to frost damage [32].

The sprinkler irrigation system effectively achieved frost protection for tea crops, with a recommended application rate of rainfall capacity 2–4 mm/h to enhance protective efficacy. To ensure optimal performance, the system should remain operational throughout the entire frost night and continue for an additional half hour after sunrise [35]. The temperature in the irrigated area was approximately 2.8 °C higher than that in the non-irrigated area, and the former exhibited a slower temperature increase after sunrise, which contributed positively to frost protection [36]. When ambient temperatures reached −4.0 °C, the microsprinklers were used to irrigate three designated areas. Each sprinkler operated on a cycle of 3.0 min of spraying followed by 6.0 min of rest, successfully maintaining canopy temperatures at approximately 1.0 °C [37].

Installing PV modules in tea fields can significantly alter the surface energy exchange process by reducing convective and radiative heat loss from the tea-plucking table to the atmosphere. This technology establishes a distinct microclimate within the PVtea field, demonstrating particular effectiveness during early spring nights that are prone to frost. The present study shows that the surface temperature of the plucking table in PVtea was 0.5–2 °C higher than that in the control group. This temperature increase effectively suppresses the occurrence of SFD. One mechanism underlying this effect is the radiation barrier function provided by the PV modules. Their physical coverage reduces long-wave radiation emitted from the plucking table to the cold night sky, thereby decreasing ground emissivity and limiting radiative cooling. A second mechanism involves the suppression

 https://doi.org/10.3390/agriculture16010125

of air convection. The structure of the PV array acts as a local barrier, restricting the downward movement of cold air masses near the plucking table and reducing their mixing with upper layers, which, in turn, minimizes heat loss due to turbulence and convection. Radiation-type frost formation primarily occurs under clear and calm nighttime conditions when the plucking table rapidly loses heat, cooling below the dew point. By reducing net radiative loss and lowering effective long-wave radiation emission to the atmosphere, PV modules weaken convective heat dissipation and slow the cooling rate. As a result, the nighttime temperature of the plucking table in PV-covered areas remains above the CTT, significantly reducing both the likelihood and severity of frost formation and protecting tea shoots from SFD. The PVtea model leverages PV technology to improve the microclimate of tea fields, offering an effective physical strategy for enhancing the safety and sustainability of tea growing.

4.2.2. Protecting Tea Plants Against Heat Damage

Global warming increases the exposure of tea plants to elevated heat conditions during the summer plucking period, and the most daunting issue of global climate change is the deleterious impact of extreme temperatures on tea productivity and quality, which has resulted in a quest among researchers and growers [38,39]. Investigation showed that the average heat damage (HD) ratios were 40–50% of tea-producing areas in China during typical HD years [40]. The critical land surface temperature (LST) threshold triggering HD in tea plants is 30.1 °C [41]. HD is always related to drought damage (DD), and it causes a range of biochemical, physiological, and chemical variations, resulting in membrane damage and loss in the functions of the cell and, finally, a decrease in the tea growth and leaf yield. A significantly greater negative impact of heat stress and drought stress was observed, as indicated by reductions in water retention, chlorophyll content, and oxidation potential, as well as an increase in membrane damage [42], resulting in a decrease in total polyphenol and water extract, which are closely related to the quality of tea [43]. To mitigate the adverse effects of high temperatures and drought on tea yield and quality, researchers and producers have implemented strategies such as cultivating heat- and drought-resistant tea cultivars or applying chemical protectants, including fulvic acid (FA) [44,45]. The present study shows that installing PV modules in tea fields to harness partial sunlight for power generation can significantly lower the temperature on the plucking table during summer days, thereby effectively mitigating HD to tea plants. These findings suggest that PV-integrated tea plantation represents a sustainable cultivation model capable of mitigating the adverse effects of global climate change on both tea productivity and quality.

Furthermore, from 2018 to 2024, the total electricity generation of the two experimental blocks (288 m^2 each) was 121,202 kWh and 119,985 kWh, respectively, equivalent to an average annual generation of 20,098.9 kWh per block or 697,878.5 kWh per hectare. The installed capacity of a pure photovoltaic power station on barren land is generally 1050 kWp per hectare. The PVtea system in this study achieves an electricity output of 656 kWp per hectare, which is 62.5% of a pure photovoltaic power station. Considering the approximately 10% tea leaf yield increase associated with PVtea, the land-use equivalence ratio (LER) of PVtea is approximately 1.7. This indicates that the PVtea system achieves a favorable balance among agricultural productivity, clean electricity generation, and climate resilience.

5. Conclusions

Photovoltaic (PV) power generation represents a sustainable method of clean energy production. However, the extensive land-use associated with PV power stations poses a significant constraint on its large-scale development. Integrating solar power technology

 https://doi.org/10.3390/agriculture16010125

with agricultural practices offers an effective solution to this land-use challenge. Tea plants are known to be shade-tolerant, and high-quality tea cultivation often requires artificial shading conditions. In this study, PV modules were installed above tea bushes at a capacity of 656 kilowatts per hectare, resulting in an average annual generation of 697,878.5 kWh per hectare. The results indicate that the annual average PAR received by the tea plants in the PVtea garden was 52.9% of that in the control (open-field tea garden). Specifically, on sunny days, the PAR was reduced to 32.0% of the control level, while on cloudy days, it remained at 49.0%. Additionally, the daytime relative humidity within PVtea was significantly increased. These changes in the microclimate of the PVtea can effectively alleviate photoinhibition caused by intense midday sunlight, maintain the high activity of leaf photosystem II (PSII), promote photosynthesis in tea plants, and ultimately contribute to a 9.3–15.3% increase in tea leaf yield. The microclimate changes in the PVtea improve the synthesis and accumulation of secondary metabolites in tea plants, with higher accumulation of amino acids in tea leaves during periods of high light and increased temperature in summer, which is beneficial for improving the quality of green tea. Additionally, the tea plants' resistance to frost damage in the spring and heat stress during the summer was also improved. It is concluded that PVtea integrates PV power generation with tea-growing practices, which not only facilitates clean energy production but also improves tea production, resulting in an increase in land-use equivalence ratio (LER) by 70%. The limitation of the study was that it was not applied to other crops.

Supplementary Materials: The following supporting information can be downloaded at: https://www.mdpi.com/article/10.3390/agriculture16010125/s1, Figure S1: PVtea at Jingshan Town, Zhejiang, China; Figure S2: PVtea at Mumen Town, Sichuan, China; Table S1: Fresh tea leaf yield in PVtea at Jingshan Town; Table S2: Fresh tea leaf yield in PVtea at Mumen Town; Table S3: Sensory assessment score of made teas.

Author Contributions: X.-Q.Z.: Conceptualization, funding acquisition, data curation, and writing—original draft; X.-H.Z.: Data curation and writing—original draft; J.-G.Z.: Funding acquisition, resources and conceptualization; R.-J.Z.: Conceptualization and resources; J.-L.L.: Methodology and validation; J.-H.Y.: Data curation; Y.-R.L.: Conceptualization, project administration, and writing—review and editing. All authors have read and agreed to the published version of the manuscript.

Funding: This study was financially supported by the Ministry of Agriculture and Rural Affairs of the People's Republic of China through the earmarked fund for CARS-Tea (CARS-19) and the 2022 Hangzhou Key Science and Technology Research and Development Program in the field of carbon peak and carbon neutrality (202204T0).

Data Availability Statement: The original contributions presented in this study are included in the article and Supplementary Materials. Further inquiries can be directed to the corresponding author.

Conflicts of Interest: Author Jian-Gao Zhang was employed by the company Zhejiang Zhongxin Energy Development Co., Ltd. The remaining authors declare that the research was conducted in the absence of any commercial or financial relationships that could be construed as a potential conflict of interest.

References

1. Li, L.; Tan, Z.F. The potential of renewable energy for the alleviation of climate change in China. *Appl. Mech. Mater.* **2013**, *278–280*, 2308–2312. [CrossRef]
2. Heremans, G.; Trompoukis, C.; Daems, N.; Bosserez, T.; Vankelecom, I.F.J.; Martens, J.A.; Rongé, J. Vapor-fed solar hydrogen production exceeding 15% efficiency using earth abundant catalysts and anion exchange membrane. *Sustain. Energ. Fuels* **2017**, *1*, 2061–2065. [CrossRef]
3. Trommsdorff, M.; Kang, J.; Reise, C.; Schindele, S.; Bopp, G.; Ehmann, A.; Weselek, A.; Högy, P.; Obergfell, T. Combining food and energy production: Design of an agrivoltaic system applied in arable and vegetable farming in Germany. *Renew. Sust. Energ. Rev.* **2021**, *140*, 110694. [CrossRef]

4. Zeddies, H.H.; Parlasca, M.; Qaim, M. Agrivoltaics increases public acceptance of solar energy production on agricultural land. *Land Use Policy* **2025**, *156*, 107604. [CrossRef]
5. Zhang, S.; Gong, J.; Xiao, C.; Yang, X.; Li, X.; Zhang, Z.; Song, L.; Zhang, W.; Dong, X.; Hu, Y. Bupleurum chinense and Medicago sativa sustain their growth in agrophotovoltaic systems by regulating photosynthetic mechanisms. *Renew. Sust. Energ. Rev.* **2024**, *189*, 114024. [CrossRef]
6. Giri, N.C.; Mohanty, R.C. Turmeric crop farming potential under Agrivoltaic system over open field practice in Odisha, India. *Environ. Dev. Sustain.* **2024**, 1–19. [CrossRef]
7. Hu, B.; Zhou, P.; Zhang, L.P. A digital business model for accelerating distributed renewable energy expansion in rural China. *Appl. Energ.* **2022**, *316*, 119084. [CrossRef]
8. Zhang, Q.; Bi, G.; Li, T.; Wang, Q.; Xing, Z.; LeCompte, J.; Harkess, R.L. Color shade nets affect plant growth and seasonal leaf quality of *Camellia sinensis* grown in Mississippi, the United States. *Front. Nutr.* **2022**, *9*, 786421. [CrossRef]
9. Mukherjee, A.; Banerjee, S.; Sarkar, S. Productivity and radiation use efficiency of tea grown under different shade trees in the plain land of West Bengal. *J. Agrometeorol.* **2018**, *10*, 146–150. [CrossRef]
10. Krahe, J.; Krahe, M.A. Optimizing the quality and commercial value of gyokuro-styled green tea grown in Australia. *Beverages* **2022**, *8*, 22. [CrossRef]
11. Mei, Y.; Liang, X. Annual Report on China's Tea Production and Sales in 2023, China Tea Marketing Association, Updated to 16 April 2024. Available online: https://mp.weixin.qq.com/s/JcL0ww7dHZKaGexvKXoiaA (accessed on 30 December 2025).
12. Jiang, C.J. *Tea Breeding*, 2nd ed.; China Agriculture Press: Beijing, China, 2011; pp. 93–95.
13. Cai, W.; Ma, J.; Guo, J.; Zhang, L. Function of ROC4 in the efficient repair of photodamaged photosystem II in Arabidopsis. *Photochem. Photobiol.* **2008**, *84*, 1343–1348. [CrossRef]
14. Cai, W.-H.; Zheng, X.-Q.; Liang, Y.-R. High-light-induced degradation of photosystem II subunits' involvement in the albino phenotype in tea plants. *Int. J. Mol. Sci.* **2022**, *23*, 8522. [CrossRef]
15. Liang, Y.R.; Wu, Y.; Lu, J.L.; Zhang, L.Y. Application of chemical composition and infusion colour difference analysis to quality estimation of jasmine-scented tea. *Int. J. Food Sci. Technol.* **2007**, *42*, 459–468. [CrossRef]
16. Zheng, X.-Q.; Dong, S.-L.; Li, Z.-Y.; Lu, J.-L.; Ye, J.-H.; Tao, S.-K.; Hu, Y.-P.; Liang, Y.-R. Variation of major chemical composition in seed-propagated population of wild cocoa tea plant *Camellia ptilophylla* Chang. *Foods* **2023**, *12*, 123. [CrossRef] [PubMed]
17. Guo, C.F.; Sun, Y.; Zhang, M.Q. Effect of soil water stress on photosynthetic light response curve of tea plant (*Camellia sinensis*). *Chin. J. Eco-Agric.* **2008**, *16*, 1413–1418. [CrossRef]
18. Xiang, P.; Zhu, Q.; Tukhvatshin, M.; Cheng, B.; Tan, M.; Liu, J.; Wang, X.; Huang, J.; Gao, S.; Lin, D.; et al. Light control of catechin accumulation is mediated by photosynthetic capacity in tea plant (*Camellia sinensis*). *BMC Plant Biol.* **2021**, *21*, 478. [CrossRef] [PubMed]
19. Mohotti, A.J.; Lawlor, D.W. Diurnal variation of photosynthesis and photoinhibition in tea: Effects of irradiance and nitrogen supply during growth in the field. *J. Exp. Bot.* **2002**, *53*, 313–322. [CrossRef] [PubMed]
20. Xiang, P.; Marat, T.; Huang, J.; Cheng, B.; Liu, J.; Wang, X.; Wu, L.; Tan, M.; Zhu, Q.; Lin, J. Response of photosynthetic capacity to ecological factors and its relationship with EGCG biosynthesis of tea plant (*Camellia sinensis*). *BMC Plant Biol.* **2025**, *25*, 199. [CrossRef]
21. Aoki, S. Interaction of light and low temperature in depression of photosynthesis in tea leaves. *Jpn. J. Crop. Sci.* **1986**, *55*, 496–503. [CrossRef]
22. Shen, R.H.; Ma, L.F.; Yang, X.D.; Fang, L. Effects of nitrogen form and weak light stress on tea plant growth and metabolism. *J. Tea Sci.* **2023**, *43*, 349–355. Available online: https://www.tea-science.com/EN/10.13305/j.cnki.jts.2023.03.01 (accessed on 30 December 2025). (In Chinese)
23. Kotikot, S.M.; Flores, A.; Griffin, R.E.; Nyaga, J.; Case, J.L.; Mugo, R.; Sedah, A.; Adams, E.; Limaye, A.; Irwin, D.E. Statistical characterization of frost zones: Case of tea freeze damage in the Kenyan highlands. *Int. J. Appl. Earth Obs. Geoinf.* **2020**, *84*, 101971. [CrossRef]
24. Lou, W.; Zhao, Y.; Huang, X.; Zhu, T.; Yang, M.; Deng, S.; Zhou, Z.; Zhang, Y.; Sun, Q.; Chen, S. Frost risk assessment based on the frost-induced injury rate of tea buds: A case study of the Yuezhou Longjing tea production area, China. *Eur. J. Agron.* **2023**, *147*, 126839. [CrossRef]
25. Lou, W.; Ji, Z.; Sun, K.; Zhou, J. Application of remote sensing and GIS for assessing economic loss caused by frost damage to tea plantations. *Precis. Agric.* **2013**, *14*, 606–620. [CrossRef]
26. Wang, P.; Tang, J.; Ma, Y.; Wu, D.; Yang, J.; Jin, Z.; Huo, Z. Mapping threats of spring frost damage to tea plants using satellite-based minimum temperature estimation in China. *Remote Sens.* **2021**, *13*, 2713. [CrossRef]
27. Wang, P.; Ma, Y.; Tang, J.; Wu, D.; Chen, H.; Jin, Z.; Huo, Z. Spring frost damage to tea plants can be identified with daily minimum air temperatures estimated by MODIS land surface temperature products. *Remote Sens.* **2021**, *13*, 1177. [CrossRef]
28. Omae, H.; Takeda, Y. Selection of early lines without frost damage based on parameters of accumulated temperature for flushing and opening speed of new leaves in tea. *JARQ—Jpn. Agric. Res. Q.* **2003**, *37*, 183–188. [CrossRef]

29. Kimura, K.; Yasutake, D.; Nakazono, K.; Kitano, M. Dynamic distribution of thermal effects of an oscillating frost protective fan in a tea field. *Biosyst. Eng.* **2017**, *164*, 98–109. [CrossRef]
30. Ni, Y.; Pan, Z.; Huang, Z.; Liu, W. Vortex characteristics downstream a fan and its effect on the heat transfer of the temperature inversion under late spring coldness. *Adv. Mech. Eng.* **2017**, *9*, 1687814017704353. [CrossRef]
31. Xu, T.; Pan, Q.; Lu, Y. Heat transfer process of the tea plant under the action of air disturbance frost protection. *Agronomy* **2024**, *14*, 959. [CrossRef]
32. Pan, Q.; Lu, Y.; Hu, Y. Influences of sprinkler frost protection on air and soil temperature and chlorophyll fluorescence of tea plants in tea gardens. *Agriculture* **2024**, *14*, 2302. [CrossRef]
33. Kimura, K.; Kudo, K.; Maruyama, A. Spatiotemporal distribution of the potential risk of frost damage in tea fields from 1981-2020: A modeling approach considering phenology and meteorology. *J. Agric. Meteorol.* **2021**, *77*, 224–234. [CrossRef]
34. Tang, J.; Wang, P.; Li, X.; Yang, J.; Wu, D.; Ma, Y.; Li, S.; Jin, Z.; Huo, Z. Disaster event-based spring frost damage identification indicator for tea plants and its applications over the region north of the Yangtze River, China. *Ecol. Indic.* **2023**, *146*, 109912. [CrossRef]
35. Hu, Y.; Zhao, C.; Liu, P.F.; Amoah, A.E.; Li, P. Sprinkler irrigation system for tea frost protection and the application effect. *Int. J. Agric. Biol. Eng.* **2016**, *9*, 17–23. [CrossRef]
36. Lu, Y.; Hu, Y.; Zhao, C.; Snyder, R.L. Modification of water application rates and intermittent control for sprinkler frost protection. *Trans. ASABE* **2018**, *61*, 1277–1285. [CrossRef]
37. Pan, Q.; Hu, Y. Reasons of sprinkler freezing and rotational irrigation for frost protection in tea plantations. *Irrig. Drain.* **2025**, *74*, 139–147. [CrossRef]
38. Chen, S.; Gao, Y.; Fan, K.; Shi, Y.; Luo, D.; Shen, J.; Ding, Z.; Wang, Y. Prediction of drought-induced components and evaluation of drought damage of tea plants based on hyperspectral imaging. *Front. Plant Sci.* **2021**, *12*, 695102. [CrossRef]
39. Wang, H.; Hou, J.; Zhou, B.; Han, X. Effects of water supply mode on nitrogen transformation and ammonia oxidation microorganisms in a tea garden. *Agronomy* **2023**, *13*, 1279. [CrossRef]
40. Wang, P.; Li, X.; Tang, J.; Wu, D.; Pang, L.; Zhang, Y. Critical threshold-based heat damage evolution monitoring to tea plants with remotely sensed LST over mainland China. *Remote Sens.* **2024**, *16*, 1784. [CrossRef]
41. Wang, P.; Li, X.; Tang, J.; Yang, J.; Ma, Y.; Wu, D.; Huo, Z. Determining the critical threshold of meteorological heat damage to tea plants based on MODIS LST products for tea planting areas in China. *Ecol. Inform.* **2023**, *77*, 102235. [CrossRef]
42. Seth, R.; Maritim, T.K.; Parmar, R.; Sharma, R.K. Underpinning the molecular programming attributing heat stress associated thermotolerance in tea (*Camellia sinensis* (L.) O. Kuntze). *Hortic. Res.* **2021**, *8*, 99. [CrossRef]
43. Chaeikar, S.S.; Marzvan, S.; Khiavi, S.J.; Rahimi, M. Changes in growth, biochemical, and chemical characteristics and alteration of the antioxidant defense system in the leaves of tea clones (*Camellia sinensis* L.) under drought stress. *Sci. Hortic.* **2020**, *265*, 109257. [CrossRef]
44. Nalina, M.; Saroja, S.; Chakravarthi, M.; Rajkumar, R.; Radhakrishnan, B.; Chandrashekara, K.N. Water deficit-induced oxidative stress and differential response in antioxidant enzymes of tolerant and susceptible tea cultivars under field condition. *Acta Physiol. Plant* **2021**, *43*, 10. [CrossRef]
45. Sun, J.; Qiu, C.; Ding, Y.; Wang, Y.; Sun, L.; Fan, K.; Gai, Z.; Dong, G.; Wang, J.; Li, X.; et al. Fulvic acid ameliorates drought stress-induced damage in tea plants by regulating the ascorbate metabolism and flavonoids biosynthesis. *BMC Genom.* **2020**, *21*, 411. [CrossRef]

 https://doi.org/10.3390/agriculture16010125

agriculture

MDPI

Article

Effect of Nitrogen Sources on the Phenological Phases of Italian Zucchini Under Salt Stress

Gleydson de Freitas Silva [1], Tayd Dayvison Custódio Peixoto [2], Miguel Ferreira Neto [1], Antônio Gustavo de Luna Souto [1], Ricardo André Rodrigues Filho [1], Kariolania Fortunato de Paiva Araújo [1], Jussiara Sonally Jácome Cavalcante [1], Kleane Targino Oliveira Pereira [1], Rômulo Carantino Lucena Moreira [1], Pedro Dantas Fernandes [3], Nildo da Silva Dias [1], Josinaldo Lopes Araújo Rocha [3], Alberto Soares de Melo [4], Alex Álvares da Silva [5] and Francisco Vanies da Silva Sá [5,*]

[1] Department of Agronomic and Forest Sciences, Universidade Federal Rural do Semi-Árido, Mossoró 59625-900, RN, Brazil; freitas.gleydson@gmail.com (G.d.F.S.); miguel@ufersa.edu.br (M.F.N.); gustavo.luna@ufersa.edu.br (A.G.d.L.S.); ricardo.filho45389@alunos.ufersa.edu.br (R.A.R.F.); kariolania.araujo@alunos.ufersa.edu.br (K.F.d.P.A.); jussiara_sonally@hotmail.com (J.S.J.C.); kleane_rn@hotmail.com (K.T.O.P.); romulocarantino@gmail.com (R.C.L.M.); nildo@ufersa.edu.br (N.d.S.D.)

[2] Center of Agrarian and Biological Sciences, Universidade Estadual Vale do Acaraú, São Benedito 62370-000, CE, Brazil; tayd_custodio@uvanet.br

[3] Academic Unit of Agricultural Engineering, Universidade Federal de Campina Grande, Campina Grande 58429-900, PB, Brazil; pedrodantasfernandes@gmail.com (P.D.F.); josinaldo.lopes@professor.ufcg.edu.br (J.L.A.R.)

[4] Department of Biological Sciences, Universidade Estadual da Paraíba, Campina Grande 58429-900, PB, Brazil; alberto.melo@servidor.uepb.edu.br

[5] Department of Agrarian and Exact, Universidade Estadual da Paraíba, Catolé do Rocha 58884-000, PB, Brazil; alex.alvares@visitante.uepb.edu.br

* Correspondence: vanies_agronomia@hotmail.com; Tel.: +55-84-99651-3164

Abstract

Salt stress is one of the most significant abiotic factors limiting plant growth and crop productivity worldwide, especially in arid and semiarid regions. We aimed to investigate nitrogen fertilization strategies using nitrate and ammoniacal sources during different phenological phases of Italian zucchini cv. Caserta to alleviate salt stress. The experiment was conducted in a greenhouse using a randomized block design with four replications. The treatments were as follows: T1 = entire cycle with nitrate nitrogen + 0.50 dS m^{-1} (control); T2 = entire cycle with nitrate nitrogen + 4.5 dS m^{-1} (salt stress); T3 = 50% nitrate nitrogen + 50% ammoniacal nitrogen + 4.50 dS m^{-1}; T4 = ammoniacal nitrogen during the vegetative phase + nitrate nitrogen during the reproductive phase + 4.50 dS m^{-1}; T5 = nitrate nitrogen during the vegetative phase + ammoniacal nitrogen during the reproductive phase + 4.50 dS m^{-1}; T6 = entire cycle with ammoniacal nitrogen + 4.50 dS m^{-1}. Under salt stress conditions, Italian zucchini cv. Caserta showed a leaf area of 5783 cm^2 compared to an average of 4521 cm^2 under salt stress. Similarly, production per plant reached 1361 g in the control, while under salt stress it averaged only 442 g. However, under salt stress, T2 resulted in higher production compared with T3, T4, T5 and T6, although it was still lower than T1. The use of ammoniacal nitrogen throughout the cycle or during the reproductive phase caused flower abortion. Under salt stress, the application of ammoniacal nitrogen during the vegetative phase (T4) or a 1:1 ammonium–nitrate ratio throughout the cycle (T3) resulted in yields that were comparable to those achieved with nitrate-only fertilization (T2).

Keywords: *Cucurbita pepo* L.; salinity; leaf gas exchange; chlorophyll fluorescence; growth; production

1. Introduction

Zucchini (*Cucurbita pepo* L.) is a Cucurbitaceae of enormous importance for agricultural production in Brazil, with a stronger presence in family farming systems [1]. This crop is also one of the most economically and nutritionally valuable vegetables, being rich in calcium, phosphorus, iron, and fiber [2].

According to Guerra et al. [3], the cultivation of Italian zucchini remains more prominent in Brazil's Central-Southern region, while the semiarid Northeast stands out as a major hub for the cultivation of pumpkins (*Cucurbita moschata*) and squash (*Cucurbita maxima*). This scenario highlights the need for further research into the responses of Italian zucchini to salinity and water stress, as well as the growth and development conditions in these regions, considering both water quality and mineral nutrition for plants.

In 2018, Brazil's zucchini production was estimated at 228,500 tons, with approximately 75% of this total concentrated in the Southeast region [4]. Guerra et al. [3] observed a lack of studies on zucchini cultivation in the Northeast, either in open fields or protected environments. The Northeast is characterized by high temperatures and low annual rainfall, prompting farmers to seek alternative water sources for vegetable production. However, these water sources often have inferior quality due to high levels of soluble salts [2].

Salt stress is one of the leading abiotic factors limiting plant growth and crop yields globally, particularly in arid and semiarid regions [5,6]. In such areas, the natural process of soil salinization, driven by prevailing environmental conditions, is further exacerbated by the need for irrigation, often carried out with water containing high concentrations of soluble salts [7,8].

The use of saline water in agricultural production in arid and semiarid regions is primarily due to challenges in capturing and storing freshwater for irrigation. As a result, saline water from shallow wells, which is more cost-effective to extract than water from deep wells, is commonly used. However, the electrical conductivity of these water sources exceeds 4.0 dS m^{-1} [9]. For Italian zucchini, Sousa et al. [10] determined that the tolerance threshold for irrigation water salinity is 3.1 dS m^{-1}.

Studies by Munns & Tester [11], Syvertsen & Garcia-Sanchez [12], and Sá et al. [13] have shown that the excessive accumulation of salts, particularly sodium chloride (NaCl), in plants can hinder their growth and yield due to ionic toxicity and nutrient imbalances. To address these issues, various studies have been conducted to develop management strategies that enhance plant performance in saline environments, especially strategies aimed at improving nutrient absorption and assimilation [14]. Among these strategies, nitrogen nutrition using specific nitrogen (N) forms has proven promising for acclimating plants to salinity. The interaction between salinity and nitrogen nutrition has been shown to improve plant physiology, growth, and yield under salt stress conditions [15].

With regard to nitrogen, Bazzo [16] emphasizes that this nutrient is positively correlated with high crop yields, primarily because nitrogen (N) is one of the key elements directly involved in plant growth and development. The inorganic nitrogen sources most absorbed by plants are ammonium (NH_4^+) and nitrate ions (NO_3^-), with the latter being considered the primary source of mineral nitrogen for plant growth [17]. According to Cordeiro, Rodrigues, and Echer [18], urea is more easily commercialized and widely used among farmers. However, it has a high potential for reduced efficiency due to its rapid

 https://doi.org/10.3390/agriculture16010062

volatilization and leaching, leading to economic losses for producers and less efficient fertilizer use by plants.

The effects of nitrogen fertilization on the different phenological stages of Italian zucchini are still not well understood, and there is limited research on how the nitrogen sources and their proportions affect crop performance throughout the growth cycle. Since nitrogen is crucial for plant growth and yield, managing its application may help reduce the negative impacts of salinity. We hypothesized that using different nitrogen sources and proportions at specific phenological stages can alleviate salt stress and improve physiological responses, growth, and yield in Italian zucchini. In this context, we assessed fluorescence, chlorophyll content, gas exchange, growth, and yield of Italian zucchini cv. Caserta under conditions of salt stress while employing different nitrogen fertilization strategies.

2. Materials and Methods

2.1. Location and Experimental Design

The experiment was conducted in a greenhouse covered with transparent material to minimize interference with light and solar radiation, with its sides protected with 50%, at the Agricultural Sciences Center of the Federal Rural University of the Semi-Arid Region (UFERSA) in Mossoró-RN, Brazil, located at 5°11′ S and 37°20′ W, at an altitude of 18 m, from november to december 2022. The Mossoró region, where the study was carried out, has a BSh climate type, classified as a hot semi-arid climate according to Köppen's classification system [19]. During the experimental period, the maximum and minimum temperatures recorded were 44.2 and 20.4 °C, with maximum and minimum humidity of 86 and 22%, respectively.

The research was conducted in a randomized block design with six treatments: T1 = complete cycle with nitric nitrogen (NO_3^-) under low salinity (0.50 dS m^{-1}; control); T2 = complete cycle with NO_3^- under saline stress (4.50 dS m^{-1}); T3 = complete cycle using a 50% NO_3^- and 50% ammoniacal nitrogen (NH_4^+) ratio under saline stress; T4 = NH_4^+ in the vegetative phase and nitric nitrogen in the reproductive phase under saline stress; T5 = nitric nitrogen in the vegetative phase and ammoniacal nitrogen in the reproductive phase under saline stress; T6 = complete cycle with NH_4^+ under saline stress, with four replications.

2.2. Plant Material and Soil Characteristics

The sowing of Italian zucchini cv. Caserta was carried out in 12-dm^3 plastic pots filled with soil (Figure 1). At the bottom of the pots, a 0.02 m layer of gravel and a piece of geotextile fabric were added to prevent soil loss. Initially, the soil was irrigated with water according to the treatments, bringing it close to its maximum water-holding capacity. Subsequently, four seeds were sown per pot at a depth of 0.03 m. Approximately 15 days after sowing, when the seedlings had developed two true leaves, thinning was performed, leaving one plant per pot. Each experimental unit consisted of two pots, each containing one plant, totaling two plants per replication and 48 plants in the entire experiment (Figure 1).

The soil used was an Oxisol (Latossolo), collected from the 0.00–0.30 m layer at the Rafael Fernandes Experimental Farm of UFERSA, located in the Alagoinha district, Mossoró, Brazil. Soil samples were collected, crumbled, sieved (2 mm), and characterized for physical and chemical attributes related to fertility, following the methodology of EMBRAPA [20], as shown in Table 1.

https://doi.org/10.3390/agriculture16010062

Figure 1. Italian zucchini cv. Caserta conducted in a plant nursery.

Table 1. Chemical and physical analysis of the soil used in the experiment.

pH	**OM (%)**	**P**	**K^+**	**Na^+**	**Ca^{2+}**	**Mg^{2+}**	**Al^{3+}**	**H + Al**	**SB**	**CEC**	**V**	**ESP**
		---------- **($cmol_c$ dm^{-3})** ----------									--- % ---	
5.3	1.7	2.1	0.13	0.05	2.7	0.9	0.0	1.8	3.78	5.58	68	2.0
ECse $dS\ m^{-1}$	**ds $kg\ dm^{-3}$**			**Sand**				**Silt**		**Clay**		
		---------- **($g\ kg^{-1}$)** ----------										
0.1	1.6			820				30		150		

OM—organic matter; SB—sum of bases; V—base saturation; ESP—Exchangeable Sodium Percentage; ECse—electrical conductivity of the soil saturation extract; Ds—soil density; CEC—cation exchange capacity.

After the physical and chemical characterization, soil acidity was corrected using calcium hydroxide ($Ca(OH)_2$), containing 54% calcium, to raise the base saturation (V%) to 90%. Fertilization was performed by applying 300 mg of phosphorus pentoxide (P_2O_5), 150 mg of potassium oxide (K_2O), and 100 mg of nitrogen (N) per dm^3 of soil [21] divided into six applications throughout the cycle. The phosphorus dose was applied at planting in the form of monosodium phosphate (NaH_2PO_4). Nitrogen (N) and potassium (K) were applied via fertigation in six fractionated doses: three during the vegetative phase and three during the reproductive phase, which began with the appearance of the first flower buds, at 30 DAS (days after sowing). Applications were carried out at 10, 17, and 24 DAS (vegetative applications) and 31, 38, and 45 DAS (reproductive applications) (Table 2).

https://doi.org/10.3390/agriculture16010062

Table 2. Amount and sources of fertilizers used for each development phase of the Italian zucchini cv. Caserta.

Treatments	Dose per Plant	
	Vegetative Phase	Reproductive Phase
T1	2.01 g KNO_3 + 2.31 g $CaNO_3$	2.01 g KNO_3 + 2.31 g $CaNO_3$
T2	2.01 g KNO_3 + 2.31 g $CaNO_3$	2.01 g KNO_3 + 2.31 g $CaNO_3$
T3	2.01 g KNO_3 + 0.66 g Urea + 0.39 g $CaNO_3$	2.01 g KNO_3 + 0.66 g Urea + 0.39 g $CaNO_3$
T4	1.32 g Urea + 1.5 g KCl	2.01 g KNO_3 + 2.31 g $CaNO_3$
T5	2.01 g KNO_3 + 2.31 g $CaNO_3$	1.32 g Urea + 1.5 g KCl
T6	1.32 g Urea + 1.5 g KCl	1.32 g Urea + 1.5 g KCl

T1 = Nitrate fertilization throughout the cycle + 0.50 dS m^{-1} (Control); T2 = Nitrate fertilization throughout the cycle + 4.50 dS m^{-1} (ECw of 4.50 dS m^{-1}, salt stress); T3 = 50% nitrate fertilization + 50% ammoniacal fertilization throughout the cycle + 4.50 dS m^{-1}; T4 = Ammoniacal fertilization during the vegetative phase + nitrate fertilization during the reproductive phase + 4.50 dS m^{-1}; T5 = Nitrate fertilization during the vegetative phase + ammoniacal fertilization during the reproductive phase + 4.50 dS m^{-1}; T6 = Ammoniacal fertilization throughout the cycle + 4.50 dS m^{-1}.

2.3. Fertilization and Irrigation Management

Urea was used as the ammoniacal N source for nitrogen fertilization, whereas calcium nitrate (14% N and 16% Ca) and potassium nitrate (13% N and 44% K_2O) were used as nitrate N sources. Potassium nitrate and potassium chloride (60% K_2O) served as potassium sources. Calcium chloride dihydrate was used to equalize the calcium dose applied via calcium nitrate. In treatment T6, a nitrification inhibitor (Dicyandiamide—DCD) at 10% of the ammoniacal N dose was added to the solution to prevent the nitrification of NH_4^+. Micronutrient fertilization was carried out via foliar application with the fertilizer Liqui-Plex Fruit® (5% N, 1% Ca, 5.3% S, 1% B, 0.05% Cu, 5% Mn, 0.1% Mo, and 5% Zn). This fertilizer was applied once during the vegetative phase and once during the reproductive phase in all treatments, at a rate of 3 mL per liter of solution, following the manufacturer's recommendations.

The plants were irrigated with low-salinity water (electrical conductivity—ECw = 0.50 dS m^{-1}) from the local water supply and moderately saline water (ECw = 4.5 dS m^{-1}) obtained by mixing saline reject and local supply water, to prepare 100 L of solution at 4.5 dS m^{-1}, we used 44.7 L of saline reject water + 54.3 L of local water supply. The saline reject (ECw = 9.50 dS m^{-1}) was collected from a desalination system treating artesian well water at the Jurema settlement, located in the rural area of Mossoró, Brazil (Table 3).

Table 3. Physicochemical characterization of the water sources used in the experiment.

Water Sources	Parameters									
	pH	EC	K^+	Na^+	Mg^{2+}	Ca^{2+}	Cl^-	CO_3^{2-}	HCO_3^-	SAR
		dS m^{-1}				mmol$_c$ L^{-1}				(mmol$_c$ L^{-1})$^{0.5}$
1	7.57	0.50	0.31	3.79	1.20	0.83	2.40	0.60	3.20	3.76
2	7.10	9.50	0.83	54.13	24.20	37.80	116.00	0.00	3.40	9.70

1—local supply water; 2—saline waste; EC—electrical conductivity; SAR—sodium adsorption rate.

Irrigation was carried out starting from sowing, twice a day—early in the morning and late in the afternoon. The amount of water applied during each irrigation session was calculated as the difference between the volume applied during the day and the amount leached (lysimetry) in a pot designated for this purpose in each treatment (Figure 1). A drip irrigation system was used, equipped with self-compensating drippers with a flow rate of 1.4 L h^{-1}. The irrigation volume was applied according to the treatment throughout the entire crop cycle and is detailed in Table 4.

 https://doi.org/10.3390/agriculture16010062

Table 4. Water and salt amounts applied via irrigation to the Italian zucchini cv. Caserta.

Treatments	Irrigation Volume	Salts Applied via Irrigation
	L per Plant	g per Plant
T1	12.50	4.00
T2	12.68	36.51
T3	12.61	36.32
T4	12.85	37.00
T5	12.73	36.67
T6	12.34	35.55

T1 = Nitrate fertilization throughout the cycle + 0.50 dS m^{-1} (Control); T2 = Nitrate fertilization throughout the cycle + 4.50 dS m^{-1} (ECw of 4.50 dS m^{-1}, salt stress); T3 = 50% nitrate fertilization + 50% ammoniacal fertilization throughout the cycle + 4.50 dS m^{-1}; T4 = Ammoniacal fertilization during the vegetative phase + nitrate fertilization during the reproductive phase + 4.50 dS m^{-1}; T5 = Nitrate fertilization during the vegetative phase + ammoniacal fertilization during the reproductive phase + 4.50 dS m^{-1}; T6 = Ammoniacal fertilization throughout the cycle + 4.50 dS m^{-1}.

2.4. Parameters Evaluated

2.4.1. Leaf Gas Exchange, Chlorophyll Fluorescence, Electrolyte Leakage, and Photosynthetic Pigments

Fifty days after sowing (DAS), the third leaf from the apex was selected for the determination of leaf gas exchange between 7:00 a.m. and 11:00 a.m. This was performed using an infrared gas analyzer (IRGA) ("LCPro+"—ADC Bio Scientific Ltd, Hertfordshire, UK.), operating with a temperature set to 25 °C, an irradiation of 1200 μmol photons m^{-2} s^{-1}, and an airflow rate of 200 mL min^{-1} under atmospheric CO_2 levels. The following variables were measured: CO_2 assimilation rate (AN), leaf transpiration rate (E), stomatal conductance (gs), internal CO_2 concentration (Ci), and leaf temperature (Tl).

Subsequently, chlorophyll *a* fluorescence was assessed in the same leaves used for gas exchange measurements. The plants were evaluated using a pulse-modulated fluorometer (model OS5p, Opti Science) under the Fv/Fm protocol for dark-adapted conditions. Under these conditions, the following variables were determined: initial fluorescence (Fo), maximum fluorescence (Fm), variable fluorescence (Fv = Fm − Fo), and maximum quantum efficiency of PSII (Fv/Fm). The protocol was conducted after dark-adapting the leaves for 30 min, with readings taken between 7:00 a.m. and 11:00 a.m. using the device's clip to ensure all acceptors were oxidized, allowing the reaction centers to remain open.

Evaluations under light conditions were also conducted using the Yield protocol. An actinic light source with a multi-flash saturating pulse, coupled to a clip for measuring photosynthetically active radiation (PAR-Clip), was used to determine the following variables: initial fluorescence before the saturation pulse (F'), maximum fluorescence after light adaptation (Fm'), electron transport rate (ETR), and quantum efficiency of photosystem II (Y (II)). Additional calculations included minimal fluorescence in illuminated tissues (Fo') [22], photochemical quenching coefficient using the lake model (qL), quantum yield of regulated photochemical quenching (Y (NPQ)), and quantum yield of non-regulated photochemical quenching (Y (NO)) [23].

During the same period, membrane integrity was determined by measuring electrolyte leakage (EE), expressed as a percentage (%). This analysis was performed on selected leaf samples following the methodology described by Singh et al. [24]. From these same leaves, 0.625 cm diameter leaf disks were extracted to determine chloroplastic pigments, measured in μm cm^{-2}: chlorophyll a (*Chl a*), chlorophyll b (*Chl b*), and carotenoids (Carot). Pigment extraction was performed using a dimethyl sulfoxide solution saturated with $CaCO_3$, as described by Wellburn [25].

https://doi.org/10.3390/agriculture16010062

2.4.2. Growth and Production

At 50 DAS, during the reproductive phase, plant growth parameters were evaluated by measuring plant height (PH), stem diameter (SD), and the number of leaves (NL). Plant height, in cm, was measured from the base to the apical bud using a measuring tape. The stem diameter, in mm, was measured with a digital caliper 2 cm above the base. The number of leaves was counted, considering fully expanded leaves with a minimum length of 5 cm.

At 50 DAS, the leaf area (LA) was also determined. To estimate the leaf area, the width of all leaves with a minimum width of 5 cm was measured. The individual leaf area was calculated using Equation (1) suggested by Fialho et al. [26], and these values were summed to determine the total leaf area.

$$AF = 47.3647 + 0.6211L^2 \quad (1)$$

where

AF—leaf area, cm^2.

L—leaf width, cm.

The shoot dry mass (SDM) was measured at 75 days after sowing (DAS). Plants were placed in labeled paper bags and dried in a forced-air circulation oven at 65 °C for 72 h until a constant mass was reached. Afterward, the samples were weighed on a semi-analytical digital scale (precision < 0.001 g).

Crop production was evaluated from 50 to 75 DAS, as the growth cycle of the Italian zucchini typically spans this period. Production was assessed by determining the number of fruits per plant (NFP), obtained by counting the total fruits produced by each plant. The average fruit mass (AFM) was measured using a semi-analytical balance, with values expressed in grams. The longitudinal (LD) and transversal (TD) fruit diameters were measured with a measuring tape, with results expressed in centimeters. The transversal diameter was determined as the average of three measurements per fruit: the base, the middle, and the top. Production per plant (PP) was calculated as the multiplication of the number of fruits per plant by the average fruit mass per plant, with results expressed in grams.

2.5. Statistical Analysis

Significant differences between treatments were determined using analysis of variance (ANOVA) with the F-test at 5% and 1% significance levels. When a significant treatment effect was detected ($p \leq 0.05$), means were compared using Tukey's test at the 5% significance level. All analyses were conducted using SISVAR statistical software, version 5.6 [27].

3. Results

3.1. Leaf Gas Exchange, Electrolyte Leakage, Photosynthetic Pigments, and Chlorophyll Fluorescence

A significant treatment effect was observed for the CO_2 assimilation rate ($p \leq 0.05$) and stomatal conductance ($p \leq 0.05$). However, no significant effects ($p > 0.05$) were found for internal CO_2 concentration, transpiration, leaf temperature, or the maximum quantum efficiency of photosystem II due to saline irrigation and nitrogen fertilization at different phenological stages of the Italian zucchini cultivar 'Caserta' (Table 5).

https://doi.org/10.3390/agriculture16010062

Table 5. Summary of the analysis of variance (F-values) and Tukey's test for the gas exchange variables and quantum yield efficiency of PSII in the Italian zucchini cultivar 'Caserta' irrigated with saline water and fertilized with nitrogen sources at different phenological phases.

Sources of Variation	A_N (µmol CO_2 m^{-2} s^{-1})	Gs (mol CO_2 m^{-2} s^{-1})	Ci (µmol CO_2 mol^{-1})	E (mmol m^{-2} s^{-1})	Tl (°C)	Fv/Fm
	F-Test (*p*-Value)					
Blocks	0.0116 ^{ns}	0.0000 **	0.0002 **	0.0048 **	0.00001 **	0.2761 ^{ns}
Treatments	0.0140 *	0.0400 *	0.4502 ^{ns}	0.0751 ^{ns}	0.2942 ^{ns}	0.6749 ^{ns}
	Means ± EP					
T1	17.43 ± 1.08 a	0.24 ± 0.03 a	215.75 ± 24.27 a	2.94 ± 0.33 a	32.90 ± 1.28 a	0.807 ± 0.001 a
T2	11.79 ± 1.59 b	0.19 ± 0.05 ab	250.50 ± 26.75 a	2.43 ± 0.33 a	32.53 ± 1.07 a	0.807 ± 0.002 a
T3	12.47 ± 1.79 ab	0.22 ± 0.05 ab	258.00 ± 28.25 a	2.64 ± 0.33 a	32.20 ± 1.07 a	0.811 ± 0.004 a
T4	11.27 ± 1.41 b	0.16 ± 0.05 ab	227.50 ± 40.63 a	2.08 ± 0.32 a	32.28 ± 1.38 a	0.808 ± 0.004 a
T5	13.31 ± 2.35 ab	0.18 ± 0.04 ab	242.25 ± 20.83 a	2.34 ± 0.28 a	31.98 ± 1.32 a	0.810 ± 0.002 a
T6	10.41 ± 0.82 b	0.13 ± 0.02 b	228.25 ± 19.96 a	2.05 ± 0.13 a	32.75 ± 1.08 a	0.802 ± 0.008 a
DMS	5.58	0.09	73.18	0.98	1.38	0.01

**, * and ^{ns} = significant at 1%, 5%, and not significant, respectively, by the F-test; T1 = entire cycle with nitrate-N + 0.50 dS m^{-1} (control); T2 = entire cycle with nitrate-N + 4.5 dS m^{-1} (salt stress); T3 = 50% nitrate-N + 50% ammonium-N + 4.50 dS m^{-1}; T4 = ammonium-N in the vegetative phase + nitrate-N in the reproductive phase + 4.50 dS m^{-1}; T5 = nitrate-N in the vegetative phase + ammonium-N in the reproductive phase + 4.50 dS m^{-1}; T6 = entire cycle with ammonium-N + 4.50 dS m^{-1}. Identical letters in the column do not differ by the Tukey test at a 5% probability level. Standard Error (SE), *n* = 4. Means followed by identical letters in the column do not differ by the Tukey test at 5% probability.

Irrigation with water at 0.5 dS m^{-1} and 100% nitrogen in nitrate form (T1) resulted in the highest CO_2 assimilation rate, with an average value of 17.43 µmol CO_2 m^{-2} s^{-1}. However, this did not differ statistically from treatments T3 (12.47 µmol CO_2 m^{-2} s^{-1}) and T5 (13.31 µmol CO_2 m^{-2} s^{-1}). The lowest assimilation rate (AN) was observed in treatment T6 (irrigation with water at 4.5 dS m^{-1} and 100% ammoniacal nitrogen), which showed a value of 10.41 µmol m^{-2} s^{-1}. This represented a 40.27% reduction compared to the control (T1) and did not differ statistically from the other saline stress treatments (Table 5).

For stomatal conductance, treatment T1 showed the highest mean value (0.24 mol CO_2 m^{-2} s^{-1}), but it was not significantly different from T2, T3, T4, and T5 (Table 5). Conversely, treatment T6 showed the lowest mean value (0.13 mol CO_2 m^{-2} s^{-1}), representing a 45.83% reduction compared to the control, but it was not significantly different from T2, T3, T4, and T5 (Table 5).

A significant effect of treatments was observed on electrolyte leakage (EL) ($p \leq 0.001$), as well as on chlorophyll *a* ($p \leq 0.01$), chlorophyll *b* ($p \leq 0.001$), and carotenoid levels ($p \leq 0.001$) (Table 6).

Electrolyte leakage was highest in treatment T6 (66.72%), though this did not differ significantly from T2 (54.00%) and T5 (52.99%). The lowest EL value was observed in T1 (22.61%), indicating less cellular membrane damage in Italian zucchini plants (Table 6). Under saline stress, treatment T3 (47.65%) showed the lowest EL value, with a 28.58% reduction in membrane damage compared to T6.

The chlorophyll *a* (*Chl a*) content was highest in the control treatment (T1, 11.01 µm cm^{-2}). The lowest value was observed in T6 (6.14 µm cm^{-2}), which represented a 44.23% reduction compared to the control. Treatments 2, 3, 4, 5 and 6 did not show significant differences among themselves (Table 6).

The chlorophyll *b* (*Chl b*) content was also highest in T1 (2.59 µm cm^{-2}), but it was not significantly different from T2 (1.92 µm cm^{-2}) and T5 (1.91 µm cm^{-2}). The lowest value was recorded in T3 (0.94 µm cm^{-2}), although it did not differ statistically from T4 (1.66 µm cm^{-2}) or T6 (1.17 µm cm^{-2}). Treatment T3 resulted in a 29.52% reduction in *chl b* compared to T1 (Table 6).

https://doi.org/10.3390/agriculture16010062

Table 6. Summary of variance analysis by F-values and Tukey's test for electrolyte leakage (EL) and leaf chlorophyll (*Chl a* and *Chl b*) and carotenoid (Carot) content in Italian zucchini cv. Caserta, irrigated with saline water and nitrogen sources at different phenological phases.

		F-Test (p-Value)		
Sources of Variation	**EL (%)**	***Chl a* (µm cm^{-2})**	***Chl b* (µm cm^{-2})**	**Carot (µm cm^{-2})**
Blocks	0.3507 ^{ns}	0.3445 ^{ns}	0.1214 ^{ns}	0.3888 ^{ns}
Treatments	0.0000 **	0.0012 **	0.0001 **	0.0007 **
		Means ± SE		
T1	22.61 ± 1.25 c	11.01 ± 0.58 a	2.59 ± 0.12 a	3.23 ± 0.25 a
T2	54.00 ± 3.63 ab	7.73 ± 0.33 b	1.93 ± 0.28 ab	2.12 ± 0.07 b
T3	47.65 ± 1.63 b	5.97 ± 0.37 b	0.94 ± 0.13 c	2.00 ± 0.02 b
T4	49.02 ± 1.71 b	6.88 ± 0.45 b	1.66 ± 0.18 bc	2.21 ± 0.12 b
T5	52.99 ± 4.70 ab	7.69 ± 1.14 b	1.91 ± 0.22 ab	2.30 ± 0.22 b
T6	66.72 ± 6.30 a	6.14 ± 0.90 b	1.17 ± 0.11 bc	1.86 ± 0.22 b
DMS	16.74	3.14	0.77	0.79

** and ^{ns} = significant at 1% and not significant, respectively, by the F-test; T1 = entire cycle with nitrate-N + 0.50 dS m^{-1} (control); T2 = entire cycle with nitrate-N + 4.5 dS m^{-1} (salt stress); T3 = 50% nitrate-N + 50% ammonium-N + 4.50 dS m^{-1}; T4 = ammonium-N in the vegetative phase + nitrate-N in the reproductive phase + 4.50 dS m^{-1}; T5 = nitrate-N in the vegetative phase + ammonium-N in the reproductive phase + 4.50 dS m^{-1}; T6 = entire cycle with ammonium-N + 4.50 dS m^{-1}. Identical letters in the column do not differ by the Tukey test at a 5% probability level. Standard Error (SE), n = 4. Means followed by identical letters in the column do not differ by the Tukey test at 5% probability.

Carotenoids were highest in T1 (3.23 µm cm^{-2}), and the lowest value was observed in T6 (1.86 µm cm^{-2}). However, T6 did not differ significantly from the other saline stress treatments. The reduction in carotenoid content between T1 and T6 was approximately 42.41% (Table 6).

No significant effect ($p > 0.05$) was observed between the treatments tested on the quantum efficiency of PSII, electron transport rate, initial fluorescence in the light phase, photochemical quenching coefficient, regulated photochemical extinction quantum yield, or non-regulated photochemical extinction quantum yield was observed for the Italian zucchini cultivar 'Caserta' (Table 7).

Table 7. Summary of variance analysis by F-values and Tukey's test for PSII quantum efficiency, electron transport rate, initial leaf fluorescence during the light phase, photochemical quenching coefficient, quantum yield of regulated photochemical quenching, and quantum yield of non-regulated photochemical quenching in Italian zucchini cv. Caserta, irrigated with saline water and nitrogen sources at different phenological phases.

		F-Test (p-Value)	
Sources of Variation	**Y**	**ETR (µmol e$^-$ m^{-2} s^{-1})**	**Fo′**
Blocks	0.4065 ^{ns}	0.2127 ^{ns}	0.1110 ^{ns}
Treatments	0.7472 ^{ns}	0.0797 ^{ns}	0.2371 ^{ns}
		Means ± SE	
T1	0.74 ± 0.01 a	38.55 ± 5.49 a	1.77 ± 0.05 a
T2	0.72 ± 0.02 a	18.75 ± 3.01 a	1.56 ± 0.07 a
T3	0.68 ± 0.04 a	25.13 ± 6.10 a	1.50 ± 0.06 a
T4	0.72 ± 0.02 a	16.20 ± 1.66 a	1.64 ± 0.12 a
T5	0.70 ± 0.05 a	27.83 ± 9.09 a	1.70 ± 0.11 a
T6	0.72 ± 0.02 a	20.73 ± 3.72 a	1.69 ± 0.10 a
DMS	0.13	23.56	0.37

https://doi.org/10.3390/agriculture16010062

Table 7. *Cont.*

Sources of Variation	F-Test (p-Value)		
	qL	Y_{NPQ}	Y_{NO}
Blocks	0.0555 ^{ns}	0.5115 ^{ns}	0.1607 ^{ns}
Treatments	0.2323 ^{ns}	0.7969 ^{ns}	0.4812 ^{ns}
	Means ± SE		
T1	0.0097 ± 0.0006 a	0.2203 ± 0.0040 a	0.0424 ± 0.0021 a
T2	0.0072 ± 0.0005 a	0.2259 ± 0.0131 a	0.0505 ± 0.0031 a
T3	0.0063 ± 0.0012 a	0.2614 ± 0.0321 a	0.0612 ± 0.0114 a
T4	0.0081 ± 0.0010 a	0.2340 ± 0.0192 a	0.0493 ± 0.0042 a
T5	0.0088 ± 0.0018 a	0.2516 ± 0.0379 a	0.0501 ± 0.0089 a
T6	0.0088 ± 0.0013 a	0.2287 ± 0.0141 a	0.0494 ± 0.0053 a
DMS	0.0045	0.1082	0.0285

^{ns} = not significant by the F-test; T1 = entire cycle with nitrate-N + 0.50 dS m^{-1} (control); T2 = entire cycle with nitrate-N + 4.5 dS m^{-1} (salt stress); T3 = 50% nitrate-N + 50% ammonium-N + 4.50 dS m^{-1}; T4 = ammonium-N in the vegetative phase + nitrate-N in the reproductive phase + 4.50 dS m^{-1}; T5 = nitrate-N in the vegetative phase + ammonium-N in the reproductive phase + 4.50 dS m^{-1}; T6 = entire cycle with ammonium-N + 4.50 dS m^{-1}. Identical letters in the column do not differ by the Tukey test at a 5% probability level. Standard Error (SE), n = 4. Means followed by identical letters in the column do not differ by the Tukey test at a 5% probability.

3.2. Growth and Production

The treatments had a significant effect on plant height (PH, $p \leq 0.05$), stem diameter (SD, $p \leq 0.001$), number of leaves (NL, $p \leq 0.05$), shoot dry mass (SDM, $p \leq 0.01$) and leaf area (LA, $p \leq 0.01$) by F-test. However, no significant differences were observed in PH among treatments in the Tukey test ($p \leq 0.05$) (Table 8).

Table 8. Summary of variance analysis using F-values and Tukey test for biometric variables of Italian zucchini, cultivar Caserta, irrigated with saline water and fertilized with nitrogen sources at different phenological stages.

Sources of Variation	F-Test (p-Value)				
	PH (cm)	SD (mm)	NL (unit)	SDM (g)	LA (cm^2)
Blocks	0.1235 ^{ns}	0.0405 *	0.8804 ^{ns}	0.3030 ^{ns}	0.2787 ^{ns}
Treatments	0.0372 *	0.0002 **	0.0156 *	0.0015 **	0.0003 **
	Means ± SE				
T1	18.75 ± 1.11 a	18.08 ± 0.78 a	21.50 ± 1.32 a	38.66 ± 3.87 a	5783.15 ± 452.79 a
T2	17.13 ± 1.05 a	14.18 ± 0.34 bc	19.00 ± 1.08 ab	29.72 ± 0.73 ab	4181.51 ± 130.17 b
T3	15.25 ± 0.85 a	15.60 ± 1.05 ab	17.25 ± 0.63 ab	28.56 ± 2.63 ab	3702.79 ± 247.33 b
T4	15.00 ± 1.08 a	14.43 ± 0.53 bc	18.00 ± 1.78 ab	26.58 ± 2.35 b	3457.96 ± 241.10 b
T5	14.75 ± 1.18 a	15.58 ± 0.35 ab	15.00 ± 0.41 b	27.40 ±1.98 b	3526.85 ± 485.78 b
T6	15.13 ± 0.43 a	12.90 ± 0.67 c	16.00 ± 0.41 b	20.05 ± 1.06 b	3216.46 ± 141.05 b
DMS	4.10	2.57	5.25	10.49	1.402.82

**, * and ^{ns} = significant at 1%, 5%, and not significant, respectively, by the F-test; T1 = entire cycle with nitrate-N + 0.50 dS m^{-1} (control); T2 = entire cycle with nitrate-N + 4.5 dS m^{-1} (salt stress); T3 = 50% nitrate-N + 50% ammonium-N + 4.50 dS m^{-1}; T4 = ammonium-N in the vegetative phase + nitrate-N in the reproductive phase + 4.50 dS m^{-1}; T5 = nitrate-N in the vegetative phase + ammonium-N in the reproductive phase + 4.50 dS m^{-1}; T6 = entire cycle with ammonium-N + 4.50 dS m^{-1}. Identical letters in the column do not differ by the Tukey test at a 5% probability level. Standard Error (SE), n = 4. Means followed by identical letters in the column do not differ by the Tukey test at a 5% probability level.

The highest mean SD (18.08 mm) was recorded in the control treatment using water with 0.5 dS m^{-1} electrical conductivity and 100% nitrate-based fertilization (T1). However, this result did not differ statistically from treatments T3 and T5. The lowest stem diameter

 https://doi.org/10.3390/agriculture16010062

(12.90 mm) was observed in the treatment using water with 4.5 dS m^{-1} electrical conductivity and 100% ammonium-based fertilization (T6), representing a 28.65% reduction compared to the control, T6 showed no statistically significant difference compared to treatments T2 and T4 (Table 8).

For the NL, treatment T1 exhibited the highest mean, with approximately 21 leaves. However, this value did not differ statistically from treatments T2, T3, and T4. The NL in T1 was significantly higher than in treatments T5 and T6 ($p \leq 0.05$). Percentagewise, the NL in T1 exceeded T5 and T6 by 30.23% and 25.58%, respectively (Table 8).

For shoot dry mass (SDM), the highest value was recorded in T1, with an average of 38.66 g, which did not differ statistically from treatments T2 and T3 ($p \leq 0.05$). Plants in treatment T6 showed the lowest SDM values (20.05 g), reflecting a 48.14% reduction compared to the control, but without statistical differences from treatments T2, T3, T4 and T5 ($p \leq 0.05$) (Table 8).

With regard to leaf area (LA), the control treatment recorded the highest value, 5783.15 cm^2, higher than all other treatments. Conversely, all treatments exhibited reductions in LA compared to the control, regardless of the nitrogen fertilization strategy adopted (Table 8).

The treatments significantly affected ($p \leq 0.001$) the number of fruits per plant (NFP), average fruit mass (AFM), longitudinal fruit diameter (LD), transverse fruit diameter (TD), and production per plant (PP) (Table 9).

Table 9. Summary of variance analysis using F-values and Tukey test for production variables of Italian zucchini, cultivar Caserta, irrigated with saline water and fertilized with nitrogen sources at different phenological stages.

	F-Test (*p*-Value)				
Sources of Variation	**NFP (Unit)**	**AFM (g)**	**LD (cm)**	**TD (cm)**	**PP (g)**
Blocks	0.4199 ^{ns}	0.8003 ^{ns}	0.4106 ^{ns}	0.9409 ^{ns}	0.3749 ^{ns}
Treatments	0.00001 **	0.00001 **	0.00001 **	0.00001 **	0.00001 **
	Means ± SE				
T1	3.25 ± 0.25 a	421.20 ± 15.88 a	22.00 ± 0.35 a	5.63 ± 0.21 a	1360.98 ± 74.69 a
T2	2.00 ± 0.00 b	254.80 ± 13.05 b	18.00 ± 1.15 b	5.10 ± 0.06 a	509.60 ± 26.10 b
T3	2.00 ± 0.00 b	199.28 ± 2.36 c	16.75 ± 0.43 b	4.43 ± 0.17 b	398.55 ± 4.72 b
T4	2.00 ± 0.00 b	209.40 ± 1.45 c	17.20 ± 0.41 b	4.35 ± 0.15 b	418.80 ± 2.89 b
T5	0.00 ± 0.00 c	0.00 ± 0.00 d	0.00 ± 0.00 c	0.00 ± 0.00 c	0.00 ± 0.00 c
T6	0.00 ± 0.00 c	0.00 ± 0.00 d	0.00 ± 0.00 c	0.00 ± 0.00 c	0.00 ± 0.00 c
DMS	0.46	41.27	2.52	0.63	1470.43

** and ^{ns} = significant at 1% and not significant, respectively, by the F-test; T1 = entire cycle with nitrate-N + 0.50 dS m^{-1} (control); T2 = entire cycle with nitrate-N + 4.5 dS m^{-1} (salt stress); T3 = 50% nitrate-N + 50% ammonium-N + 4.50 dS m^{-1}; T4 = ammonium-N in the vegetative phase + nitrate-N in the reproductive phase + 4.50 dS m^{-1}; T5 = nitrate-N in the vegetative phase + ammonium-N in the reproductive phase + 4.50 dS m^{-1}; T6 = entire cycle with ammonium-N + 4.50 dS m^{-1}. Identical letters in the column do not differ by the Tukey test at a 5% probability level. Standard Error (SE), *n* = 4. Means followed by identical letters in the column do not differ by the Tukey test at a 5% probability level.

The Italian zucchini plants (cv. Caserta) showed no fruit production in treatments T5 and T6, which involved ammonium-based fertilization. Toxicity symptoms were observed, particularly during the reproductive phase (Table 9). The highest number of fruits was recorded in treatment T1, with an average of three fruits per plant. Treatments T2, T3, and T4 produced two fruits per plant, reflecting a reduction of approximately 38.46% compared to T1.

The highest average fruit mass was observed in T1, with a value of 421.20 g. Among the treatments that produced fruits, T3 showed the lowest average fruit mass (199.28 g),

https://doi.org/10.3390/agriculture16010062

which was not statistically different from T4. This represents a reduction of approximately 52.69% compared to the control (Table 9).

Salt stress reduced fruit size, as evidenced by decreases in both longitudinal and transverse fruit diameters for Italian zucchini cv. Caserta (Table 9). For longitudinal fruit diameter, T1 (control) achieved the highest value (22.00 cm), while T3 recorded the lowest (16.75 cm). However, T3 was not statistically different from treatments T2 and T4, showing a 23.86% reduction compared to T1 (Table 9). Regarding transverse fruit diameter, T1 had the highest average (5.63 cm), which was not statistically different from T2. Treatment T4 had the lowest average (4.35 cm), not differing statistically from T3. This represents a 22.73% reduction between the largest and smallest transverse fruit diameters (Table 9).

In terms of PP, treatment T1 showed the highest value, 1360.98 g (Table 9) compared to the other treatments ($p \leq 0.05$). Treatments T2, T3, and T4, which were equal to each other ($p \leq 0.05$), showed an average reduction of 67.50% compared to the control, while T5 and T6 showed a 100% reduction as these treatments did not reach the production target.

4. Discussion

Excess salt impairs cellular expansion in plants by reducing water absorption, adversely affecting growth, development, and crop yield [8,28]. Based on the results, nitrogen management strategies that consider the proportions of nitrate- and ammonium-based fertilizers may improve the responses of Italian zucchini (cv. Caserta) under salt stress. Ammonium-based fertilization negatively affects Italian zucchini production. However, applying this type of fertilization during specific growth stages may mitigate adverse effects. It has been observed that the application of ammonium-based nitrogen during the reproductive stage of zucchini induces flower abortion and should therefore be avoided. However, when ammonium and nitrate nitrogen are applied in a 1:1 ratio, their use is permissible throughout the crop cycle. The exclusive application of nitrate-based nitrogen poses no restrictions at any stage of crop development.

Zucchini plants in treatments T3 and T5 had photosynthetic rates similar to those in the control (Table 5). However, leaf area and yield per plant were reduced under salt stress (Table 8). These results suggest that nitrogen fertilization, despite improving photosynthesis, was insufficient to counteract the effects of salinity on growth and yield. This outcome reflects the stress conditions imposed on the plants, which likely redirected energy reserves toward acclimation, including osmotic adjustment [29].

Ionic homeostasis under salt stress involves sodium exclusion from leaves and roots, compartmentalization of sodium and chloride in vacuoles, production of compatible solutes for osmotic adjustment, and regulation of K^+, Na^+, Cl^-, and nitrogen forms (NO_3^- and NH_4^+) [28]. Energy consumption for homeostasis, though essential, reduces crop growth and productivity, especially when photosynthetic activity is suppressed by salt stress. This phenomenon has been reported for various crops, including Italian zucchini [30–34]

With regard to electrolyte leakage, treatment T6 exhibited the highest value at 66.72%, not differing significantly from treatments T2 and T5 (Table 6). Conversely, treatment T1 presented the lowest value at 22.61%, significantly differing from treatments T3 and T4. The percentage of cellular electrolyte leakage can serve as an indicator of cell membrane stability, and elevated levels may indicate cell death [35]. Under salt stress, treatments with exclusive use of nitrogen in the ammoniacal form caused greater damage to membrane integrity, showing no statistical difference from treatments using nitrate fertilization exclusively, corroborating Fernandes et al. [34].

Irrigation with moderately saline water (4.5 dS m^{-1}) inhibited the photosynthetic pigments (chlorophyll and carotenoids) in zucchini squash cv. Caserta (Table 6). Over time, the excessive accumulation of salts in young leaves can adversely affect the photosynthetic

 https://doi.org/10.3390/agriculture16010062

process, leading to losses of vital photosynthetic pigments essential for converting light energy into chemical energy [8]. Under irrigation conditions with water at an EC of 4.5 dS m^{-1}, despite no significant differences among nitrogen application methods, T2 (irrigation with 4.5 dS m^{-1} water + 100% nitrate fertilization) exhibited the highest levels of photosynthetic pigments. This benefit is attributed to the nitrate applied as $Ca(NO_3)_2$ and KNO_3, which provides calcium (Ca^{2+}) and potassium (K^+) in addition to nitrogen (N), reducing sodium (Na^+) absorption through competitive inhibition at the same non-selective transport sites used by calcium and potassium [36,37].

Treatment T3, on the other hand, showed the lowest value of photosynthetic pigments compared to other treatments, whereas treatments T4 and T6 did not differ significantly (Table 6). These results indicate higher concentrations of chlorophyll b, even under saline water conditions, when nitrate-based fertilization was exclusively used, differing from the results of Fernandes et al. [34], who evaluated nitrate and ammonium fertilization in zucchini squash cv. Caserta grown hydroponically under salt stress. They observed increased foliar chlorophyll b levels when ammonium nitrogen was used as the nitrogen source. Regarding foliar carotenoid levels, the control treatment (T1) had the highest result, while treatment T6 recorded the lowest, though it did not differ from treatments T2, T3, T4, and T5 (Table 8).

The reduction in foliar carotenoid content in zucchini squash cv. Caserta is associated with salt stress caused by irrigation with 4.5 dS m^{-1} water, which affects carotenoid synthesis and, consequently, modulates plant growth under such conditions. Under stress, carotenoids act as structural pigments of the antenna complex, serving as photoprotectors of the photosynthetic apparatus against excessive light [38]. They dissipate energy in a way that neutralizes reactive oxygen species to maintain the stability of cellular structures [39]. These findings differ from those reported by Fernandes et al. [34], who found no significant effects of ECw on chlorophyll *a* and carotenoid levels.

Although no significant effects were observed on Fv/Fm, Y, ETR, Fo′, qL, YNPQ, and YNO in zucchini squash under salt stress, reductions in CO_2 assimilation rates were noted. These results suggest that the decrease in zucchini photosynthesis under salt stress was not solely due to stomatal effects, as only treatment T6 exhibited gs values lower than those of the control treatment. Therefore, the reduction in CO_2 assimilation rates observed in this study corroborates the observed decreases in chlorophyll *a* and carotenoid levels, along with increased damage to plasma membranes. Oxidative stress reduced the amount of photosynthetic pigments in light-harvesting systems, such as photosystem II (P680) and photosystem I (P700), thereby lowering the CO_2 assimilation rate in zucchini squash cv. Caserta under salt stress. The lack of differences in Fv/Fm and ETR indicates that the PSII photochemical apparatus was functioning normally. However, the decline in CO_2 assimilation suggests limitations in the later stages of photosynthesis, such as CO_2 diffusion or biochemical processes like Rubisco activity and RuBP regeneration. Thus, despite preserved electron flow and no photoinhibition, the plant could not convert energy into carbon fixation, leading to a disconnect between photochemical variables and CO_2 assimilation.

Fertilization exclusively with the ammoniacal source under irrigation conditions with moderately saline water (4.5 dS m^{-1}) negatively affected the growth of Italian zucchini, leading to reductions in stem diameter, number of leaves, and shoot dry mass. These reductions were particularly significant in treatments where ammonium was applied alongside irrigation with 4.5 dS m^{-1} water during the reproductive phase and throughout the crop cycle (Table 8). Urea in the soil can undergo hydrolysis, releasing ammonium ions (NH_4^+), which may compete with other essential cations, such as calcium, magnesium, and sulfur, causing nutrient imbalances and exacerbating the deleterious effects of toxic salts

https://doi.org/10.3390/agriculture16010062

on plants [40]. Moreover, excessive urea application can contribute to soil acidification, and when converted to ammonia (NH_3^+), it may become toxic to plants, damaging the root system and interfering with nutrient uptake [40].

The low leaf area of zucchini under salt stress (Table 8) reflected its reduced capacity for synthesizing photoassimilates due to decreased photosynthetic rates (Table 5), which negatively impacted fruit production per plant (Table 9). This pattern of reduced growth, photosynthesis, and production of Italian zucchini under saline conditions has been reported by Souza et al. [41], Fernandes et al. [34,42], and Sousa et al. [10].

The literature demonstrates that ionic and biochemical homeostasis and the production of zucchini fertilized with nitrate surpass those fertilized with ammoniacal sources when irrigated with water at 6.5 dS m^{-1} [34,43–45]. These studies identified flower abortion, which prevented fruit formation in plants fertilized with NH_4^+. According to Fernandes et al. [42], the excessive use of NH_4^+ can cause toxicity in Italian zucchini plants. This toxicity can manifest in various symptoms, including reduced growth, leaf chlorosis, and marginal leaf burn. The underlying reason for these symptoms is an alteration in the plant's ionic balance, as excessive NH_4^+ inhibits the absorption of other cations such as K^+, Mg^{2+}, and Ca^{2+}. The study also found that using NH_4^+ as the sole fertilizer is toxic to zucchini plants, particularly during the flowering stage. Consequently, only plants fertilized with NO_3^- were able to reach the fruiting stage.

When salt concentrations exceed the tolerance threshold, salinity becomes detrimental to agricultural crops, resulting in inhibited growth, as observed in Italian zucchini, and leading to yield losses [46]. The deleterious effect of saline/water stress on the growth of zucchini, cultivar Caserta, has been reported by Sousa et al. [10], showing reductions in height, leaf area, number of leaves, and stem diameter when irrigated with water at 2.5 dS m^{-1}. Souza et al. [41] observed that the use of saline water resulted in reduced zucchini productivity at levels starting from 5.0 dS m^{-1}. Biomass accumulation in zucchini is negatively affected under saline stress starting at an ECw of 2.5 dS m^{-1} [47].

Treatments with salt stress and ammoniacal fertilization during the reproductive phase, as well as salt stress and 100% ammoniacal fertilization throughout the entire cycle, did not result in fruit production due to ammonium toxicity during the reproductive phase of zucchini (Table 6). Fernandes et al. [34] reported flower abortion and the absence of fruit formation in Italian zucchini plants fertilized exclusively with an NH_4^+ source. Similarly, this study found that the exclusive use of ammoniacal fertilization during the reproductive phase of zucchini causes flower abortion and prevents fruit formation. However, the application of ammoniacal nitrogen only during the vegetative phase or in combination with nitrate sources did not result in plant toxicity.

When evaluating the use of biostimulants to improve the productivity and quality of zucchini fruits under salt stress, Souza et al. [41] observed no significant difference in the number of fruits per plant (NFP) between irrigation water salinities of 0.5 and 5.0 dS m^{-1}. However, they reported significant differences in the average fruit mass and the transverse diameter of the fruits. Fernandes et al. [42] found that increasing ECw from 0.5 to 6.5 dS m^{-1} reduced the number of fruits per plant by 71.5%, even in treatments fertilized with both nitrate and ammonium.

The highest average fruit mass (AFM) was observed in the control treatment. Among the treatments producing fruits under salt stress, treatment T2 had the highest AFM (Table 6). Under salt stress conditions, a reduction in average fruit mass is commonly observed, as Souza et al. [41] noted in Italian zucchini, reporting a 25.15% decrease in AFM in the treatment subjected to salt stress (EC = 5.0 dS m^{-1}). Similarly, Fernandes et al. [34] reported a 57.60% reduction in AFM in Italian zucchini under salt stress compared to the control.

https://doi.org/10.3390/agriculture16010062

A reduction in the longitudinal diameter of the fruits was observed by Fernandes et al. [42] in Italian zucchini under salt stress, with average losses of 43.70% compared to the control treatment. The transverse diameter of zucchini fruits was also affected by nitrogen fertilization under saline water stress, showing the lowest values. However, in the treatment with exclusive nitrate fertilization (T2), this reduction did not differ statistically from the control. Comparing treatments with moderately saline water (EC = 5.0 dS m^{-1}) to those with low salinity water (EC = 0.5 dS m^{-1}) in Italian zucchini production, Souza et al. [41] reported an 8.93% decrease in fruit transverse diameter. Similarly, Fernandes et al. [42] observed a 38.20% reduction in transverse diameter in treatments with saline water.

Exclusive nitrate fertilization in Italian zucchini cv. Caserta, when irrigated with moderately saline water (T2), resulted in higher yields compared to ammoniacal fertilization (Table 6). However, the findings of the present study reveal that under saline conditions, it is feasible to use 50% nitrogen fertilization in the ammoniacal form and 50% in the nitrate form, or 100% ammoniacal nitrogen during the vegetative phase, without causing toxicity issues. This strategy allowed the plants to complete their growth cycle and produce yields similar to those fertilized exclusively with nitrate nitrogen.

Although the present study did not evaluate enzyme activity and oxidative stress, studies evaluating these processes have demonstrated that high NH_4^+ supply saturates the activity of glutamine synthetase, the enzyme responsible for incorporating NH_4^+ into glutamate for amino acid and protein synthesis [48], in addition to increasing cytoplasmic pH through the overproduction of OH^- ions [49]. The reproductive phase requires substantial energy for bud differentiation and flower formation, which may limit the metabolic capacity necessary for NH_4^+ assimilation by enzymes such as glutamine synthetase, glutamate synthase, and glutamate dehydrogenase [49].

Despite fertilization, zucchini exhibited a less effective physiological response under saline conditions compared with the control treatment [50–52]. This reduced performance can be explained by the osmotic and ionic constraints imposed by salinity. High salt concentrations decrease the soil osmotic potential, limiting water uptake by the roots and inducing plant water stress. As a consequence, stomatal closure occurs, restricting gas exchange processes and reducing photosynthetic efficiency [8]. In parallel, plants subjected to salinity accumulate potentially toxic ions such as Na^+ and Cl^-, which impair membrane integrity and disrupt nutrient uptake pathways in the roots. Together, these osmotic and ionic effects compromise plant metabolism and help explain the lower effectiveness of zucchini fertilization under saline stress.

5. Conclusions

The findings of this study indicate that under salt stress conditions (ECw 4.5 dS m^{-1}), the CO_2 assimilation, growth, and yield of Italian zucchini cv. Caserta were adversely affected, regardless of the nitrogen source used. Under salt stress, cv. Caserta showed an average decrease in leaf area (5783 to 4521 cm^2) and production per plant (1361 to 442 g). Nitrate fertilization resulted in higher yields compared to ammonium. The continuous application of ammonium nitrogen, or its use during the reproductive phase, led to flower abortion. However, under salinity, supplying ammonium only during the vegetative stage (T4) or using a 1:1 ammonium-to-nitrate ratio throughout the growth cycle (T3) produced yields that were similar to those achieved with nitrate-only fertilization (T2). While salt stress did not affect photochemical efficiency, it significantly reduced the levels of photosynthetic pigments. These findings are specific to the salinity conditions studied, and future research should explore how various nitrogen sources, ratios, salinity levels, and growth stages influence the performance of Italian zucchini cv. Caserta.

 https://doi.org/10.3390/agriculture16010062

Author Contributions: Conceptualization, methodology, investigation, data curation, writing—original draft preparation, G.d.F.S., M.F.N., A.G.d.L.S., T.D.C.P. and F.V.d.S.S.; writing—review and editing, formal analysis, visualization, methodology, investigation, R.A.R.F., K.F.d.P.A., J.S.J.C., K.T.O.P., R.C.L.M., N.d.S.D., P.D.F., J.L.A.R., A.S.d.M. and A.Á.d.S. Supervision, project administration, funding acquisition, M.F.N., N.d.S.D. and F.V.d.S.S. All authors have read and agreed to the published version of the manuscript.

Funding: This study was financed in part by the Coordenação de Aperfeiçoamento de Pessoal de Nível Superior—Brasil (CAPES), Finance Code 001.

Data Availability Statement: All data are presented in the paper.

Conflicts of Interest: The authors declare no conflicts of interest.

References

1. Costa, I.B.C.; Bonfim, F.P.G.; Pasa, M.C. Agricultura familiar: Estudo de caso do sistema produtivo em comunidade rural do estado do mato grosso, brasil. *Braz. J. Dev.* **2023**, *9*, 26754–26765. [CrossRef]
2. Dantas, M.V.; Lima, G.S.; Gheyi, H.R.; Pinheiro, F.W.A.; Silva, P.C.C.; Soares, L.A.A. Gas exchange and hydroponic production of zucchini under salt stress and H_2O_2 application. *Rev. Caatinga* **2022**, *35*, 436–449. [CrossRef]
3. Guerra, A.M.N.M.; Silva, M.G.M.; Evangelista, R.S.; Santos, E.B.; Nogueira, W.P. Produção de cultivares de abobrinha italiana a pleno sol e sombreada no nordeste brasileiro. *Agropecuária Técnica* **2020**, *41*, 1–7. [CrossRef]
4. IBGE—Instituto Brasileiro de Geografia e Estatística. *Número de Estabelecimentos Agropecuários e Quantidade Produzida por Produtos da Horticultura—Resultados Preliminares 2018*; IBGE: Rio de Janeiro, Brazil, 2019.
5. Fadl, M.E.; Jalhoum, M.E.M.; Abdelrahman, M.A.E.; Ali, E.A.; Zahra, W.R.; Abuzaid, A.S.; Fiorentino, C.; D'antonio, P.; Belal, A.A.; Scopa, A. Soil salinity assessing and mapping using several statistical and distribution techniques in arid and semi-arid ecosystems, Egypt. *Agronomy* **2023**, *13*, 583. [CrossRef]
6. Wang, N.; Chen, S.; Huang, J.; Frappart, F.; Taghizadeh, R.; Zhang, X.; Wigneron, J.; Xue, J.; Xiao, Y.; Peng, J.; et al. Global soil salinity estimation at 10 m using multi-source remote sensing. *J Remote Sens.* **2024**, *4*, 0130. [CrossRef]
7. Cirillo, C.; Rouphael, Y.; Caputo, R.; Raimondi, G.; Sifola, M.I.; de Pascale, S. Effects of high salinity and the exogenous of an osmolyte on growth, photosynthesis and mineral composition in two ornamental shrubs. *J. Hortic. Sci. Biotechnol.* **2016**, *91*, 14–22. [CrossRef]
8. Acosta-Motos, J.R.; Ortuño, M.F.; Vicente, A.B.; Vivancos, P.D.; Blanco, M.J.S.; Hernandez, J.A. Plant responses to salt stress: Adaptive mechanisms. *Agronomy* **2017**, *7*, 18. [CrossRef]
9. Medeiros, J.F.; Terceiro Neto, C.P.C.; Dias, N.S.; Gheyi, H.R.; Silva, M.V.T.; Loiola, A.T. Salinidade e ph de um argissolo irrigado com água salina sob estratégias de manejo. *Rev. Bras. Agric. Irrig.* **2017**, *11*, 1407–1419. [CrossRef]
10. Sousa, H.C.; Sousa, G.G.; Cambissa, P.B.C.; Lessa, C.I.N.; Goes, G.F.; Silva, F.D.B.; Abreu, F.S.; Viana, T.V.A. Gas exchange and growth of zucchini crop subjected to salt and water stress. *Rev. Bras. Eng. Agrícola Ambient.* **2022**, *26*, 815–822. [CrossRef]
11. Munns, R.; Tester, M. Mechanism of salinity tolerance. *Annu. Rev. Plant Biol.* **2008**, *59*, 651–681. [CrossRef]
12. Syvertsen, J.P.; Garcia-Sanchez, F. Multiple abiotic stresses occurring with salinity stress in citrus. *Environ. Exp. Bot.* **2014**, *103*, 128–137. [CrossRef]
13. Sá, F.V.S.; Brito, M.E.B.; Silva, L.A.; Moreira, R.C.L.; Fernandes, P.D.; Figueiredo, L.C. Fisiologia da percepção do estresse salino em híbridos de tangerineira—Sunki comum sob solução hidropônica salinizada. *Comun. Sci.* **2015**, *6*, 463–470. [CrossRef]
14. Sá, F.V.S.; Gheyi, H.R.; Lima, G.S.; Paiva, E.P.; Lacerda, C.F.; Fernandes, P.D. Saline water, nitrogen and phosphorus on water relations and physiological aspects of west indian cherry. *Comun. Sci.* **2018**, *9*, 430–437. [CrossRef]
15. Santos, J.B.; Gheyi, H.R.; Lima, G.S.; Xavier, D.A.; Cavalcante, L.F.; Centeno, C.R.M. Morfofisiologia e produção do algodoeiro herbáceo irrigado com águas salinas e adubado com nitrogênio. *Comun. Sci.* **2016**, *7*, 86–96. [CrossRef]
16. Bazzo, J.H.B.; Riede, C.R.; Arruda, K.M.A.; Cardoso, C.P.; Franzoni, I.; Fonseca, I.C.B.; Zucareli, C. Performance of white oat cultivars in response to nitrogen fertilization and trinexapac-ethyl. *Semin. Ciências Agrárias* **2019**, *40*, 2121–2135. [CrossRef]
17. Blanger, B.R.; Lobo, F.A.; Genuncio, G.C.; Possamai, A.C.S. Avaliação de um método para estimar a absorção radicular do nitrato em arroz. *Rev. Ibero Am. Ciências Ambient.* **2021**, *12*, 1–19. [CrossRef]
18. Cordeiro, C.F.S.; Rodrigues, D.R.; Echer, F.R. Cover crops and controlled-release urea decrease need for mineral nitrogen fertilizer for cotton in sandy soil. *Field Crop. Res.* **2022**, *276*, 108387. [CrossRef]
19. Alvares, C.A.; Stape, J.L.; Sentelhas, P.C.; de Moraes Gonçalves, J.L.; Sparovek, G. Köppen's climate classification map for Brazil. *Meteorol. Z.* **2013**, *22*, 711–728. [CrossRef]
20. Embrapa—Empresa Brasileira de Pesquisa Agropecuária. *Manual de Análises Químicas de Solos, Plantas e Fertilizantes*, 2nd ed.; Embrapa Informação Tecnológica: Brasília, Brazil, 2009.

https://doi.org/10.3390/agriculture16010062

21. Novais, R.F.; Neves, J.C.L.; Barros, N.F. Ensaio em ambiente controlado. In *Métodos de Pesquisa em Fertilidade do Solo*; Oliveira, A.J., Garrido, W.E., Araújo, J.D., Lourenço, S.; Embrapa-Sea: Brasília, Brazil, 1991; pp. 189–254.
22. Oxborough, K.; Baker, N.R. An instrument capable of imaging chlorophyll a fluorescence from leaves at very low irradiance and at cellular and subcellular levels of organization. *Plant Cell Environ.* **1997**, *20*, 1473–1483. [CrossRef]
23. Kramer, D.M.; Johnson, G.; Kiirats, O.; Edwads, G.E. New fluorescence parameters for the determination of qa redox state and excitation energy fluxes. *Photosynth. Res.* **2004**, *79*, 209–218. [CrossRef]
24. Singh, H.P.; Batish, D.R.; Kohli, R.K.; Arora, K. Arsenic-induced root growth inhibition in mung bean (*phaseolus aureus* roxb.) is due to oxidative stress resulting from enhanced lipid peroxidation. *Plant Growth Regul.* **2007**, *53*, 65–73. [CrossRef]
25. Wellburn, A.R. The spectral determination of chlorophyll a and b, as well as total carotenoids, using various solvents with spectrophotometers of different resolution. *J. Plant Physiol.* **1994**, *144*, 307–313. [CrossRef]
26. Fialho, G.S.; Dalvi, L.P.; Dalvi, N.B.C.; Kuhlcamp, K.T.; Effgen, E.M. Prediction of the leaf area in zucchini fruit: A nondestructive, exact, simple, fast and practical method. *Rev. Bras. Agropecuária Sustentável* **2011**, *1*, 59–63.
27. Ferreira, D.F. Sisvar: A computer analysis system to fixed effects split plot type designs. *Rev. Bras. Biom.* **2019**, *37*, 529–535. [CrossRef]
28. Munns, R.; Day, D.A.; Fricke, W.; Watt, M.; Arsova, B.; Barkla, B.J.; Bose, J.; Byrt, C.S.; Chen, Z.; Foster, K.J.; et al. Energy costs of salt tolerance in crop plants. *New Phytol.* **2020**, *225*, 1072–1090. [CrossRef]
29. Lira, E.; Souza, J.; Galdino, L.; Macêdo, C.; Silva, A.; Melo, Y.; Santos, I.; Arriel, M.; Meneses, C.; Maia, J. Changes in reserve mobilization caused by salinity could interfere in the initial growth of *Jatropha curcas*. *Sustainability* **2021**, *13*, 7446. [CrossRef]
30. Sá, F.V.S.; Gueyi, H.R.; Lima, G.S.; Paiva, E.P.; Silva, L.A.; Moreira, R.C.L.; Fernandes, P.D.; Dias, A.S. Ecophysiology of west indian cherry irrigated with saline water under phosphorus and nitrogen doses. *Biosci. J.* **2019**, *35*, 211–221. [CrossRef]
31. Sá, F.V.S.; Gheyi, H.R.; Lima, G.S.; Pinheiro, F.W.A.; Paiva, E.P.; Moreira, R.C.L.; Silva, L.A.; Fernandes, P.D. The right combination of n-p-k fertilization may mitigate salt stress in custard apple (*annona squamosa* L.). *Acta Physiologiae Plantarum* **2021**, *43*, 59. [CrossRef]
32. Silva, F.H.; Morais, P.L.D.; Albuquerque, C.C.; Albuquerque, M.T.; Dias, N.S. Physiological aspects of melon (*cucumis melo* l.) as a function of salinity. *J. Plant Growth Regul.* **2021**, *40*, 1298–1314. [CrossRef]
33. Praxedes, S.S.C.; Ferreira Neto, M.; Loiola, A.T.; Santos, F.J.Q.; Umbelino, B.F.; Silva, L.A.; Moreira, R.C.L.; Melo, A.S.; Lacerda, C.F.; Fernandes, P.D.; et al. Photosynthetic responses, growth, production, and tolerance of traditional varieties of cowpea under salt stress. *Plants* **2022**, *11*, 1863. [CrossRef]
34. Fernandes, C.S.; Sá, F.V.S.; Ferreira Neto, M.; Dias, N.S.; Reges, L.B.L.; Gheyi, H.R.; Paiva, E.P.; Silva, A.A.; Melo, A.S. Ionic homeostasis, biochemical components and yield of Italian zucchini under nitrogen forms and salt stress. *Braz. J. Biol.* **2022**, *82*, e233567. [CrossRef] [PubMed]
35. Hatsugai, N.; Katagiri, F. Quantification of plant cell death by electrolyte leakage assay. *Bio-Protocol* **2018**, *5*, e2758. [CrossRef] [PubMed]
36. Raddatz, N.; Morales de Los Ríos, L.; Lindahl, M.; Quintero, F.J.; Pardo, J.M. Coordinated transport of nitrate, potassium, and sodium. *Front. Plant Sci.* **2020**, *11*, 247. [CrossRef]
37. Pathak, J.; Ahmed, H.; Kumari, N.; Pandey, A.; Rajneesh, R.P. Role of calcium and potassium in amelioration of environmental stress in plant. In *Protective Chemical Agents in the Amelioration of Plant Abiotic Stress*; John Wiley & Sons, Ltd.: Chichester, UK, 2020; pp. 535–562. [CrossRef]
38. Leiva-Ampuero, A.; Agurto, M.; Matus, J.T.; Hoppe, G.; Huidobro, C.; Inostroza-Blancheteau, C.; Reyes-Díaz, M.; Stange, C.; Canessa, P.; Vega, A. Salinity impairs photosynthetic capacity and enhances carotenoid-related gene expression and biosynthesis in tomato (*Solanum lycopersicum* L. cv. micro-tom). *PeerJ* **2020**, *8*, e9742. [CrossRef]
39. Kopsell, D.A.; Kopsell, D.E.; Curran-Celentano, J. Carotenoid pigments in kale are influenced by nitrogen concentration and form. *J. Sci. Food Agric.* **2007**, *87*, 900–907. [CrossRef]
40. Han, J.; Shi, J.; Zeng, L.; Xu, J.; Wu, L. Effects of nitrogen fertilization on the acidity and salinity of greenhouse soils. *Environ. Sci. Pollut. Res. Int.* **2015**, *22*, 2976–2986. [CrossRef]
41. Souza, M.W.L.; Torres, S.B.; Oliveira, F.A.; Marques, I.C.S.; Pereira, K.T.O.; Guimarães, T.T. Saline-water irrigation and plant growth regulator application on zucchini fruit yield and quality. *Rev. Bras. Eng. Agrícola Ambient.* **2020**, *24*, 679–684. [CrossRef]
42. Fernandes, C.S.; Ferreira Neto, M.; Dias, N.S.; Reges, L.B.L.; Silva, L.A.; Moreira, R.C.L.; Silva, A.Á.; Paiva, E.P.; Fernandes, P.D.; Sá, F.V.S. The appropriate source of nitrogen for italian zucchini under salt stress conditions. *J. Soil Sci. Plant Nutr.* **2022**, *22*, 560–570. [CrossRef]
43. Nobre, R.G.; Gheyi, H.R.; Soares, F.A.L.; Cardoso, J.A.F. Produção de girassol sob estresse salino e adubação nitrogenada. *Rev. Bras. Ciência Solo* **2011**, *35*, 929–937. [CrossRef]
44. Lima, G.S.; Nobre, R.G.; Gheyi, H.R.; Soares, L.A.A.; Silva, A.O. Crescimento e componentes de produção da mamoneira sob estresse salino e adubação nitrogenada. *Engenharia Agrícola* **2014**, *34*, 854–866. [CrossRef]

45. Ribeiro, P.H.P.; Silva, S.; Dantas Neto, J.; Oliveira, C.S.; Chaves, L.H.G. Crescimento e componentes de produção do girassol em função da irrigação com água salina e adubação nitrogenada. *Rev. Eng. Agric.* **2015**, *23*, 48–56. [CrossRef]
46. Dantas, M.V.; Lima, G.S.; Greyi, H.R.; Pinheiro, F.W.A.; Silva, L.A.; Fernandes, P.D. Summer squash morphophysiology under salt stress and exogenous application of h_2o_2 in hydroponic cultivation. *Comun. Sci.* **2021**, *12*, e3464.
47. Canjá, J.F.; Azevedo, J.; Sousa, G.G.; Magalhaes, C.L.; Viana, T.V.A. Initial growth of zucchini irrigated with saline water in soil with biofertilizers. *Agro@mbiente On-Line* **2021**, *15*, 1–15. [CrossRef]
48. Roosta, H.R.; Schjoerring, J.K. Effects of ammonium toxicity on nitrogen metabolism and elemental profile of cucumber plants. *J. Plant Nutr.* **2007**, *30*, 1933–1951. [CrossRef]
49. Taiz, L.; Zeiger, E.; Møller, I.M.; Murphy, A. *Plant Physiology and Development*, 6th ed.; Sinauer Associates: New York, NY, USA, 2015.
50. Volkov, V.; Beilby, M.J. Salinity tolerance in plants: Mechanisms and regulation of ion transport. *Front. Plant Sci.* **2017**, *8*, 1795. [CrossRef] [PubMed]
51. Wan, Q.; Hongbo, S.; Zhaolong, X.; Jia, L.; Dayong, Z.; Yihong, H. Salinity tolerance mechanism of osmotin and osmotin-like proteins: A promising candidate for enhancing plant salt tolerance. *Curr. Genom.* **2017**, *18*, 553–556. [CrossRef]
52. Liang, W.; Ma, X.; Wan, P.; Liu, L. Plant salt-tolerance mechanism: A review. *Biochem. Biophys. Res. Commun.* **2018**, *495*, 286–291. [CrossRef]

https://doi.org/10.3390/agriculture16010062

agriculture

MDPI

Article

Detecting the Maturity of Red Strawberries Using Improved YOLOv8s Model

Shengyi Zhao [1,2,3,4], Chen Fang [1,3,4], Tianzheng Hua [1,2,3,4] and Yong Jiang [1,2,3,4,*]

1 School of Agricultural Engineering, Jiangsu University, Zhenjiang 212013, China
2 National Digital Agricultural Equipment (Artificial Intelligence and Agricultural Robotics) Innovation Sub-Centre, Jiangsu University, Zhenjiang 212013, China
3 Key Laboratory of Modern Agricultural Equipment and Technology, Ministry of Education, Jiangsu University, Zhenjiang 212013, China
4 Key Laboratory for Theory and Technology of Intelligent Agricultural Machinery and Equipment, Jiangsu University, Zhenjiang 212013, China
* Correspondence: 1000006668@ujs.edu.cn

Abstract: Strawberry picking relies primarily on manual labor, making it the most labor-intensive stage in strawberry cultivation. Harvesting robots have become essential for strawberry production, and fruit ripeness detection models are critical for picking operations. This study collected strawberry ripeness photographs under various natural environments and enhanced feature expression through diverse image enhancement techniques. Considering practical deployment on harvesting robots, the low-parameter, high-accuracy YOLOv8s was selected as the base model. Leveraging the ease of integration of the Global Attention Mechanism (GAM) within the YOLO model, we incorporated GAM before the SPFF module to enhance the extraction capabilities of both global and local features. Experimental results demonstrate that the improved YOLOv8s achieves excellent performance, with a mAP of 91.5% for three maturity classes and a frame rate of 53 FPS. Compared with other mainstream models, the improved YOLOv8s presented in this paper demonstrates superior detection performance, achieving mAP improvements of 12.1%, 8.0%, 6.1%, 4.6%, and 3.1% over YOLOv3, YOLOv5s, YOLOv7s, YOLOv8s, and CBAM-YOLOv8s, respectively. It also exhibits robust detection capabilities under varying lighting conditions and occlusions, meeting the demands for high precision and rapid performance during harvesting operations.

Keywords: strawberries; maturity assessment; object detection; attention mechanism

1. Introduction

Originating from Europe and East Asia, strawberries are a delicious and tasty fruit. They are widely loved by consumers for their enticing aroma, tender and smooth texture, and vibrant red color [1,2]. In addition, strawberries can adapt to diverse environments, making them widely cultivated worldwide and one of the most popular fruits globally [3]. As cultivation areas continue to expand, economic benefits have significantly increased. Strawberries hold substantial economic importance across the globe and possess immense potential as an export commodity [4].

According to the latest data from the United Nations Food and Agriculture Organization (FAO), China is the world's largest producer of strawberries [5,6]. Its cultivation area and output have consistently been in the lead worldwide, accounting for approximately one-third of the world's total production. After strawberries begin to set fruit, they typically require 20 to 30 days to ripen. The degree of ripeness is a key factor affecting strawberry quality and flavor; harvesting them too early or too late will significantly reduce their quality [7]. Currently, strawberry harvesting relies primarily on manual picking, making it the most labor-intensive task in strawberry cultivation. During peak seasons, rising labor costs indirectly increase the overall production expenses of strawberry farming [5]. Additionally, manual picking can easily cause surface damage and lead to misjudgments about fruit ripeness, failing to ensure strawberries are harvested at their optimal picking time.

With the rapid advancement of modern computer technology, the integration of traditional agriculture with modern computing has emerged as a significant research focus [8]. Introducing robots into strawberry harvesting not only liberates labor and reduces the workload for farmers but also enhances picking efficiency during peak ripening periods [9]. Therefore, conducting research on automated strawberry ripeness detection in natural environments holds substantial scientific significance and practical value. It not only provides crucial technical support for strawberry production processes but also lays the foundation for developing intelligent harvesting equipment [10].

In recent years, computer vision technology has been widely applied in the agricultural sector, enabling the detection of ripeness in fruits such as apples, bananas, grapes, and tomatoes [11–13]. Numerous scholars have employed traditional image segmentation techniques to extract features such as the size, color, and contours of target objects. Cho et al. [14] selected spectral data correlated with tomato ripeness and established a ripeness classification model using support vector classification (SVC) and snapshot hyperspectral imaging. Arefi et al. [15] employed threshold analysis for background removal in the RGB color space, followed by mature tomato extraction using HSI, YIQ, and RGB color spaces, achieving an accuracy rate of 96.36%. Fadhel et al. [16] employed color thresholding and k-means clustering to identify unripe strawberries. The results demonstrate that color thresholding outperformed k-means in terms of accuracy, effectiveness, and speed in code implementation. Mo et al. [17] developed a banana fruit ripeness discrimination model based on genetic algorithms and SVM, utilizing manually measured banana rhombic features as indicators of ripeness. The final model achieved an average prediction accuracy of 86.20%. Khojastehnazhand et al. [18] employed RGB values converted to HSV for classification, focusing on the HSV color space and Naive Bayes method to classify dragon fruit ripeness, achieving an accuracy rate of 86.6%. Malik et al. [19] proposed a novel detection method based on an improved HSV color space and watershed segmentation algorithm, utilizing low-cost and readily available RGB cameras to assist robots in automatically picking ripe red tomatoes.

Traditional image processing-based maturity detection has achieved certain results but exhibits several significant drawbacks: (1) Algorithms are developed for specific crops and environments, requiring predefined parameters such as thresholds and filter kernel sizes, lacking dynamic adjustment capabilities. (2) Feature extraction parameter settings rely on subjective human experience, making it difficult to adapt to complex and variable real-world scenarios, resulting in poor universality. (3) Due to the excessive complexity of crop feature extraction, the results obtained are easily affected by environmental noise, light changes, overlap, and other surrounding disturbances. This leads to low detection efficiency and poor robustness and generalization capabilities, making it difficult to meet field detection requirements.

In recent years, deep learning technology has garnered significant attention. By leveraging neural networks to learn more abstract and advanced features from raw images, it enables automatic feature learning and extraction without human intervention [20]. This technology effectively analyzes large datasets and identifies replicable features, substantially enhancing performance in image classification, segmentation, and detection [21]. It currently represents a major research focus in the field of classification and recognition.

Zhou et al. [22] employed the YOLOv3 model to detect the maturity levels of strawberry flowers and fruits at different stages using both aerial and ground-level imaging, achieving excellent results with average accuracies of 88.0% and 89.0%. MIAO et al. [23] modified the original model backbone to MobileNetV3 and incorporated a Global Attention Mechanism into the feature fusion network to address the challenges of subtle maturity differences among cherry tomatoes and frequent mutual occlusion between fruits. The improved YOLOv7 model achieved a mAP of 98.2% on the test set while reducing the model size by 55.7 MB. Qiu et al. [24] proposed an algorithm based on an improved YOLOv4 model, utilizing MobilenetV3, separable convolutions, and the SeNet attention mechanism for precise grape ripeness detection and spatial localization within orchards, achieving an accuracy of 93.52%. Fan et al. [25] proposed an algorithm combining YOLOv5 with dark channel enhancement based on the three standard criteria of strawberry maturity: ripe, near-ripe, and unripe. Huang et al. [26] proposed a production line-based mango ripeness detection method using an enhanced YOLOv8s model. By modifying the C2F module within the YOLOv8s model's neck structure and incorporating a channel attention mechanism, the approach significantly improves detection efficiency. Wang et al. [27] proposed an artificial intelligence algorithm framework for locating lychee picking points and detecting obstacles, achieving 88.1% accuracy in segmenting lychee fruits from branches and an 88% success rate in identifying picking points.

In summary, although deep learning-based fruit ripeness detection technology has made certain progress, related methods still require improvement in terms of robustness and practical application [28]. Strawberry fruits exhibit the following characteristics: (1) Strawberry canopies are dense, and the fruit bodies are relatively small, often overlapping and shading each other. (2) Due to varying light conditions during growth, strawberry fruits do not ripen simultaneously. Even on the same plant, the degree of ripeness among individual fruits is not uniform.

This study collected photographs of strawberry ripeness across diverse natural environments and enhanced the expressive power of the dataset's feature information through various image enhancement techniques. Subsequently, this study selected the YOLOv8s model—known for its minimal parameters and high accuracy—as the base model. By incorporating a Global Attention Mechanism (GAM) into the backbone network, the model effectively enhances its extraction of key information regarding strawberry ripeness. Finally, through multiple comparative validations, this paper demonstrates the effectiveness of the improved model, ensuring its suitability for subsequent application in harvesting robots.

2. Materials and Methods

2.1. Building Strawberry Maturity Dataset

Deep learning methods have been successfully used to grade and identify strawberry ripeness. Beyond demanding high-quality detection models, this approach relies on ample datasets for model training. Commonly used public datasets such as ImageNet, Open Images Dataset, COCO, and PASCALVOC feature diverse object categories but lack images depicting strawberries at different stages of ripeness. Therefore, this study necessitates the construction of a strawberry ripeness dataset for model training.

2.1.1. Image Collection

All datasets used for training were manually collected at the No. 1 Farm Strawberry Garden in Changzhou City, Jiangsu Province. The strawberry variety selected was Hongyan, with samples collected between November 2024 and March 2025. To ensure consistency between training data features and those encountered in the subsequent target recognition tasks of the harvesting robot, data were acquired using an Intel Realsense D415 (Intel, Santa Clara, CA, USA). With a Python (V3.6) program, images were automatically captured every 500 milliseconds at a resolution of 640 × 640 pixels. During collection, the camera maintained a distance of 5~15 cm from the target. The lighting conditions and shooting angles were varied as much as possible throughout the process. Finally, the collected images underwent manual screening to remove blurry images or those lacking strawberry fruit objects. The collected strawberry image samples are shown in Figure 1.

Figure 1. Strawberry Maturity Sample.

This dataset comprises 1500 images capturing diverse lighting conditions, including front lighting, backlighting, overcast skies, and sunny weather. The strawberry data further encompasses various growth states such as single fruits, clusters of fruits, overlapping fruits, and foliage obstruction. This comprehensive approach ensures the trained model can effectively handle complex scenarios encountered in real-world strawberry cultivation environments.

2.1.2. Image Enhancement

To ensure the efficiency of the strawberry detection model, a rich dataset is required to guarantee the model's strong generalization ability and stability. Sufficient data enhance the model's adaptability to diverse scenarios. Furthermore, due to the complex growing environment of strawberries, factors such as lighting and shadows during photography are difficult to control. By simulating variations in natural lighting conditions, angular deviations, and imaging blur, we expanded the diversity of the strawberry ripeness dataset to enhance the model's generalization capability in real-world scenarios. To achieve this, three data augmentation techniques were employed to enrich the features of strawberry samples within the dataset:

(1) To ensure the model can effectively recognize targets under varying lighting conditions, this study adjusted the exposure levels of captured raw images to accommodate detection requirements in both intense sunlight and dim environments.

(2) To enable the model to adapt to blurry photos caused by the camera's inability to focus, this study employed a Gaussian blur method to simulate the out-of-focus effect.

(3) To simulate the randomness of imaging angles in natural environments, this study enhanced the dataset using image flipping techniques.

After processing the strawberry images as described above. We applied the three image enhancement techniques described above to each original image; the dataset size was expanded from 1500 images to 6000 images (Figure 2).

(**a**) Original image (**b**) Brightness (**c**) Blurred (**d**) Flip

Figure 2. Image Enhancement Sample.

During our research, we surveyed over a dozen strawberry farms in Jiangsu Province and found that farmers determine strawberry ripeness based on the extent of red coloration covering the fruit during harvest. Through extensive sample analysis, strawberry ripeness was categorized into three stages: unripe, half-ripe, and ripe. Unripe strawberries exhibit no red coloration. Half-ripe strawberries display red coverage between 30% and 70% of the fruit's surface area. Strawberries with over 70% red coverage are classified as ripe. The LabelImg annotation tool was used to manually label each sample according to the three stages. Then, text labels were generated in the YOLO format (Table 1).

Table 1. Strawberry Maturity Classification.

Classification	Image	Features
unripe		The red-colored area is less than 30%.
half-ripe		The red-colored area ranges from 30% to 70%.
ripe		The red-colored area exceeds 70%.

2.2. *YOLOv8s Model*

As a newer iteration within the mature YOLO framework, YOLOv8 focuses on achieving an optimal balance between accuracy and speed [29]. Building upon the success of YOLOv5, YOLOv8 introduces new features and enhancements, resulting in significantly improved accuracy. Based on CNNs, YOLOv8 employs an end-to-end object detection approach, directly feeding images into the model to output bounding boxes for target locations and category predictions [30]. YOLOv8 stands as one of the most advanced models in the YOLO series, offering both superior performance and cutting-edge architecture. Its optimized training mechanism, mixed-precision support, and multi-format export capabilities enable flexible and efficient deployment, making it the current benchmark within the YOLO family for balancing performance and ease of use.

For different detection tasks, five YOLOv8 models of varying sizes have been designed. YOLOv8s strikes a good balance between detection speed and accuracy. Compared to YOLOv8n, it achieves higher detection accuracy, while outperforming YOLOv8m and

YOLOv8l in detection speed [31]. This makes it highly suitable for robotic harvesting tasks. This study selected YOLOv8s as the base model and optimized its architecture based on the characteristics of strawberry fruit images.

As shown in Figure 3, the YOLOv8s network architecture consists of four components: the input module, backbone network, neck network, and detection head. The backbone network is responsible for extracting features from the input image, primarily composed of CBS (Convolution-BN-SiLU), C2f (Cross-Stage Local Fusion), and SPPF (Spatial Pyramid Pooling-Fast) modules. The C2f module enhances the model's feature representation capabilities by optimizing feature propagation paths. This significantly improves overall object detection accuracy and inference speed while maintaining efficient computational resource utilization.

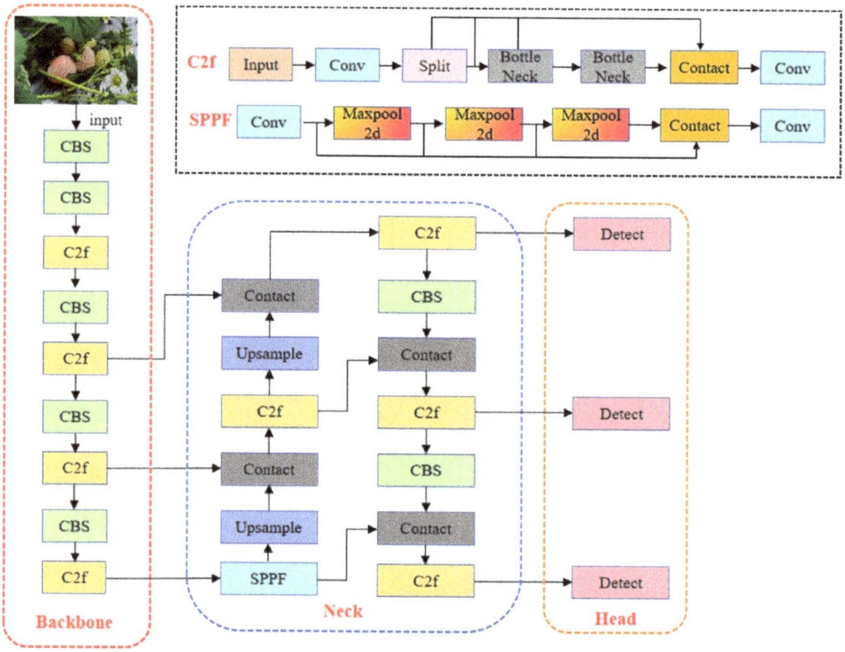

Figure 3. YOLOv8s Overall Network Framework.

The neck section employs a Rep-PAN (Reparametrized Path Aggregation Network) architecture. During training, it utilizes a multi-branch topology to extract features, effectively integrating characteristics from different levels. At inference time, structural reparameterization techniques merge these into a lightweight, computationally efficient graph, perfectly balancing model representational power with inference efficiency. The detection head receives multi-scale feature maps from Rep-PAN, introduces Dynamic Label Allocation (DLA) to reasonably adjust the distribution of positive and negative samples, and adopts an Anchor-Free strategy in the improved prediction head to reduce computational complexity.

2.3. Global Attention Mechanism

In recent years, numerous researchers have incorporated attention mechanisms into their models to achieve improvements in detection accuracy. The core concept of the attention mechanism is to simulate the focusing capabilities of the human visual system, enabling the model to selectively focus on different parts of the input when processing complex data. This enhances both the model's performance and interpretability. Currently, attention mechanisms are primarily categorized into channel-based and spatial-based approaches. By dynamically adjusting feature channel weights by analyzing the importance of each channel, the network can focus more on information from critical channels. Spatial attention enhances the model's focus on key regions by calculating the distribution of importance across spatial dimensions.

CBAM combines channel and spatial attention mechanisms, first adjusting the channel dimension and then optimizing the spatial dimension [32]. This enables the network to focus more intently on task-relevant regions when processing images. However, the drawback of CBAM lies in its reliance on local information processing, which makes it difficult to capture global contextual information and results in poor performance in complex agricultural scenarios.

In this paper, the Global Attention Module (GAM) [33] replaces CBAM. GAM amplifies global-level interaction features while minimizing information dispersion. By computing global context, it effectively captures global information within gesture features, enhancing the model's focus on critical features. Figure 4 shows the channel and spatial attention modules. The calculation method is as follows:

$$F_2 = M_C(F_1) \otimes F_1;\ F_3 = M_S(F_2) \otimes F_2 \tag{1}$$

where F_1 is the input feature, F_2 is the intermediate state, and F_3 is the output feature.

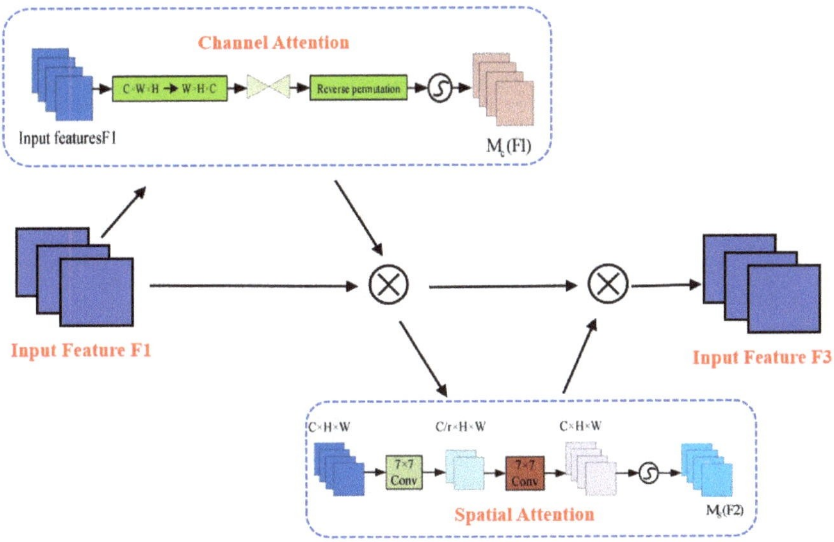

Figure 4. GAM Network Framework.

The channel attention submodule employs multilayer perceptrons to preserve information across channels, spatial width, and height, thereby enhancing global interaction features. The spatial attention submodule fuses spatial information through convolutional layers. GAM addresses CBAM's limitation of insufficient global feature representation when processing images by simultaneously focusing on interactions across both channel and spatial dimensions, thereby strengthening the expression capability of global features.

The channel attention submodule preserves information across three dimensions through a three-dimensional arrangement. A two-layer multilayer perceptron (MLP) is employed to extend cross-dimensional channel-spatial dependencies. This MLP adopts an encoder–decoder architecture with a compression ratio of r (rate = 4).

In the spatial attention submodule, two convolutional layers are employed to fuse spatial features and focus on spatial information. First, a convolution with a 7×7 kernel reduces the number of channels to decrease computational load. Subsequently, another 7×7 convolution increases the channel count to maintain consistency, and the padding value is set to 3. Finally, the output passes through a *Sigmoid* function.

2.4. Fusion of GAM and YOLOv8s

As shown in Figure 5, the GAM module is integrated into the backbone network of the YOLOv8 algorithm to enhance the extraction of both global and local features. This approach preserves finer-grained details while suppressing irrelevant noise, thereby improving the detection of occluded objects. A 4×4-pixel detection head was added to the head section, along with a fourth output layer P_2, to enhance the detection of small targets like strawberry fruits. This design not only preserves YOLOv8's efficiency but also significantly improves the model's ability to detect small objects, achieving higher detection accuracy and robustness.

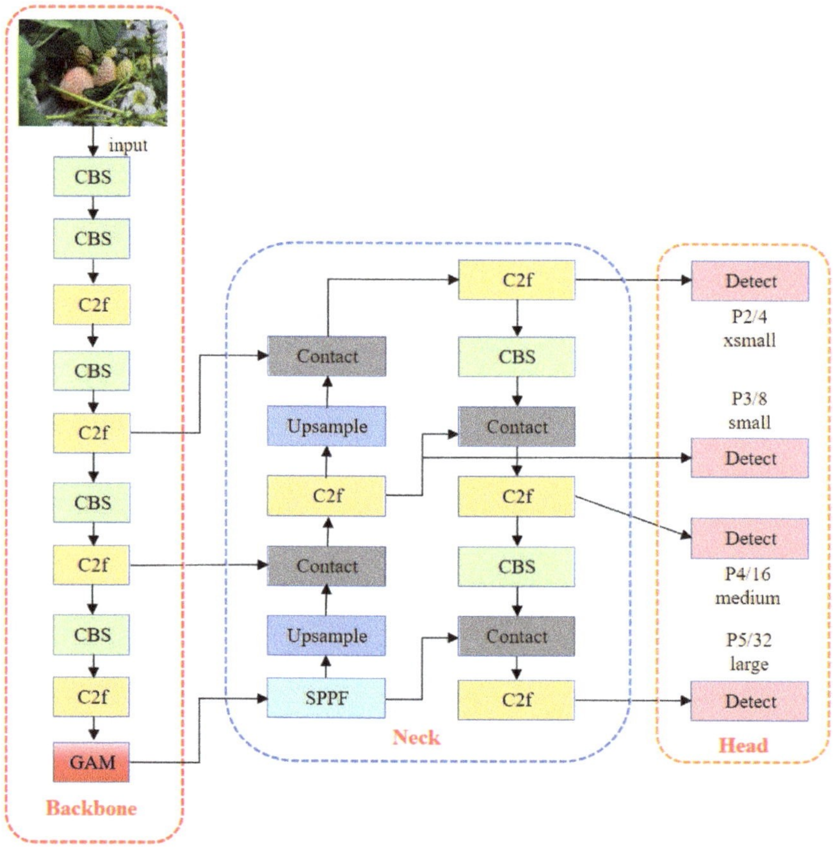

Figure 5. GAM+ YOLOv8s Network Framework.

2.5. Settings and Data Description

Through stratified random sampling, the 6000 images were divided into training, validation, and test sets in an 8:1:1 ratio. The division strictly maintained consistent proportions across the three categories: unripe, semiripe, and ripe. The training set contained

4800 images covering all scenarios without extreme interference. It was used for iterative model parameter training and gradient descent optimization, with model weights saved every 10 iterations. The validation set comprised 600 images with identical scene distributions to the training set. It was used for hyperparameter optimization, overfitting monitoring, and optimal model selection. Early stopping was triggered if validation loss increases for 15 consecutive iterations. The test set contained 600 images representing extreme scenarios not covered in the training or validation sets, enabling unbiased evaluation of the model's generalization capability.

The experimental hardware utilized a Dell T7920 workstation (Dell, American) equipped with an Intel Xeon Gold 6248 CPU, NVIDIA Tesla V100 GPU, 64GB of memory, and 2TB NVMe SSD storage. This configuration provides sufficient computational power for efficient deep learning model training. Training parameters were set based on preliminary experimental results, with the final parameters detailed in the table below. Key parameter selections are outlined in Table 2.

Table 2. Model Training Parameter Settings.

Parameter Name	Value
Epoch	150
Batch Size	8
Algorithm Optimizer	SGD
Initial Learning Rate	0.01
Weight Decay	0.0005
Bounding Box Loss	CIoU Loss
Image Dimensions	640 × 640

3. Results and Discussion

The experimental results and discussion focus on the effectiveness of the YOLOv8s model enhanced with a Global Attention Mechanism (GAM) for strawberry ripeness detection. The model's performance advantages are systematically quantified across five dimensions: standardization of experimental baseline conditions, validation of core improvement modules, cross-comparison with multiple algorithms, robustness testing in complex scenarios, and analysis of the underlying mechanisms behind the results.

3.1. *Performance Enhancement of YOLOv8s Through GAM Module*

To validate the effectiveness of GAM in enhancing the YOLOv8s model, this study employed a controlled variable method. The sole variable tested was "whether the GAM module is introduced," while all other experimental conditions remained identical. We compare the convergence characteristics, detection accuracy, and feature extraction capabilities between the original YOLOv8s and the improved YOLOv8s, with a particular focus on verifying GAM's advantages in complex scenarios.

Model convergence characteristics serve as key indicators for evaluating model stability and training efficiency. By monitoring the trends in training loss (Train-Loss), validation loss (Val-Loss), and validation set mAP, we compared the convergence processes of the two model groups. The results are shown in Figure 6.

By monitoring the trends in training loss, validation loss, and validation set mAP, we observed that both model groups entered the convergence phase after 40 epochs. However, the improved YOLOv8s converged faster, with loss values below 0.3 after 30 epochs, whereas the original model required 45 epochs. The final training loss of the improved model stabilized at 0.23 ± 0.05, and the validation loss stabilized at 0.38 ± 0.02, representing reductions of 0.13 and 0.14, respectively, compared to the original model. Regarding validation set mAP, the improved model reached 88.5% at 50 epochs and stabilized at 91.5% by epoch 120, whereas the original model achieved only 86.9% mAP at epoch 120.

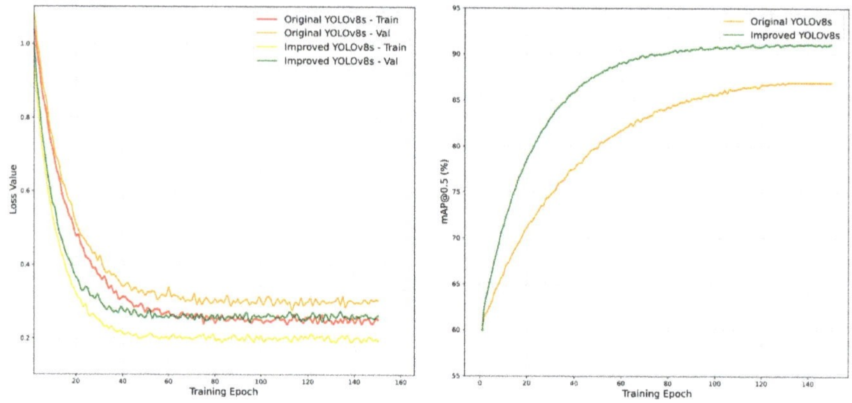

(**a**) Training loss versus validation loss curve (**b**) Validation set mAP@0.5 change curve

Figure 6. Comparison of Convergence Curves Between Original and Improved YOLOv8s.

The core reason for this discrepancy lies in the original YOLOv8s backbone network's reliance solely on local feature aggregation, making it prone to learning irrelevant features like leaf patterns and soil backgrounds. This leads to slow loss reduction and overfitting. In contrast, the GAM module employs a dual-branch interaction of "channel attention–spatial attention" to focus on effective features such as the red areas and contour textures of strawberry fruits. It suppresses the weights of irrelevant features, thereby accelerating convergence and enhancing generalization stability.

3.2. Performance Comparison of Different Models in Strawberry Maturity Assessment

To validate the competitiveness of the improved YOLOv8s among comparable algorithms, we selected traditional YOLO variants, attention-enhanced models, and other common detection models for comparison. A comprehensive evaluation was conducted across three dimensions: accuracy, speed, and lightweight performance. All models were retrained using the training dataset from this study to ensure consistent conditions.

The comparison models include YOLOv3, YOLOv5s, and YOLOv7s; the attention-enhanced model CBAM-YOLOv8s; and the proposed GAM-YOLOv8s. YOLOv3 serves as the baseline model for early strawberry ripeness detection, based on the Darknet53 backbone network; YOLOv5s represents lightweight models, based on CSPDarknet53; YOLOv7s represents medium-to-high-accuracy models, based on the ELAN backbone network; CBAM-YOLOv8s replaces GAM with the traditional CBAM attention mechanism, functioning as the attention mechanism comparison group.

All models were retrained using the training set from this study (under consistent training conditions). After training, various performance metrics were calculated on the test set, and the results are shown in Table 3.

Table 3. Performance Comparison of Different Models in Strawberry Maturity Assessment.

Model	IoU (%)	Precision (%)	Recall (%)	AP Unripe (%)	AP Half-Ripe (%)	AP Ripe (%)	mAP@0.5 (%)	FPS
YOLOv3	72.3	81.5	79.8	78.6	76.2	83.5	79.4	28
YOLOv5s	76.5	85.2	84.1	82.3	80.5	87.8	83.5	42

Table 3. *Cont.*

Model	IoU (%)	Precision (%)	Recall (%)	AP Unripe (%)	AP Half-Ripe (%)	AP Ripe (%)	mAP@0.5 (%)	FPS
YOLOv7s	78.1	86.7	85.9	84.1	82.8	89.2	85.4	38
YOLOv8s	80.2	88.3	87.5	85.7	84.3	90.6	86.9	55
CBAM-YOLOv8s	82.5	89.6	88.8	87.2	86.1	91.8	88.4	51
GAM-YOLOv8s (our model)	85.7	92.1	91.3	90.5	89.8	94.2	91.5	53

The test results indicate that, among the traditional YOLO series, YOLOv3 exhibits relatively low overall performance, achieving only 79.4% mAP@0.5, with an inference speed of 28FPS. Both accuracy and real-time capability fall short. YOLOv5s improves accuracy to 83.5% and reaches an inference speed of 42FPS, demonstrating greater lightweight performance, though recognition accuracy remains insufficient. YOLOv7s achieves further accuracy improvement to 85.4%, but its speed drops to 38FPS, resulting in slightly poorer balance. As a new-generation base model, YOLOv8s enhances mAP@0.5 to 86.9% while significantly boosting inference speed to 55FPS, delivering overall performance superior to its predecessor. CBAM-YOLOv8s, which incorporates an attention mechanism based on YOLOv8s, achieves an even higher accuracy of 88.4%, though its inference speed slightly decreases to 51FPS. This demonstrates that the attention mechanism positively contributes to accuracy.

The proposed GAM-YOLOv8s model achieves the best performance across all metrics, with a mAP@0.5 of 91.5%. This represents improvements of 12.1%, 8.0%, 6.1%, 4.6%, and 3.1% over YOLOv3, YOLOv5s, YOLOv7s, YOLOv8s, and CBAM-YOLOv8s, respectively. Meanwhile, the inference speed remains at 53FPS, which is only slightly lower than that of the baseline YOLOv8s. This demonstrates that the model maintains high real-time performance and lightweight advantages while significantly improving detection accuracy. Its comprehensive performance stands out, making it highly valuable for practical applications.

We further analyzed the feature heatmaps, which are shown in Figure 7. We compared the activation focus of the base YOLOv8s (Figure 7a) and the proposed GAM-YOLOv8s (Figure 7c). By measuring the pixel intensities in the feature maps, we observed that the GAM-YOLOv8s model concentrated significantly more activation energy (approximately 85–90%) within the ground-truth bounding box of the strawberry, compared to the more dispersed activations of the base model. This difference in feature concentration demonstrates that GAM effectively filters background noise and directs the model's attention. This focused learning mechanism ensures more efficient and relevant gradient updates, resulting in faster convergence and lower, more stable loss values.

3.3. Verification of Model Robustness Improvement Across Different Scenarios

In the natural environment of strawberry cultivation, variations in light intensity and fruit occlusion represent the two primary core interference factors affecting detection accuracy. To validate the robustness of the enhanced YOLOv8s model, this study designed "Light Robustness Tests" and "Occlusion Robustness Tests" to compare performance differences between the improved model and the baseline YOLOv8s. The test dataset was categorized into four lighting conditions: "front-lit," "back-lit," "cloudy," and "dusk low-light," with 150 images per category. The mAP comparisons between the two models are presented in the table below (Table 4).

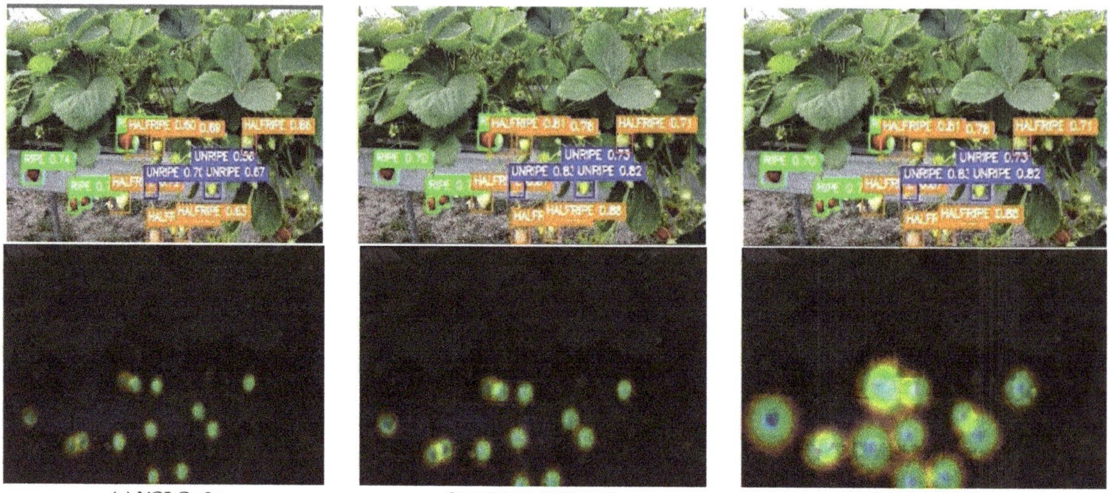

(**a**) YOLOv8s (**b**) CBAM-YOLOv8s (**c**) GAM-YOLOv8s

Figure 7. Feature Heatmap Comparison Between Original YOLOv8s and Enhanced YOLOv8s.

Table 4. Performance Comparison of Different Models in Strawberry Maturity Assessment.

Lighting Scenario	Illuminance (Lux)	YOLOv8s	GAM-YOLOv8s	Improvement Rate
Front lighting	8000–12,000	90.2	93.5	3.3
Backlighting	1000–3000	78.5	87.2	8.7
Overcast	500–1500	85.6	90.1	4.5
Low light at dusk	200–500	72.3	82.8	10.5

Comparisons of mAP@0.5 under varying lighting conditions reveal that both the baseline YOLOv8s and the improved GAM-YOLOv8s exhibit detection performance that fluctuates with light intensity. However, the enhanced model demonstrates superior overall performance and greater robustness. In front-lit scenes (8000–12,000 lux), the base model achieved a mAP of 90.2%, while GAM-YOLOv8s improved to 93.5%, representing a 3.3 percentage point increase. In back-lit conditions (1000–3000 lux), the baseline model drops to 78.5%, while the improved model maintains 87.2%, an increase of 8.7 percentage points. Under cloudy conditions (500–1500 lux), the baseline model achieves 85.6%, and the improved model reaches 90.1%, an increase of 4.5 percentage points. In low-light conditions at dusk (200–500 lux), the baseline model achieved only a mAP of 72.3%, a performance close to that of our defined "practicality threshold," which is set at 70% mAP@0.5. Given our operational requirements for harvesting robots, a detection accuracy below 70% would result in missed detections of ripe fruit or a high false detection rate of unripe fruit. In contrast, GAM-YOLOv8s maintained a mAP of 82.8%, well above this survival threshold, achieving a maximum improvement of 10.5 percentage points. Overall, as illumination decreased from bright to dim, the baseline model's mAP declined by 19.8%, while the improved model only decreased by 10.7%. Moreover, the improvement effect exhibited an increasing trend as lighting conditions deteriorated.

Occlusion Robustness Testing categorizes the dataset into three scenarios based on fruit occlusion levels, with 200 images per category and 67–68 samples of each maturity stage across all three categories. Strawberry leaves serve as occluding objects to simulate canopy shading in natural cultivation. The unobstructed scenario features fruit surfaces without leaf coverage, preserving complete features. The partially obstructed scenario shows 30–50% leaf coverage on fruit surfaces, with partial features visible. The severely obstructed scenario exhibits 50–70% leaf coverage, revealing only edge or localized features. Test results using the AP and average miss rate as core metrics for each category are presented in Table 5.

Table 5. Model Performance Comparison at Different Occlusion Levels.

Occlusion Level	Occlusion Area Ratio	Model	AP Unripe (%)	AP Half-Ripe (%)	AP Ripe (%)	Average Miss Rate (%)
Unobstructed	0%	YOLOv8s	89.2	88.5	93.8	3.2
		GAM-YOLOv8s	92.5	91.8	95.6	1.8
Partially obstructed	30–50%	YOLOv8s	81.5	79.8	86.2	8.5
		GAM-YOLOv8s	88.6	87.2	92.3	4.2
Severely obstructed	50–70%	YOLOv8s	71.8	69.5	78.6	15.6
		GAM-YOLOv8s	85.6	83.5	90.2	6.8

Table 5 shows that, as the strawberry occlusion area increases from 0% to 50–70%, both YOLOv8s and GAM-YOLOv8s exhibit declining detection accuracy (AP). However, the improved model demonstrates a significant advantage that grows more pronounced with increasing occlusion. Without occlusion, YOLOv8s achieved AP values of 89.2%, 88.5%, and 93.8% for the three strawberry categories, with an average miss rate of 3.2%. GAM-YOLOv8s improved these to 92.5%, 91.8%, and 95.6%, while reducing the miss rate to 1.8%. Under partial occlusion (30–50%), YOLOv8s' AP dropped to 81.5–86.2% with a false negative rate rising to 8.5%, while GAM-YOLOv8s achieved an AP of 87.2–92.3% and a false negative rate of 4.2%, demonstrating a 7.1–7.4% AP improvement and a clear advantage. Under severe occlusion (50–70%), YOLOv8s' AP dropped to 71.8–78.6% with a false negative rate of 15.6%. GAM-YOLOv8s maintained an AP of 83.5–90.2% and a false negative rate of 6.8%, achieving an AP improvement of 13.8–14%. The results demonstrate that GAM-YOLOv8s significantly mitigates performance degradation in occlusion scenarios, with the most pronounced effects observed under moderate to severe occlusion conditions.

3.4. Compared with Existing Research

In recent years, many scholars have used various deep learning models to study strawberry maturity (Table 6). Tao et al. [34] propose a strawberry maturity recognition algorithm named YOLOv5s-BiCE, which replaces the upsampling algorithm with the CARAFE module structure to achieve multi-scale feature fusion. Compared to the original YOLOv5s model, YOLOv5s-BiCE achieves a 2.8% improvement in mean average precision and a 7.4% increase in accuracy. Chen et al. [35] propose a CES-YOLOv8 network architecture that replaces some C2f modules in the YOLOv8 backbone with ConvNeXt V2 modules to enhance feature capture capabilities for strawberries at different ripeness levels. The introduction of the ECA attention mechanism further improves feature representation. Experimental results demonstrate that the CES-YOLOv8 model achieves an accuracy of 88.20%, a recall rate of 89.80%, a mAP50 of 92.10%, and an F1 score of 88.99% in complex environments.

Table 6. Comparison with Existing Research.

Author	Year	Method	mAP
Tao et al. [34]	2024	YOLOv5s-BiCE	96.1%
Chen et al. [35]	2024	CES-YOLOv8	88.99%

Table 6. *Cont.*

Author	Year	Method	mAP
Cai et al. [36]	2022	Improved DeepLabV3+	90.90%
Tamrakar et al. [37]	2024	YOLOv5s-CGhostnet	91.7%
Our model	2025	YOLOv8s-GAM	91.5%

Cai et al. [36] introduced an attention mechanism into the main network and the hollow space pyramid pooling module of the DeepLabV3+ network to enhance the feature information of strawberry images. Experimental results demonstrate that this method can accurately segment strawberry images at different stages of ripeness, achieving a model mean pixel accuracy of 90.9% and a mean intersection-over-union ratio of 83.05%, with a frame rate of 7.67 frames per second. Tamrakar et al. [37] proposed a lightweight improved version of the YOLOv5s-CGhostnet model to enhance strawberry detection performance. Compared to the original YOLOv5 model, the model size was significantly reduced by 85.09%, and the GFLOPS computational load decreased by 88.5%.

In terms of detection results, our model outperforms those of Chen et al. [35] and Cai et al. [36], with 2.51% and 0.6% higher mAP values, respectively. Compared with Tao et al. [34] and Tamrakar et al. [37], our model has a slightly lower mAP. The main reason is that the image samples used in this paper are closer to the natural environment, with more complex backgrounds and lighting. Regarding model parameters, Tao et al. [34] and Tamrakar et al. [37] selected YOLOv5s as their base model due to its lower number of parameters. This paper opts for YOLOv8s, which features fewer parameters and higher accuracy. Its computational speed outperforms Chen et al. [35] and Cai et al. [36], enabling precise strawberry maturity detection with minimal computational effort. Through comparisons with models in previous research, the model proposed in this paper is effective and has good detection effects in natural environments.

4. Conclusions and Future Work

To achieve precise detection of red strawberry ripeness, this study collected photographs of strawberry maturity under various natural environments and enhanced the expressiveness of feature information in the dataset through different image enhancement methods. Considering the application requirements for picking robots, this study selected the YOLOv8s model as the base model due to its low parameter count and high accuracy. Leveraging the advantages of attention mechanisms, we integrated a Global Attention Mechanism (GAM) into the backbone network. This effectively enhances the model's extraction of key information related to strawberry ripeness while suppressing the learning of irrelevant information. Experimental results demonstrate that the improved YOLOv8s achieves excellent performance, with a mAP of 91.5% for three maturity levels and a frame rate of 53 FPS. Compared with other mainstream YOLO models, our model delivers optimal performance. It also exhibits robust detection capabilities under varying lighting conditions and occlusions, meeting the requirements for subsequent deployment on harvesting robots.

In subsequent tasks, we will further expand the strawberry ripeness dataset to include variations in weather conditions and growth stages, enhancing the model's generalization capabilities under natural environments. We will also deploy this model onto strawberry-harvesting robots for field validation and application. Based on actual test results, we will optimize the model's performance. Concurrently, leveraging the harvesting robot's hardware architecture, we will develop autonomous learning capabilities to enable simultaneous model training and sample collection. The strawberry ripeness detection model developed in this research holds significant application potential and will contribute to the advancement of agricultural modernization.

Author Contributions: Conceptualization, S.Z., C.F., T.H. and Y.J.; methodology, S.Z., C.F., T.H. and Y.J.; software, S.Z.; validation, S.Z., C.F. and Y.J.; formal analysis, S.Z., C.F., T.H. and Y.J.; investigation, S.Z. and Y.J.; resources, C.F.; data curation, Y.J. and T.H.; writing—original draft preparation, S.Z.; writing—review and editing, S.Z. and Y.J. All authors have read and agreed to the published version of the manuscript.

Funding: The work was supported by the Young Scientists Fund of the National Natural Science Foundation of China (No. 32401693), Jiangsu Province Modern Agricultural Machinery Equipment and Technology Demonstration and Promotion Project (No. NJ2023-23), and Priority Academic Program Development of Jiangsu Higher Education Institutions (No. PAPD2023-87), and Changzhou Agricultural Science and Technology Innovation and Demonstration Extension Project (No. KCSF(2023)11).

Institutional Review Board Statement: Not applicable.

Informed Consent Statement: Not applicable.

Data Availability Statement: The original contributions presented in this study are included in the article. Further inquiries can be directed to the corresponding author.

Conflicts of Interest: The authors declare no conflicts of interest.

References

1. Skrovankova, S.; Sumczynski, D.; Mlcek, J.; Jurikova, T.; Sochor, J. Bioactive compounds and antioxidant activity in different types of berries. *Int. J. Mol. Sci.* **2015**, *16*, 24673–24706. [CrossRef] [PubMed]
2. Wang, Z.; Cang, T.; Qi, P.; Zhao, X.; Xu, H.; Wang, X.; Zhang, H.; Wang, X. Dissipation of four fungicides on greenhouse strawberries and an assessment of their risks. *Food Control* **2015**, *55*, 215–220. [CrossRef]
3. Elhariri, E.; El-Bendary, N.; Saleh, S.M. Strawberry-DS: Dataset of annotated strawberry fruits images with various developmental stages. *Data Brief* **2023**, *48*, 109165. [CrossRef]
4. Zhao, S.; Liu, J.; Hua, T.; Jiang, Y. Improved UNet Recognition Model for Multiple Strawberry Pests Based on Small Samples. *Agronomy* **2025**, *15*, 2252. [CrossRef]
5. Liu, J.; Wu, S. Research Progress and Prospect of Strawberry Whole-process Farming Mechanization Technology and Equipment. *Trans. Chin. Soc. Agric. Mach.* **2021**, *52*, 1–16.
6. Zhao, S.; Liu, J.; Wu, S. Multiple disease detection method for greenhouse-cultivated strawberry based on multiscale feature fusion Faster R_CNN. *Comput. Electron. Agric.* **2022**, *199*, 107176. [CrossRef]
7. Tang, C.; Chen, D.; Wang, X.; Ni, X.; Liu, Y.; Liu, Y.; Mao, X.; Wang, S. A fine recognition method of strawberry ripeness combining Mask R-CNN and region segmentation. *Front. Plant Sci.* **2023**, *14*, 1211830. [CrossRef]
8. Liu, J.; Jiang, Y. Industrialization trends and multi-arm technology direction of harvesting robots. *Trans. Chin. Soc. Agric. Mach.* **2024**, *55*, 1–17.
9. Xiong, Y.; Peng, C.; Grimstad, L.; From, P.J.; Isler, V. Development and field evaluation of a strawberry harvesting robot with a cable-driven gripper. *Comput. Electron. Agric.* **2019**, *157*, 392–402. [CrossRef]
10. Yu, Y.; Zhang, K.; Liu, H.; Yang, L.; Zhang, D. Real-time visual localization of the picking points for a ridge-planting strawberry harvesting robot. *IEEE Access* **2020**, *8*, 116556–116568. [CrossRef]
11. Liu, Y.; Wei, C.; Yoon, S.-C.; Ni, X.; Wang, W.; Liu, Y.; Wang, D.; Wang, X.; Guo, X. Development of multimodal fusion technology for tomato maturity assessment. *Sensors* **2024**, *24*, 2467. [CrossRef]
12. Wang, F.; Zhao, C.; Yang, H.; Jiang, H.; Li, L.; Yang, G. Non-destructive and in-site estimation of apple quality and maturity by hyperspectral imaging. *Comput. Electron. Agric.* **2022**, *195*, 106843. [CrossRef]
13. Gupta, A.K.; Pathak, U.; Tongbram, T.; Medhi, M.; Terdwongworakul, A.; Magwaza, L.S.; Mditshwa, A.; Chen, T.; Mishra, P. Emerging approaches to determine maturity of citrus fruit. *Crit. Rev. Food Sci. Nutr.* **2022**, *62*, 5245–5266. [CrossRef]
14. Cho, B.-H.; Kim, Y.-H.; Lee, K.-B.; Hong, Y.-K.; Kim, K.-C. Potential of snapshot-type hyperspectral imagery using support vector classifier for the classification of tomatoes maturity. *Sensors* **2022**, *22*, 4378. [CrossRef]
15. Arefi, A.; Motlagh, A.M.; Mollazade, K.; Teimourlou, R.F. Recognition and localization of ripen tomato based on machine vision. *Aust. J. Crop Sci.* **2011**, *5*, 1144–1149.
16. Fadhel, M.A.; Hatem, A.S.; Alkhalisy, M.A.E.; Awad, F.H.; Alzubaidi, L. Recognition of the unripe strawberry by using color segmentation techniques. *Int. J. Eng. Technol.* **2018**, *7*, 3383–3387.
17. Mo, S.-T.; Dong, T.; Zhao, X.-X.; Kan, J.-M. Discriminant model of banana fruit maturity based on genetic algorithm and SVM. *J. Fruit Sci.* **2022**, *39*, 2418–2427.

18. Khojastehnazhand, M.; Mohammadi, V.; Minaei, S. Maturity detection and volume estimation of apricot using image processing technique. *Sci. Hortic.* **2019**, *251*, 247–251. [CrossRef]
19. Malik, M.H.; Zhang, T.; Li, H.; Zhang, M.; Shabbir, S.; Saeed, A. Mature tomato fruit detection algorithm based on improved HSV and watershed algorithm. *IFAC Pap.* **2018**, *51*, 431–436. [CrossRef]
20. Chen, M.; Shi, X.; Zhang, Y.; Wu, D.; Guizani, M. Deep feature learning for medical image analysis with convolutional autoencoder neural network. *IEEE Trans. Big Data* **2017**, *7*, 750–758. [CrossRef]
21. Manakitsa, N.; Maraslidis, G.S.; Moysis, L.; Fragulis, G.F. A review of machine learning and deep learning for object detection, semantic segmentation, and human action recognition in machine and robotic vision. *Technologies* **2024**, *12*, 15. [CrossRef]
22. Zhou, X.; Lee, W.S.; Ampatzidis, Y.; Chen, Y.; Peres, N.; Fraisse, C. Strawberry maturity classification from UAV and near-ground imaging using deep learning. *Smart Agric. Technol.* **2021**, *1*, 100001. [CrossRef]
23. Miao, R.; Li, Z.; Wu, J. Lightweight maturity detection of cherry tomato based on improved YOLO v7. *Trans. Chin. Soc. Agric. Mach.* **2023**, *54*, 225–233.
24. Qiu, C.; Tian, G.; Zhao, J.; Liu, Q.; Xie, S.; Zheng, K. Grape maturity detection and visual pre-positioning based on improved YOLOv4. *Electronics* **2022**, *11*, 2677. [CrossRef]
25. Fan, Y.; Zhang, S.; Feng, K.; Qian, K.; Wang, Y.; Qin, S. Strawberry maturity recognition algorithm combining dark channel enhancement and YOLOv5. *Sensors* **2022**, *22*, 419. [CrossRef] [PubMed]
26. Huang, Y.; Jiang, X.; Zhou, C.; Zhuo, X.; Xiong, J.; Zhang, M. Study on mango ripeness detection on production line based on improved YOLOv8s. *J. Food Meas. Charact.* **2025**, *19*, 768–780. [CrossRef]
27. Wang, C.; Li, C.; Han, Q.; Wu, F.; Zou, X. A performance analysis of a litchi picking robot system for actively removing obstructions, using an artificial intelligence algorithm. *Agronomy* **2023**, *13*, 2795. [CrossRef]
28. Tang, Y.; Qiu, J.; Zhang, Y.; Wu, D.; Cao, Y.; Zhao, K.; Zhu, L. Optimization strategies of fruit detection to overcome the challenge of unstructured background in field orchard environment: A review. *Precis. Agric.* **2023**, *24*, 1183–1219. [CrossRef]
29. Sohan, M.; Sai Ram, T.; Rami Reddy, C.V. A review on yolov8 and its advancements. In Proceedings of the International Conference on Data Intelligence and Cognitive Informatics, Tirunelveli, India, 18–20 November 2024; pp. 529–545.
30. Yang, G.; Wang, J.; Nie, Z.; Yang, H.; Yu, S. A lightweight YOLOv8 tomato detection algorithm combining feature enhancement and attention. *Agronomy* **2023**, *13*, 1824. [CrossRef]
31. Wang, A.; Qian, W.; Li, A.; Xu, Y.; Hu, J.; Xie, Y.; Zhang, L. NVW-YOLOv8s: An improved YOLOv8s network for real-time detection and segmentation of tomato fruits at different ripeness stages. *Comput. Electron. Agric.* **2024**, *219*, 108833. [CrossRef]
32. Woo, S.; Park, J.; Lee, J.-Y.; Kweon, I.S. Cbam: Convolutional block attention module. In Proceedings of the European conference on computer vision (ECCV), Munich, Germany, 8–14 September 2018; pp. 3–19.
33. Liu, Y.; Shao, Z.; Hoffmann, N. Global attention mechanism: Retain information to enhance channel-spatial interactions. *arXiv* **2021**, arXiv:2112.05561. [CrossRef]
34. Tao, Z.; Li, K.; Rao, Y.; Li, W.; Zhu, J. Strawberry maturity recognition based on improved YOLOv5. *Agronomy* **2024**, *14*, 460. [CrossRef]
35. Chen, Y.; Xu, H.; Chang, P.; Huang, Y.; Zhong, F.; Jia, Q.; Chen, L.; Zhong, H.; Liu, S. CES-YOLOv8: Strawberry maturity detection based on the improved YOLOv8. *Agronomy* **2024**, *14*, 1353. [CrossRef]
36. Cai, C.; Tan, J.; Zhang, P.; Ye, Y.; Zhang, J. Determining strawberries' varying maturity levels by utilizing image segmentation methods of improved deeplabv3+. *Agronomy* **2022**, *12*, 1875. [CrossRef]
37. Tamrakar, N.; Karki, S.; Kang, M.Y.; Deb, N.C.; Arulmozhi, E.; Kang, D.Y.; Kook, J.; Kim, H.T. Lightweight Improved YOLOv5s-CGhostnet for Detection of Strawberry Maturity Levels and Counting. *AgriEngineering* **2024**, *6*, 962–978. [CrossRef]

agriculture

MDPI

Article

Microclimate Condition Influence on the Physicochemical Properties and Antioxidant Activity of Pomegranate (*Punica granatum* L.): A Case Study of the East Adriatic Coast

Mira Radunić [1,2,*], Maja Jukić Špika [2,3,*], Jelena Gadže [4], Smiljana Goreta Ban [2,5], Juan Carlos Díaz-Pérez [6] and Dan MacLean [7]

1 Department of Plant Sciences, Institute for Adriatic Crops and Karst Reclamation, Put Duilova 11, 21000 Split, Croatia
2 Centre of Excellence for Biodiversity and Molecular Plant Breeding, Svetošimunska 25, 10000 Zagreb, Croatia; smilja@iptpo.hr
3 Department of Applied Science, Institute for Adriatic Crops and Karst Reclamation, Put Duilova 11, 21000 Split, Croatia
4 Department for Pomology, Division of Horticulture and Landscape Architecture, Faculty of Agriculture, University of Zagreb, Svetošimunska 25, 10000 Zagreb, Croatia; jgadze@agr.hr
5 Institute of Agriculture and Tourism Poreč, Karla Huguesa 8, 52440 Poreč, Croatia
6 Department of Horticulture, University of Giorgia, 2360 Rainwater Road, Tifton, GA 31793, USA; jcdiaz@uga.edu
7 AgroFresh, 510-530 Walnut Street, Suite 1350, Philadelphia, PA 19106, USA
* Correspondence: mira.radunic@krs.hr (M.R.); maja.jukic.spika@krs.hr (M.J.Š.)

Abstract: The pomegranate cultivar Barski slatki, the most widely cultivated on the Eastern Adriatic coast, was evaluated over one growing season across four growing areas to assess its pomological and chemical properties and antioxidant activity. Results showed that location significantly influenced fruit weight, volume, number of arils per fruit, and both total and individual aril weight, with the Kaštela (CRO) site producing the largest fruits and highest aril yields. Climatic factors, such as precipitation during bud differentiation, flowering, and early fruit development, were found to impact fruit set, aril number, and fruit size. Aril and juice yields, however, remained relatively stable across sites. Notable differences were observed in total soluble solids, titratable acidity, pH, total phenolic content, and anthocyanin profiles. Location with higher rainfall occurring during fruit growth favored enhanced phenolic accumulation. Although total anthocyanin content remained consistent among locations, significant variation occurred in aril coloration and composition of individual anthocyanins. In conclusion, microclimatic factors, particularly rainfall distribution, temperature, and altitude, play a decisive role in shaping the physical, chemical, and visual attributes of 'Barski slatki'. Despite being cultivated under similar Mediterranean conditions, the observed differences across sites highlight the strong adaptability of this cultivar to diverse agroecological environments, while maintaining stable quality.

Keywords: pomegranate; meteorological factors; climate change; fruit size; anthocyanin; phenolics; antioxidant activity; super fruit; fruit quality

1. Introduction

The pomegranate (*Punica granatum* L.) has been cultivated throughout the Mediterranean region since ancient times, particularly in South Europe and North Africa, as well as in parts of Asia, especially Persia. Countries such as Turkey, Spain, Italy, Greece, and

Morocco, which are characterized by a favorable Mediterranean climate and long-standing agricultural traditions, play a central role in global pomegranate production [1]. In recent decades, pomegranate cultivation has expanded to new subtropical and tropical regions in the Americas, South Africa, and Australia, driven by the increasing recognition of the pomegranate as a functional food and a rich source of nutraceuticals, including bioactive compounds such as anthocyanins, phenolic acids, flavonoids, tannins, and vitamins [2].

Pomegranates thrive in climatic conditions characterized by hot, dry summers and mild, rainy winters [3,4]. The Mediterranean region, with its varied topography and microclimates, ranging from coastal plains to inland valleys, offers optimal conditions for cultivating different pomegranate cultivars. Despite their tolerance to low temperatures (as low as −10 °C) and moderate water requirements (rainfall or equivalent irrigation of around 500 mm annually) [4], pomegranate trees are prone to physiological fruit disorders caused by weather and microclimate fluctuations [5]. Optimal fruit development is achieved under a regime of summer drought, followed by winter rains that replenish soil moisture. This climatic combination plays a crucial role in achieving the desired balance of sweetness, acidity, and juiciness, while fruit quality can also be objectively evaluated using modern non-destructive methods [6].

Fruit quality, both internal and external, is determined by a combination of factors, including genotype, growing region, climate, and cultivation practices [7,8]. While genotype plays a fundamental role in determining physicochemical traits of fruit peel and arils (weight and red color), environmental variables such as temperature, micro location, and altitude have a decisive influence on the expression of traits such as fruit size, skin coloration, and biochemical composition [2,9,10]. In particular, temperature has a pronounced influence on the physiological and biochemical development of different fruit species [11]. Although warmer temperatures can enhance sugar metabolism in the fruit, high temperatures are generally associated with a reduction in fruit size and sugar content, as well as changes in skin pigmentation and firmness [12,13]. During key phenological stages, elevated temperatures can alter fruit morphology, sugar accumulation, firmness, and coloration [11]. In addition to temperature, other microclimatic elements such as humidity, UV light intensity, PAR (Photosynthetically Active Radiation), day–night temperature differences, precipitation, and soil properties also exert a cumulative influence, albeit to a lesser extent [14,15].

On the eastern Adriatic coast, pomegranate cultivation remains rooted in tradition and is mostly limited to small orchards and home gardens. Numerous cultivars are successfully grown in these areas, but the predominant local cultivar is 'Barski slatki', which is highly valued for its sensory and nutritional properties. Nevertheless, there is a notable gap in the scientific literature regarding the influence of specific environmental parameters on the physicochemical and antioxidant properties of this cultivar. This study aims to fill this gap by investigating how different microclimatic conditions on the eastern Adriatic coast affect the physical, chemical, and antioxidant properties of 'Barski slatki' pomegranates. Understanding these microclimatic interactions is essential not only for improving cultivation practices and postharvest handling, but also for advancing varietal selection and increasing the market potential of fruit destined for both fresh consumption and value-added processing.

2. Materials and Methods

2.1. Plant Material

Fruit samples of pomegranate 'Barski slatki' were collected from four orchards along the Eastern Adriatic coast (Kaštela and Metković in Croatia; Mostar, Bosnia and Herzegovina, and Bar, Montenegro). The 12-year-old trees grown on their own roots were trained into a shrub with 3–5 main trunks and planted at a spacing of 3 × 4.5 m. The trees were in good physiological condition, exhibiting optimal vigor and productivity. The orchards were drip-irrigated, and other cultivation practices, such as plant protection, pruning, and fertilization, were applied in accordance with accepted commercial practices [16]. The main soil types at the study sites were as follows: Kaštela—anthrosol developed from flysch, karst synclines, and colluvial deposition; Metković—alluvial soil; Mostar—eutric cambisol; and Bar—eutric brown soil. Flowering at the study sites occurred as follows: Kaštela—12 May; Metković—14 May; Mostar—10 May; and Bar—17 May. Fruits were harvested at the time of harvest maturity (6–10 October), which was determined according to standard local practices, when the fruit skin began to show superficial cracking. At each site, fruits were collected from three trees and pooled to form one composite sample representing a single biological replicate. Three such composite samples were obtained per site. From each tree, three fruits of uniform size were randomly collected around the canopy.

2.2. Climatic Condition

The sampling locations (Figure 1) are situated in the Mediterranean climate, which is defined as a Csa climate type according to the Köppen–Geiger climate classification [17].

Figure 1. Sampling locations: Kaštela and Metković (Croatia), Mostar (Bosnia and Herzegovina), and Bar (Montenegro). The inset in red in the top right corner shows the position of the study area within Europe.

The climatic conditions shown in Table 1 are as follows:

1. Kaštela is situated on the coastal part of Croatia (CRO), Middle Dalmatia (43°33′05″ N, 16°20′23″ E, 9 altitude). The main climatic characteristics were an average annual temperature of 16.4 °C, an absolute minimum of −3.7 °C (December), and an absolute maximum of 37.2 °C (July). Annual rainfall in this location was 1428 mm, but during April to October, it was 712 mm (50% of the total). Average sunny hours totaled 2700.
2. Metković is situated in the lower part of the Neretva river valley, South Dalmatia, Croatia (CRO) (43°00′29″ N, 17°38′47″ E, 5 m altitude). The main climatic characteristics were a mean daily temperature of 15.6 °C, an absolute minimum of −6.4 °C (December), and an absolute maximum of 36.5 °C (August). Annual rainfall in this location was 1555.4 mm, with 646 mm (42% of total) recorded from April to October. Average sunny hours totaled 2700.

3. Mostar is situated in the upper part of the Neretva Valley, in the southern part of Bosnia and Herzegovina (BIH) (43°25′13″ N, 17°52′26″ E, 100 m altitude). Mostar has a temperate Mediterranean climate with milder but colder winters (characterized by little to no snow) and hot summers. Average annual temperature was 15.7 °C, with an absolute minimum of −7.8 °C (December) and an absolute maximum of 38.9 °C (July). Annual rainfall was 1848.9 mm, with 741.6 mm (40% of total) recorded from April to September. Annual sunny hours totaled 2285.
4. Bar is situated on the southeast coastal part of Montenegro (MNE) (42°02′22″ N, 19°09′09″ E, 143 m altitude). The climate of Bar is determined by the proximity of the Adriatic Sea and Lake Skadar and the mountain massif of Rumia. The average annual temperature of 16.9 °C characterizes it, with an absolute minimum of 0 °C (February) and an absolute maximum of 35.4 °C (August). Annual rainfall was 1826.4 mm, with 641.8 mm (35% of total) recorded during April to October. Annual sunny hours totaled 2555.

Table 1. Microclimatic variables in four pomegranate growing areas: Kaštela and Metković (Croatia), Mostar (Bosnia and Herzegovina), and Bar (Montenegro).

Location	**Kaštela, Croatia (CRO)**					**Metković, Croatia (CRO)**				
Month	**Temperature (°C)**			**Rainfall (mm)**	**Humidity (%)**	**Temperature (°C)**			**Rainfall (mm)**	**Humidity (%)**
	Mean	**Max**	**Min**			**Mean**	**Max**	**Min**		
January	8.1	15.9	−3.0	228.4	65	7.1	16.7	−4.0	243.1	68
February	7.1	15.1	−3.0	81.8	57	6.0	17.5	−3.9	71.0	63
March	10.2	19.0	2.0	88.2	58	9.8	20.7	−0.2	132.0	60
April	15.2	24.5	8.5	111.2	70	15.1	27.1	6.9	54.4	68
May	20.9	34.7	9.7	17.4	54	20.8	34.0	8.2	33.0	67
June	22.2	31.9	13.7	242.4	61	21.5	32.8	12.6	203.1	69
July	26.3	37.2	17.4	34.0	52	25.6	36.4	14.7	25.8	69
August	26.8	36.1	18.0	14.6	50	25.9	36.5	15.5	34.7	69
September	22.5	31.0	14.6	90.4	58	21.8	32.0	12.8	52.9	67
October	15.7	26.2	5.5	202.0	59	14.0	27.7	1.7	242.2	66
November	12.0	19.7	3.1	119.2	76	10.8	19.6	0.3	130.4	72
December	9.4	18.6	−3.7	198.6	66	8.3	21.0	−6.4	332.8	73
Average	16.4					15.6				
Summa				1428.2					1555.4	
Location	**Mostar, Bosnia and Herzegovina (BIH)**					**Bar, Monte Negro (MNE)**				
Month	**Temperature (°C)**			**Rainfall (mm)**	**Humidity (%)**	**Temperature (°C)**			**Rainfall (mm)**	**Humidity (%)**
	Mean	**Max**	**Min**			**Mean**	**Max**	**Min**		
January	6.3	16.3	−3.6	336.9	63	9.6	17.2	1.0	351.1	69
February	5.9	18.0	−3.0	117.4	55	8.1	20.1	0.0	141.9	54
March	9.5	20.0	1.6	173.6	60	10.8	22.2	2.7	188.7	61
April	15.6	26.4	7.3	68.1	65	15.5	25.2	8.2	52.5	72
May	21.3	35.0	7.6	35.7	55	20.5	30.8	10.7	39.7	66
June	22.0	36.1	12.0	190.0	62	22.3	32.0	14.7	175.8	68
July	26.3	38.9	15.0	15.8	55	25.5	33.8	17.7	27.0	66
August	26.8	37.9	18.1	57.8	51	25.8	35.4	19.9	11.8	66
September	22.3	34.8	14.2	127.8	56	22.8	30.8	15.6	103.1	62
October	14.8	28.6	3.6	246.6	62	16.9	26.5	6.7	231.9	65
November	10.4	19.4	2.9	164.2	75	13.1	21.1	5.9	236.7	79
December	7.5	18.2	−7.8	315.0	68	11.9	19.9	1.1	265.9	70

Table 1. *Cont.*

Location	Mostar, Bosnia and Herzegovina (BIH)					Bar, Monte Negro (MNE)				
Month	Temperature (°C)			Rainfall (mm)	Humidity (%)	Temperature (°C)			Rainfall (mm)	Humidity (%)
	Mean	Max	Min			Mean	Max	Min		
Average	15.7					16.9				
Summary				1848.9					1826.4	

Climate parameters registered for the studied locations were obtained from the Croatian Meteorological and Hydrological Service, Federal Hydrometeorological Institute, Federation of BIH, and Institute of Hydrometeorology and Seismology, Montenegro, respectively.

2.3. Physical Properties

The weight of the fruit (FW; g) was measured using a digital balance (Mettler Toledo AB 204, Greifensee, Switzerland) with a sensitivity of ±0.01 g. Fruit volume was determined using the liquid displacement method. The measurements of fruit length without calyx (FL; mm), equatorial fruit diameter (FD; mm), calyx length (CL; mm), calyx diameter (CD; mm), and peel thickness (PT; mm) were carried out using digital micrometers (Powerfix Profi, Neckarsüm, Germany). For fruit diameter and calyx diameter, two perpendicular measurements were taken around the horizontal (equatorial) plane, while peel thickness was measured on two opposite sides of the fruit. Fruit shape was determined by calculating the ratio of fruit length to fruit diameter. The calyx index (CI, %) was defined as the ratio of the calyx length to the total length of the fruit. Arils were manually separated from the pericarp and membrane, and the total aril weight (TAW) per fruit was recorded. The total number of arils per fruit (NoA/F) was estimated by counting the number of arils in a 100 g sample and extrapolating this number based on the total aril weight per fruit. Aril yield (AY; %) was calculated as the ratio of aril to fruit weight multiplied by 100. Juice was extracted from a 50 g aril sample by pressing it through four layers of cheesecloth. The juice yield (JY; %) was determined by calculating the volume of juice extracted (in milliliters) per 50 g sample and expressing the result as a percentage.

Pomegranate aril color was measured using a Croma meter CR-400 (Konica Minolta, Osaka, Japan). Color was estimated according to the CIE L*a*b* method and expressed as L*, a*, b*, C, and hue angle (h°) values. Thirty arils were randomly taken from each location for the color measurements. L* indicates the lightness and ranges from 0 to 100. Values a* and b* define red-greenness and blue-yellowness, respectively. Croma (C) and hue angle denote the visual color appearances. The C value defines color intensity. The hue angle represents the visual experience according to which the color is evaluated, with the following values: 0°–90° = red-purple; 90°–180° = yellow; 180°–270° = bluish-green; and 270°–360° = blue.

2.4. Chemical Analysis

2.4.1. Total Soluble Solids, Titratable Acidity, TSS/TA Ratio, pH, and Total Sugar Content

The measurements were performed on fresh aril juice extracted with cheesecloth as previously described. Each replicate included three pomegranate fruits. The total soluble solids (TSS) content was determined using a digital refractometer (Mettler Toledo, Columbus, OH, USA), with results expressed as percentage TSS at 20 °C. Titratable acidity (TA) was determined using the AOAC method [18] and expressed as a percentage of citric acid equivalents. The TSS/TA ratio was then calculated based on these values. The pH of the juice was measured at 20 °C using a Mettler Toledo MP 230 digital pH meter. The total sugar content (TSC) was quantified using the Luff–Schoorl method [18] and expressed in g/100 g.

2.4.2. Total Phenolic Content, Anthocyanin Content, and Antioxidant Activity

Extraction: Approximately 30 g of pomegranate arils was frozen at −80 °C prior to extraction. The frozen material was then freeze-dried using a FreeZone 2.5 L benchtop freeze dryer (LabConco, Kansas City, MO, USA). The lyophilized arils were then ground to a coarse powder under liquid nitrogen using a pre-cooled mortar and pestle. For extraction, 0.5 g of the powdered sample was mixed with 5 mL of an extraction solvent consisting of methanol, water, and formic acid (60:37:3, *v*/*v*/*v*), and extracted for 1 h. The mixture was homogenized for 10 min and then centrifuged at 4000 rpm for 10 min. To increase the efficiency of the extraction, the procedure was repeated under the same conditions, and the resulting supernatants were pooled. The combined extracts were concentrated to a volume of less than 1 mL using a rotary vacuum evaporator (Jouan RC10.22, Thermo Fisher Scientific, Waltham, MA, USA). The concentrate was further centrifuged in a microcentrifuge at 14,000 rpm for 15 min, and the extract was transferred to a vial. The extract was brought to a final volume of 1 mL, capped, and then stored at −20 °C until analysis.

Determination of Total Phenolic Content

Total phenolic content (TPC) was quantified using a modified version of the Folin–Ciocalteu colorimetric assay described by Singleton and Rossi [19]. Briefly, 300 µL of the extract (previously diluted to 1:10 with an extraction solvent of methanol, water, and formic acid in a 60:37:3 ratio) was combined with 1.5 mL of the 10-fold-diluted Folin–Ciocalteu reagent, followed by the addition of 1.2 mL of sodium carbonate solution (7.5% *w*/*v*). The reaction mixture was incubated at room temperature for 90 min. The absorbance was then measured at 760 nm using a Beckman Coulter DU730 UV–Vis spectrophotometer (Allegra™ 25R Centrifuge, Beckman Coulter, Brea, CA, USA) in comparison to a reagent blank. Calibration was performed using six standard solutions of gallic acid ranging from 50 to 500 mg/100 mL. The results were expressed as gallic acid equivalents (GAE).

Identification and Quantification of Anthocyanin Compounds

A 20 µL sample of the pomegranate extract was analyzed on an Agilent 1200 HPLC system (Agilent Technologies, Santa Clara, CA, USA) equipped with a photodiode array detector (DAD, G1315 D series) and an autosampler (G1329A series) and controlled by ChemStation software (Rev. B.03.01, Agilent Technologies, Santa Clara, CA, USA). The separation was performed on an Agilent Zorbax Eclipse XDB reversed-phase C18 column (150 mm × 3.5 mm; particle size 5 µm) maintained at 30 °C. The analytes were eluted using a two-solvent gradient system consisting of solvent A [formic acid and water (5:95, *v*/*v*)] and solvent B [formic acid, water, and acetonitrile (5:90:5, *v*/*v*/*v*)]. The gradient started with 5% B (0–2 min) and increased to 15% B at 19 min and then to 20% B at 20 min. From 20 to 21 min, the gradient was increased to 100% B and held until 24 min. The column was re-equilibrated to the initial conditions (5% B) from 24 to 28 min. The flow rate was 1 mL/min, and the chromatograms were recorded at 520 nm to detect the anthocyanins. Peaks were quantified using a standard curve generated by injecting analytical-grade cyanidin-3-glucoside (Indofine Chemical Company, Hillsborough, NJ, USA) at concentrations ranging from 0.2 to 500 µg/g in triplicate (R^2 = 0.9999). The results were expressed as cyanidin-3-glucoside equivalents. To identify the individual peaks, the collected fractions were concentrated to near dryness using a rotary evaporator. The concentrated samples were analyzed by the core facility using MALDI-TOF mass spectrometry on a Bruker Autoflex instrument (Bruker Daltonics, Billerica, MA, USA). The resulting mass spectra (m/z values) were compared with data from the literature to identify the compounds [20,21]. The total anthocyanin content (TAC) was determined by summing the concentrations of all individually identified anthocyanin compounds.

Determination of Antioxidant Activities

DPPH assay: The radical scavenging activity was evaluated according to the method of Brand-Williams and collaborators [22] with minor modifications. The extract (diluted to 1:10 with a mixture of extraction buffer, methanol, and water/formic acid, 60:37:3) was mixed with 0.5 mM DPPH in methanol (400 μL of extract and 200 μL of DPPH). After 30 min of incubation in the dark at room temperature, the absorbance was measured at 517 nm (Beckman Coulter DU730 UV–Vis spectrophotometer). A blank without DPPH was used for correction. The results were expressed as percent inhibition.

TOSC determination: Total antioxidant capacity was determined using the TOSC assay [23]. The extracts were dried, reconstituted in a 100 mM phosphate buffer (pH 7.4), and diluted to 1:20. An aliquot of 100 μL was added to reaction vials containing α-keto-γ-(methylthio)-butyric acid (KMBA) and a buffer; one vial also contained Trolox as a standard. Reactions were initiated by injection of 200 mM 2,2′-azobis(2-amidinopropane) dihydrochloride (ABAP), followed by incubation at 39 °C for 90 min. Ethylene production was measured using a gas chromatograph (Agilent 7890A, Agilent Technologies, CA, USA). The system was equipped with a capillary injection port with electronic pressure control (EPC), which was set to 50 °C and operated in split mode (20:1). Separation was performed using an HP-Plot Q column (15 m × 0.32 mm × 20 μm) maintained at 40 °C, while detection was performed with a flame ionization detector (FID), also equipped with EPC and set at 225 °C. The carrier gas was high-purity helium at a flow rate of 2.66 mL/min. Data acquisition was performed at a rate of 50 Hz and processed using the ChemStation software (Rev. B.03.01). Ethylene concentrations were quantified using a second-degree polynomial calibration curve (R^2 = 0.981), and antioxidant capacity was expressed as millimole Trolox equivalents (mM TE).

2.5. Statistical Analysis

The data obtained from this study were analyzed using Statistica 14.0.0.15 [24]. Analysis of variance, specifically a one-way ANOVA, was performed to determine the differences between locations in the physical and chemical characteristics of the fruit, including the % radical scavenging activity (DPPH) and Trolox equivalents (TOSC), as well as to determine differences within the individual locations in the proportion of individual anthocyanidins. Differences between means were considered significant at $p \leq 0.5$ using the LSD Test.

3. Results

3.1. Physical Properties

Properties of the pomegranate fruit 'Barski slatki' cultivar grown at four different locations (Kaštela and Metković, CRO; Mostar, BIH; and Bar, MNE) were analyzed. The location had a significant influence on some of the physical properties of the fruit and aril (Table 2). This indicates that local climatic and environmental conditions can influence fruit morphology, which is important for understanding the impact of the growing site on cultivar performance.

Fruit weight and fruit volume ranged from 386.9 g to 506.8 g and from 407.2 mL to 531.1 mL, respectively, and showed significant differences among different growing areas (Table 2). In general, fruit weight and fruit volume were higher in the northernmost-fruit-grown location (Kaštela, CRO) than in the southernmost location in Bar (MNE). While no differences were found in fruit length between growing locations, the location affected fruit diameter. At Mostar (BIH) (96.1 mm), FD was significantly larger than at Kaštela (91.5 mm) and Metković (86.4 mm) (CRO). The FL/FD ratio defines the fruit shape index. At both locations in Croatia (Kaštela and Metković), the fruits were round (0.92), whereas at the Mostar (BIH) and Bar (Montenegro) locations, they had a more flattened shape. Also, at both Croatian locations, the length of the calyx was significantly lower than in Mostar (BIH) and Bar (MNE).

Table 2. Physical properties of the fruit and aril of the pomegranate 'Barski slatki' evaluated at four different locations.

Fruit Properties	Location			
	Kaštela, CRO	**Metković, CRO**	**Mostar, BIH**	**Bar, MNE**
Fruit weight (FW; g)	506.8 * ± 18.5 a **	421.6 ± 20.09 ab	453.9 ± 26.4 ab	386.9 ± 31.9 b
Fruit volume (FV; mL)	531.1 ± 12.9 a	448.9 ± 20.9 ab	477.2 ± 27.6 ab	407.2 ± 33.8 b
Fruit length (FL; mm)	84.6 ± 0.8	79.6 ± 1.2	83.5 ± 2.2	79.7 ± 2.6
Fruit diameter (FD; mm)	91.5 ± 1.2 b	86.4 ± 0.8 c	96.1 ± 0.7 a	92.6 ± 2.5 ab
Fruit shape index (FL/FD)	0.92 ± 0.02 a	0.92 ± 0.01 a	0.87 ± 0.02 b	0.86 ± 0.02 b
Calix length (CL; mm)	17.0 ± 0.5 b	17.2 ± 0.6 b	18.9 ± 0.3 a	19.2 ± 0.3 a
Calix diameter (CD; mm)	23.9 ± 1.2	23.6 ± 0.9	26.3 ± 0.4	23.6 ± 1.1
Peel thickness (PT; mm)	4.2 ± 0.4	4.1 ± 0.4	3.5 ± 0.2	4.4 ± 0.4
Number of arils per fruit (NoA/F)	749 ± 41 a	607 ± 27 b	611 ± 46 b	561 ± 23 b
Total aril weight (TAW; g)	309.2 ± 14.8 a	261.1 ± 17.1 bc	294.3 ± 10.9 ab	235.0 ± 17.9 c
Aril weight (AW; g)	0.41 ± 0.02 b	0.42 ± 0.02 ab	0.47 ± 0.02 a	0.39 ± 0.03 b
Aril yield (AY; %)	60.8 ± 0.9	60.9 ± 1.7	58.6 ± 1.9	61.1 ± 1.4
Juice yield (JY; %)	71.8 ± 1.7	74.2 ± 0.4	73.3 ± 1.6	71.0 ± 2.6

* Values are given as mean ± SE. ** Different lowercase letters in each row indicate a significant difference between locations at $p \leq 0.05$ by LSD test.

Significant influence of growing area on the number of arils per fruit was recorded, varying from 561 to 749, and at Kaštela (CRO), it was higher than at the other investigated location. The total aril's weight and the weight of individual arils also depended on the growing region (Table 2). Total arils' weight varied between 235 and 309.2 g. The highest TAW was recorded at Kaštela (CRO) and was higher than at Bar (MNE) and Metković (CRO). Aril weight varied between 0.39 g and 0.47 g and was higher in Mostar (BIH) than in Kaštela (CRO) and Bar (MNE). Although fruit size varied between locations, no differences were found in terms of peel thickness, aril yield, and juice yield. This highlights that some fruit traits may be less sensitive to environmental variation than others.

3.2. Chemical Properties

We found a significant influence of location on TSS, TA, TSC, and pH in juice (Table 3).

Table 3. Total soluble solids (TSS), titratable acidity (TA), TSS/TA ratio, total sugar content (TSC), and pH of juice from pomegranate 'Barski slatki' grown at four locations.

Juice Properties	Location			
	Kaštela, CRO	**Metković, CRO**	**Mostar, BIH**	**Bar, MNE**
Total soluble solids (TSS; °Brix)	14.1 * ± 0.1 b **	15.0 ± 0.2 a	13.1 ± 0.0 c	12.8 ± 0.4 c
Total acidity (TA; %)	0.53 ± 0.01 bc	0.59 ± 0.03 a	0.48 ± 0.01 c	0.55 ± 0.01 ab
TSS/TA ratio	26.4 ± 0.5	25.7 ± 0.9	27.0 ± 0.3	23.2 ± 1.1
Total sugar content (TSC; %)	11.8 ± 0.4 b	12.1 ± 0.2 b	12.6 ± 0.1 b	13.9 ± 0.5 a
pH	3.57 ± 0.02 a	3.52 ± 0.05 ab	3.60 ± 0.03 a	3.42 ± 0.03 b

* Values are given as mean ± SE. ** Different lowercase letters in each row indicate a significant difference between locations at $p \leq 0.05$ by LSD test.

The total soluble solid content varied between 12.8 and 15 °Brix (Table 3). The TSS content was the highest at the Metković site (CRO) and the lowest at the Mostar (BIH) and Bar (MNE) sites. Location also had a significant influence on total acidity, with higher acidity (0.59%) found at the Metković location compared to Kaštela (CRO) (0.53%) and Mostar (BIH) (0.48%). Despite the differences in TSS and TA, the TSS/TA ratio was not found to vary by location, suggesting that overall taste balance remains relatively stable across sites despite variation in individual components.

The total sugar content at Bar (MNE) (13.9%) was higher than that at other locations. The difference in pH between the locations was also significant: at Kaštela (CRO) and Mostar (BIH), the pH was higher than at Bar (MNE).

Table 4 shows that the location has an influence on the total phenol content, which varied between 2603 and 3101.9 mg L^{-1} GAE. At the Kaštela (CRO) location, the TPC was significantly higher than at the other locations. Although no differences were found in total anthocyanidin content or antioxidant activity across locations (Table 4), the concentration of specific individual anthocyanins varied depending on location (Figure 2).

Table 4. Total phenol content (TPC), total anthocyanin content (TAC), and antioxidant activity (DPPH and TOSC) of arils of pomegranate 'Barski slatki' grown at four locations.

	Location			
Aril Properties	**Kaštela, CRO**	**Metković, CRO**	**Mostar, BiH**	**Bar, MNE**
Total phenol content (mg L^{-1} GAE)	3101.9 * ± 35.7 a **	2603.0 ± 124.1 b	2725.5 ± 150.2 b	2772.3 ± 108.3 b
Total anthocyanin content (mg/100 g)	44.9 ± 5.58	43.99 ± 4.67	51.3 ± 6.71	59.36 ± 4.03
DPPH (%)	70.5 ± 0.8	67.7 ± 1.2	69.1 ± 1.3	69.3 ± 2.5
TOSC (umol Trolox eq/kg)	414.3 ± 12.4	425.8 ± 10.0	420.3 ± 8.2	436.4 ± 10.4

* Values are given as mean ± SE. ** Different lowercase letters in each row indicate a significant difference between locations at $p \leq 0.05$ by LSD test.

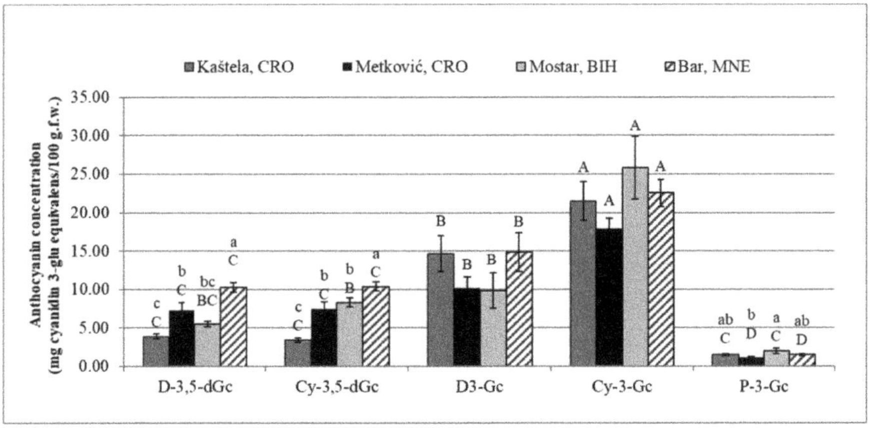

Figure 2. Anthocyanin concentration (mg cyanidin-3-glucoside equivalents per 100 g fresh weight) in the 'Barski slatki' at four locations. Values are presented as mean ± SE (n = 3). Different lowercase letters indicate significant differences among locations for each individual anthocyanidin at $p \leq 0.05$ according to the LSD test. Different uppercase letters indicate significant differences within locations in the relative proportions of individual anthocyanidins. Abbreviations: D-3,5-dGc, delphinidin-3,5-diglucoside; Cy-3-dGc, cyanidin-3-diglucoside; D-3-Gc, delphinidin-3-glucoside; Cy-3-Gc, cyanidin-3-glucoside; P-3-Gc, pelargonidin-3-glucoside.

Anthocyanins identified in this study included delphinidin-3,5-diglucoside, cyanidin-3,5-diglucoside, cyanidin-3-glucoside, pelargonidin-3-glucoside, and delphinidin-3-glucoside. The profile was consistent across all locations studied. The most prevalent anthocyanin pigment at all locations was cyanidin-3-glucoside, the content of which did not differ significantly between the investigated locations. The second most abundant anthocyanin, delphinidin-3-glucoside, also showed no significant differences between locations. The remaining three anthocyanins showed variability depending on the cultivation site. The content of cyanidin-3,5-diglucoside was highest at the Bar location (10.34 mg/100 g f.w) and lowest in Kaštela, with no significant differences between Metković and Mostar. The highest content of delphinidin-3,5-diglucoside (10.20 mg/100 g f.w) was found in Bar, while the lowest was in Kaštela (3.88 mg/100 g f.w), with no significant differences between Kaštela and Mostar. The content of pelargonidin-3-glucoside differed only in Mostar compared to Metković.

Aril color defined by the CIELAB color space parameters L* (lightness), a* (green to red), b* (blue to yellow), C (chroma, representing color intensity), and h° (hue angle) varied significantly across different growing locations (Table 5).

Table 5. Aril color parameters (CIE L*, a*, b*, C, and h°) of pomegranate 'Barski slatki' grown at four locations.

Color Parameters	Location			
	Kaštela, CRO	**Metković, CRO**	**Mostar, BIH**	**Bar, MNE**
L*	32.95 * ± 0.57 a **	35.47 ± 2.26 a	32.91 ± 2.09 a	26.36 ± 1.03 b
a*	6.47 ± 0.41 c	8.34 ± 1.83 bc	10.66 ± 0.83 ab	11.92 ± 1.62 a
b*	12.07 ± 0.65 a	14.56 ± 3.84 a	4.20 ± 1.54 b	8.61 ± 2.18 ab
C	14.33 ± 0.31 ab	17.23 ± 3.24 a	11.62 ± 0.97 b	15.06 ± 1.05 ab
h°	58.27 ± 2.40 a	56.90 ± 9.30 a	19.96 ± 5.53 b	36.07 ± 10.42 b

* Values are given as mean ± SE; n = 10. ** Different lowercase letters in each row indicate a significant difference between locations at $p \leq 0.05$ by LSD test.

Lightness (L*) of arils from both locations in Croatia and Mostar (BIH) was higher than in Bar (MNE). L* = 0 denotes black, and L* = 100 indicates diffuse white. The results showed that the arils from fruits from Bar (MNE) were more colorful than those from other locations studied. In Bar (MNE), arils had significantly higher redness (a*; more magenta color components) than the arils from both locations in Croatia, where the content of yellow color components (b*) is higher. The (C) value represents the 'purity' of the color. A lower value, i.e., lower color purity, was recorded at the location Mostar (BIH) compared to Metković (CRO). Although the h° value was higher at both locations in CRO than in BIH and MNE, the arils had a red-purple color at all locations, suggesting that while hue remains relatively consistent, subtle differences in color parameters reflect the impact of growing site on fruit visual quality.

4. Discussion

Extensive scientific knowledge about the health-promoting properties of pomegranates, along with public interest in functional foods [25–29], has led to a significant increase in the demand for pomegranate fruit and its byproducts in recent years, particularly in Western countries [30].

The pomegranate is a Mediterranean fruit species that is best adapted to the Mediterranean climate, which is characterized by intense sunlight, mild winters with minimum temperatures not below −12 °C, and warm, dry summers without rainfall during the final stage of fruit development [31,32]. Under these conditions, the fruit can reach its optimal size, color, and sugar accumulation without the risk of skin cracking. The area under pomegranate cultivation worldwide has expanded considerably, and it is now grown in various subtropical and tropical regions with different microclimatic conditions [33].

The 'Barski slatki' cultivar is grown extensively on the eastern Adriatic coast, and this study has established the influence of four different locations on select physical and chemical fruit properties.

4.1. Microclimate Condition Influence on Physical Properties

A significant influence of location was found on physical properties (Table 2), including fruit weight, fruit diameter, fruit volume, fruit shape index, calyx length, number of arils per fruit, and total and individual aril weight.

Mditshwa et al. [34] also indicated a significant influence of growing location on fruit and aril weight, fruit length and diameter, and total aril weight. Borochov–Neori et al. [35], Schwartz et al. [32], and Ghasemi Soloklui et al. [36] confirm the influence of growing area on the fruit and aril weight, while Boussaa et al. [2] reported that location had no influence on the fruit weight, length, and diameter, but influenced the peel weight and thickness, as well as aril and juice yield. Feng et al. [37] reported that the adaptability of the cultivar to environmental conditions significantly affects the physical fruit characteristics. The quality and proper flower bud initiation, flowering, and pollination determine fruit set, as these are particularly sensitive stages of development influenced by various physiological and environmental factors, such as soil moisture, temperature, rainfall, wind, and humidity [32]. Flower bud initiation begins about 4–5 weeks before flowering [38,39] which typically occurs in April under the climatic conditions of the study area. In our study area, flowering began in May and lasted about two months. Early-flowering flowers usually have larger floral structures and ovaries, resulting in higher fruit set (up to 90%) and better fruit quality [40,41], probably due to more favorable environmental conditions during their development [42]. However, unfavorable weather conditions during the flowering period can negatively affect pollination and fruit set. At the Kaštela site, about 50% of the annual rainfall occurred during the vegetation period (April to October), which is 8–15% more than at the other sites during the same period (Table 1). This rainfall distribution was particularly favorable during the critical period of flower bud differentiation and flowering. Higher precipitation at Kaštela in April (Table 1) improved soil moisture availability and created favorable conditions for early initiation and differentiation of hermaphrodite flowers, which are known to produce larger fruits. In contrast, at the same location, lower rainfall in May probably favored better pollination and fertilization, contributing to higher fruit set and a greater number of arils per fruit. During the first two weeks after fruit set, fruit growth progresses evenly, followed by a rapid increase in fruit size (weight, length, diameter, and volume) until the end of June, after which growth gradually slows down [43]. Sufficient soil moisture combined with adequate heat is crucial for optimal fruit development and growth during this phase, which ultimately affects yield and fruit quality. Although pomegranate is considered a drought-tolerant fruit species, deficit irrigation during fruit development and ripening has an adverse effect on fruit size and total yield [44]. At the Kaštela site, microclimatic conditions, especially higher rainfall in June and July (Table 1), provided higher soil moisture and more favorable conditions for fruit growth, resulting in larger fruits at harvest. In contrast, the lowest rainfall in June was recorded at the Bar site (MNE), where the smallest fruits were also harvested.

In our study, aril yield and juice yield showed no significant differences between locations for either parameter (Table 2), which contrasts with the results of Mditshwa et al. [34] and Schwartz et al. [32]. Interestingly, despite the variation in fruit weight across locations, aril yield remained constant. This response can be attributed to climatic factors that may favor the development of internal membrane structures over aril formation [34].

According to Borochov-Neori et al. [35], the number of membrane walls influences both the number of arils and the peel thickness. Environmental conditions typical of Mediterranean and subtropical climates promote the development of these structures. This finding could also explain the relatively high juice content observed in fruits from these regions.

4.2. Microclimate Condition Influence on Chemical Properties

Previous studies have reported that genotype, agroclimatic conditions, cultivation practices, fruit maturity, harvest time, storage, and postharvest handling affect the content of bioactive compounds in pomegranate fruits [45–47]. With some experience, harvest maturity can be assessed by tapping the fruit and listening for a metallic sound, a traditional method mentioned by Mir et al. [48]. In the region where our study was conducted, harvest time is typically determined based on skin color and the growers' experience when the fruit skin begins to show superficial cracking. Our results showed that the growing location influences specific chemical characteristics (Table 3). The total soluble solids (TSS) content significantly varied. As previously described by Fawole and Opara [45], TSS content increases faster in the early stages of fruit development, which is probably due to the active hydrolysis of starch into sugars during fruit ripening [49]. At the time of harvest, climatic conditions, especially the amount and distribution of rainfall, have a significant impact on TSS content [34,36,50,51]. The relatively low TSS content observed in fruits from Bar (MNE) (Table 3) could reflect the adverse effects of higher rainfall during ripening. Fruits grown under drier conditions are often sweeter than those from humid or irrigated areas [50].

Along with total soluble solids (TSS) content, titratable acidity (TA) is a crucial quality characteristic of pomegranate juice that significantly contributes to its taste and consumer acceptance [32]. In our study, the highest TA value was measured at the Metković (CRO) site, situated in the lower reaches of the Neretva River at an altitude of 5 m in a typical Mediterranean climate zone. Conversely, the lowest TA was found in fruits from Mostar (BIH), which is located in the upper reaches of the same river at an altitude of 100 m in the inland region where the Mediterranean influence persists. This pattern is consistent with earlier findings by Shulman et al. [52], who found higher acidity in pomegranates grown in coastal regions of Israel compared to those in warmer inland valleys.

The pH value, which is directly related to the perception of the acidity of pomegranate juice, is also an important parameter [53]. Immature fruits usually have lower pH values due to the higher acidity, while the pH increases as the fruits ripen. In our study, lower pH values were observed in fruits from Bar (MNE), whereas the fruits from Kaštela (CRO) and Mostar (BIH) had a higher pH value (Table 3). These differences may reflect developmental stages as well as site-specific growing conditions. Previous research has attributed variations in TSS, TA, and pH to both genotype and regional environmental factors [45,54]. Similarly, Mditshwa et al. [34] confirmed the significant influence of location on TSS and TA, although they found no significant differences in pH.

The phenolic content of pomegranate fruit is influenced by a complex interplay of factors, including genetic background, climatic conditions, and cultivation practices [55–57]. Among the environmental variables, altitude has been identified as a factor that can significantly affect the phenolic biosynthetic pathway [2,46] by altering temperature, UV radiation, and water availability. In our study, total phenolic content (TPC) varied significantly between locations (Table 4). Fruits harvested from the lower altitude (Kaštela, at 5 m above sea level) had a higher TPC than fruit from higher altitude sites such as Mostar and Bar (100 and 143 m above sea level, respectively). This result aligns with the results reported for pomegranates [34], and similar trends were observed for olives [58,59], sweet cherries [60], and grapes [61]. Schwartz et al. [32] also reported a higher TPC in the juice of pomegranates from the Mediterranean compared to the arid desert region, underlining the importance of mesoclimatic suitability. Attanayake et al. [30] observed lower phenolic content in pomegranates grown in the colder and rainier regions of Sri Lanka, suggesting that excessive rainfall may inhibit phenolic accumulation. Some studies, however, indicated that higher altitudes can lead to higher phenolic content and antioxidant capacity, probably due to environmental stress and radiation exposure [36]. In our study, lower altitudes seem to favor higher phenolic accumulation, possibly due to better hydration, favorable temperature, and radiation exposure, although this was not consistent across all sites; Kaštela and Metković (5 and 9 m a.s.l., respectively) did not follow the same trend as the two other locations at significantly higher altitudes. This suggests that rainfall distribution may also play an important role, as a more favorable distribution of precipitation during fruit growth and development (e.g., in Kaštela) resulted in fruits with higher moisture content at the developmental stage, which in turn promoted phenolic accumulation. Overall, these results emphasize the need to consider not only altitude but also the broader climatic context, including latitude, rainfall patterns, and microclimatic conditions, when interpreting the effects of growing environment on phenolic composition.

Color is a key visual quality parameter for pomegranate fruit, playing a decisive role in its attractiveness and marketability for the consumer. The distinctive red color of pomegranate juice is primarily attributed to anthocyanins, a class of flavonoids responsible for the pigmentation of many fruits [34,52]. In our study, although the total anthocyanin content was not significantly affected by location, the growing location had a significant effect on the color of the arils (Table 5) and the concentration of individual anthocyanin compounds (Figure 2). Fruits grown in Bar (MNE) at 147 m altitude had the lightest arils and the most pronounced red hue, as reflected in the significantly higher a* values indicating a greater proportion of magenta-red color components. In contrast, the arils from Croatian areas (Kaštela and Metković) had a higher yellow hue, indicated by higher b* values. These results are in line with previous reports by Ubi [62] and Mditshwa et al. [34], which highlight an environmental modulation of pomegranate coloration.

At all locations, cyanidin-3-glucoside was the predominant anthocyanin, followed by delphinidin-3-glucoside (Figure 2), which is consistent with previous anthocyanin profiles in pomegranate. Interestingly, despite apparent differences in color intensity (a*) and lightness (L*), the concentrations of these two main anthocyanins remained relatively stable across the different locations. On the other hand, the remaining three anthocyanins showed site-specific variation. The content of cyanidin-3,5-diglucoside and delphinidin-3,5-diglucoside was highest at Bar (147 m altitude) and lowest in Kaštela (9 m altitude). Pelargonidin-3-glucoside was present in significantly different concentrations only in Mostar compared to Metković (Figure 2).

Environmental temperature is a crucial factor regulating the biosynthesis and degradation of anthocyanins. These pigments are known to accumulate in response to cooler temperatures, which enhances gene expression related to the anthocyanin biosynthetic pathway [30,63,64]. Anthocyanin accumulation is inversely correlated with harvest season temperatures, and phenolic compounds are often more abundant under cooler growing conditions. While low temperatures can stimulate anthocyanin production, excessively high temperatures can inhibit synthesis and promote pigment degradation [63,65]. Temperature not only affects the anthocyanin concentration but also influences the chemical stability of anthocyanins. Under warmer conditions, anthocyanins are predominantly glucosylated, which may enhance their structural stability [45,66]. According to our findings, the fruits from Bar (MNE) with milder temperatures showed higher levels of diglucoside forms, supporting the hypothesis that temperate climatic conditions favor both anthocyanin biosynthesis and stabilization (Figure 2). However, it is essential to note that genotype–environment interactions can result in varied responses. For example, Attanayake et al. (2018) [30] observed increased expression of anthocyanin synthase genes and higher anthocyanin accumulation in drier, warmer environments, highlighting the role of crop-specific sensitivity. Despite these nuances, a consensus in several studies confirms that cooler temperatures during ripening tend to increase anthocyanin accumulation, resulting in deeper red pigmentation of the skin and arils [45,63]. In addition, cyanidins appear to have more temperature stability, while delphinidins prefer to accumulate under cooler conditions. In summary, although the total anthocyanin content remained stable across all locations, variations in local growing conditions, including temperature, were associated with changes in the color parameters and profile of individual anthocyanins, highlighting the contribution of the environment to the visual and nutritional quality of pomegranate fruit.

5. Conclusions

Microclimatic conditions, including rainfall distribution, altitude, and temperature, significantly influenced the physical, chemical, and visual attributes of 'Barski slatki' pomegranate. The results indicated that precipitation at the location during critical phenological stages, such as bud differentiation, flowering, and early fruit development, was associated with variation in fruit set, aril number, and fruit size, although these effects are part of a complex interplay. While aril and juice yields were relatively stable across sites, notable differences were observed in total soluble solids, titratable acidity, pH, total phenolic content, and anthocyanin profiles. Locations with higher spring rainfall (Kaštela, CRO) favored enhanced fruit growth and higher phenolic accumulation. Although the total anthocyanin content remained consistent across locations, significant variation occurred in aril coloration and the relative abundance of individual anthocyanins. Fruits from Bar (MNE), the highest-altitude site, significantly exhibited the highest accumulation and concentration of diglucoside forms.

These findings highlight the importance of incorporating environmental variables into cultivar selection and orchard management strategies to maximize fruit quality for both fresh consumption and processing. Overall, pomegranate 'Barski slatki' demonstrates a notable adaptability to diverse agroecological conditions within the Mediterranean climate zone, maintaining stable fruit quality.

Author Contributions: M.R., conceptualization, methodology, sample collection, investigation, formal analyses, writing—original draft preparation, editing, data curation, and review; M.J.Š., formal analyses, software and data analyses, writing—original draft preparation, review, and editing; J.G., sample collection, formal analyses, writing—original draft preparation, and editing; S.G.B., conceptualization, review, and editing; J.C.D.-P., review and editing; D.M., formal analyses, methodology, review, and editing. All authors have read and agreed to the published version of the manuscript.

Funding: This research was funded by the Croatian National Program for Conservation and Sustainable Use of Plant Genetic Resources for Food and Agriculture 2021–2027, by the European Agricultural Fund for Rural Development (EAFRD) within the framework of Intervention 70.05 — Support for the conservation, sustainable use, and development of genetic resources in agriculture, and by the project INOMED-2I (09-207/1-23) granted by the European Union – "NextGenerationEU".

Data Availability Statement: All data are included within the article.

Acknowledgments: We especially note our appreciation of the support from Dennis Phillips of the University of Giorgia Proteomic and Mass Spectrometry (OAMS) Core Facility in the identification of pomegranate fruit chemical compounds, as well as from technical assistant Radojka Plećaš for help in the laboratory, grower Jozo Obšivač for providing the plant material, and Mate Čarija for assistance with the visualization of the sampling site.

Conflicts of Interest: Author Dan Maclean was employed by the company AgroFresh. The remaining authors declare that the research was conducted in the absence of any commercial or financial relationships that could be construed as a potential conflict of interest.

References

1. Çiçek, M.; Pakyürek, M.; Çelik, F. Determination of Morphological and Pomological Characteristics of Diyarbakır Region Pomegranates (*Punica granatum* L.). *Int. J. Agric. Environ. Food Sci.* **2019**, *3*, 197–203. [CrossRef]
2. Boussaa, F.; Zaouay, F.; Burlo-Carbonell, F.; Noguera-Artiaga, L.; Carbonell-Barrachina, A.; Melgarejo, P.; Hernandez, F.; Mars, M. Growing Location Affects Physical Properties, Bioactive Compounds, and Antioxidant Activity of Pomegranate Fruit (*Punica granatum* L. Var. Gabsi). *Int. J. Fruit Sci.* **2020**, *20*, 508–523. [CrossRef]
3. Özgüven, A.I.; Yılmaz, C. Pomegranate Growing in Turkey. In Proceedings of the Production, Processing and Marketing of Pomegranate in the Mediterranean Region: Advances in Research and Technology, Orihuela, Spain, 15–17 October 1998; Melgarejo-Moreno, P., Martinez-Nicolas, J.J., Martinez-Tome, J., Eds.; CIHEAM-IAMZ: Zaragoza, Spain, 1998; pp. 41–48.
4. Özcan, E.; Ünaldi, Ü.E. Ecology of Pomegranate and Its Economics in Turkey. In Proceedings of the International Symposium on Geography Environment and Culture in the Mediterranean Region, Marseille, France, 27–29 March 2007; Available online: https://web.deu.edu.tr/geomed2010/2007/Ozcan-Unaldi.pdf (accessed on 18 August 2025).
5. Davarpanah, S.; Tehranifar, A.; Abadía, J.; Val, J.; Davarynejad, G.; Aran, M.; Khorassani, R. Foliar Calcium Fertilization Reduces Fruit Cracking in Pomegranate (*Punica granatum* Cv. Ardestani). *Sci. Hortic.* **2018**, *230*, 86–91. [CrossRef]
6. Czieczor, L.; Bentkamp, C.; Damerow, L.; Blanke, M. Non-Invasive Determination of the Quality of Pomegranate Fruit. *Postharvest Biol. Technol.* **2018**, *136*, 74–79. [CrossRef]
7. Mirdehghan, S.H.; Rahemi, M. Seasonal Changes of Mineral Nutrients and Phenolics in Pomegranate (*Punica granatum* L.) Fruit. *Sci. Hortic.* **2007**, *111*, 120–127. [CrossRef]
8. Caleb, O.J.; Mahajan, P.V.; Linus Opara, U.; Witthuhn, C.R. Modeling the Effect of Time and Temperature on Respiration Rate of Pomegranate Arils (Cv. "Acco" and "Herskawitz"). *J. Food Sci.* **2012**, *77*, E80–E87. [CrossRef]
9. Mphahlele, R.R.; Caleb, O.J.; Fawole, O.A.; Opara, U.L. Effects of Different Maturity Stages and Growing Locations on Changes in Chemical, Biochemical and Aroma Volatile Composition of 'Wonderful' Pomegranate Juice. *J. Sci. Food Agric.* **2016**, *96*, 1002–1009. [CrossRef]
10. AL-Kalbani, B.S.; Al-Yahyai, R.A.; Al-Sadi, A.M.; Al-Mamari, A.-G.H. Physical and Chemical Fruit Quality Attributes of Two Pomegranate Cultivars Grown at Varying Altitudes of Al-Hajar Mountains in Oman. *J. Agric. Mar. Sci.* **2021**, *26*, 42–50. [CrossRef]
11. Gómez-Devia, L.; Nevo, O. Effects of Temperature Gradient on Functional Fruit Traits: An Elevation-for-Temperature Approach. *BMC Ecol. Evol.* **2024**, *24*, 94. [CrossRef] [PubMed]
12. Welles, G.W.H.; Buitelaar, K. Factors Affecting Soluble Solids Content of Muskmelon (Cucumis melo L.). *Wagening. J. Life Sci.* **1988**, *36*, 3. [CrossRef]
13. Koshita, Y. Effect of Temperature on Fruit Color Development. In *Abiotic Stress Biology in Horticultural Plants*; Kanayama, Y., Kochetov, A., Eds.; Springer: Tokyo, Japan, 2015; pp. 47–58.
14. Moretti, C.L.; Mattos, L.M.; Calbo, A.G.; Sargent, S.A. Climate Changes and Potential Impacts on Postharvest Quality of Fruit and Vegetable Crops: A Review. *Food Res. Int.* **2010**, *43*, 1824–1832. [CrossRef]
15. Fischer, G.; Parra-Coronado, A.; Balaguera-López, H.E. Altitude as a Determinant of Fruit Quality with Emphasis on the Andean Tropics of Colombia. A Review. *Agron. Colomb.* **2022**, *40*, 212–227. [CrossRef]
16. Radunić, M.; Goreta Ban, S.; Gadže, J.; MacLean, D.; Martino, K.; Scherm, H.; Horton, D. *Šipak-Pomegranate*; Institute for Adriatic Crops and Karst Reclamation Split: Split, Croatia, 2012.
17. Kottek, M.; Grieser, J.; Beck, C.; Rudolf, B.; Rubel, F. World Map of the Köppen-Geiger Climate Classification Updated. *Meteorol. Zeitschrift* **2006**, *15*, 259–263. [CrossRef]
18. AOAC. *Official Methods of Analysis*, 16th ed.; AOAC: Washington, DC, USA, 1995.
19. Singleton, V.L.; Rossi, J.A. Colorimetry of Total Phenolics with Phosphomolybdic-Phosphotungstic Acid Reagents. *Am. J. Enol. Vitic.* **1965**, *16*, 144–158. [CrossRef]
20. Mousavinejad, G.; Emam-Djomeh, Z.; Rezaei, K.; Khodaparast, M.H.H. Identification and Quantification of Phenolic Compounds and Their Effects on Antioxidant Activity in Pomegranate Juices of Eight Iranian Cultivars. *Food Chem.* **2009**, *115*, 1274–1278. [CrossRef]
21. Wu, X.; Prior, R.L. Systematic Identification and Characterization of Anthocyanins by HPLC-ESI-MS/MS in Common Foods in the United States: Fruits and Berries. *J. Agric. Food Chem.* **2005**, *53*, 2589–2599. [CrossRef]
22. Brand-Williams, W.; Cuvelier, M.E.; Berset, C. Use of a Free Radical Method to Evaluate Antioxidant Activity. *Lebensm. Wiss. Und Technol.* **1995**, *28*, 25–30. [CrossRef]
23. MacLean, D.D.; Murr, D.P.; DeEll, J.R. A Modified Total Oxyradical Scavenging Capacity Assay for Antioxidants in Plant Tissues. *Postharvest Biol. Technol.* **2003**, *29*, 183–194. [CrossRef]

24. TIBCO. *Statistica, v. 14.0.0.15*; TIBCO Software Inc.: San Ramon, CA, USA, 2020.
25. Gil, M.I.; Tomas-Barberan, F.A.; Hess-Pierce, B.; Holcroft, D.M.; Kader, A.A. Antioxidant Activity of Pomegranate Juice and Its Relationship with Phenolic Composition and Processing. *J. Agric. Food Chem.* **2000**, *48*, 4581–4589. [CrossRef]
26. Kaplan, M.; Hayek, T.; Raz, A.; Coleman, R.; Dornfeld, L.; Vaya, J.; Aviram, M. Pomegranate Juice Supplementation to Atherosclerotic Mice Reduces Macrophage Lipid Peroxidation, Cellular Cholesterol Accumulation and Development of Atherosclerosis. *J. Nutr.* **2001**, *131*, 2082–2089. [CrossRef] [PubMed]
27. Aviram, M.; Rosenblat, M.; Gaitini, D.; Nitecki, S.; Hoffman, A.; Dornfeld, L.; Volkova, N.; Presser, D.; Attias, J.; Liker, H.; et al. Pomegranate Juice Consumption for 3 Years by Patients with Carotid Artery Stenosis Reduces Common Carotid Intima-Media Thickness, Blood Pressure and LDL Oxidation. *Clin. Nutr.* **2004**, *23*, 423–433. [CrossRef]
28. de Nigris, F.; Williams-Ignarro, S.; Lerman, L.O.; Crimi, E.; Botti, C.; Mansueto, G.; D'Armiento, F.P.; De Rosa, G.; Sica, V.; Ignarro, L.J.; et al. Beneficial Effects of Pomegranate Juice on Oxidation-Sensitive Genes and Endothelial Nitric Oxide Synthase Activity at Sites of Perturbed Shear Stress. *Proc. Natl. Acad. Sci. USA* **2005**, *102*, 4896–4901. [CrossRef]
29. Bahari, H.; Pourreza, S.; Goudarzi, K.; Mirmohammadali, S.N.; Asbaghi, O.; Kolbadi, K.S.H.; Naderian, M.; Hosseini, A. The Effects of Pomegranate Consumption on Obesity Indices in Adults: A Systematic Review and Meta-analysis. *Food Sci. Nutr.* **2024**, *12*, 641–660. [CrossRef]
30. Attanayake, R.; Eeswaran, R.; Rajapaksha, R.; Weerakkody, P.; Bandaranayake, P.C.G. Biochemical Composition and Expression of Anthocyanin Biosynthetic Genes of a Yellow Peeled and Pinkish Ariled Pomegranate (*Punica granatum* L.) Cultivar Are Differentially Regulated in Response to Agro-Climatic Conditions. *J. Agric. Food Chem.* **2018**, *66*, 8761–8771. [CrossRef]
31. Levin, G.M. *Pomegranate*; Third Millennium Publishing: Tempe, AZ, USA, 2006.
32. Shwartz, E.; Glazer, I.; Bar-Ya'akov, I.; Matityahu, I.; Bar-Ilan, I.; Holland, D.; Amir, R. Changes in Chemical Constituents during the Maturation and Ripening of Two Commercially Important Pomegranate Accessions. *Food Chem.* **2009**, *115*, 965–973. [CrossRef]
33. Verma, N.; Mohanty, A.; Lal, A. Invited Mini-Review Fruit, Vegetable and Cereal Science and Biotechnology Pomegranate Genetic Resources and Germplasm Conservation: A Review. *Fruit Veg. Cereal Sci. Biotechnol* **2010**, *4*, 120–125.
34. Mditshwa, A.; Fawole, O.A.; Al-Said, F.; Al-Yahyai, R.; Opara, U.L. Phytochemical Content, Antioxidant Capacity and Physicochemical Properties of Pomegranate Grown in Different Microclimates in South Africa. *S. Afr. J. Plant Soil* **2013**, *30*, 81–90. [CrossRef]
35. Borochov-Neori, H.; Judeinstein, S.; Tripler, E.; Harari, M.; Greenberg, A.; Shomer, I.; Holland, D. Seasonal and Cultivar Variations in Antioxidant and Sensory Quality of Pomegranate (*Punica granatum* L.) Fruit. *J. Food Compos. Anal.* **2009**, *22*, 189–195. [CrossRef]
36. Ghasemi-Soloklui, A.A.; Kordrostami, M.; Gharaghani, A. Environmental and Geographical Conditions Influence Color, Physical Properties, and Physiochemical Composition of Pomegranate Fruits. *Sci. Rep.* **2023**, *13*, 15447. [CrossRef]
37. Feng, L.; Yin, Y.; Yang, X.; Tang, H. Evaluation of Physiochemical Properties of Different Pomegranate Cultivars in China. *Acta Hortic.* **2020**, *1281*, 89–96. [CrossRef]
38. Holland, D.; Hatib, K.; Bar-Ya'akov, I. Pomegranate: Botany, Horticulture, Breeding. *Hortic. Rev.* **2009**, *35*, 127–191. [CrossRef]
39. Babu, K.D. Invited Mini-Review Fruit, Vegetable and Cereal Science and Biotechnology Floral Biology of Pomegranate (*Punica granatum* L.). *Fruit Veg. Cereal Sci. Biotechnol* **2010**, *4*, 45–50.
40. Chandra, R.; Jadhav, V.T.; Sharma, J. Global Scenario of Pomegranate (*Punica granatum* L.) Culture with Special Reference to India. *Fruit Veg. Cereal Sci. Biotechnol.* **2010**, *4*, 7–18.
41. Wetzstein, H.Y.; Yi, W.; Porter, J.A.; Ravid, N. Flower Position and Size Impact Ovule Number per Flower, Fruitset, and Fruit Size in Pomegranate. *J. Am. Soc. Hortic. Sci.* **2013**, *138*, 159–166. [CrossRef]
42. Evreinoff, V.A. Contribution à l'étude Du Grenadier. *J. Agric. Trop. Bot. Appl.* **1957**, *4*, 124–138. [CrossRef]
43. Gozlekcİ, S.; Kaynak, L. Physical and Chemical Changes during Fruit Development and Flowering in Pomegranate (*Punica granatum* L.) Cultivar Hicaz Nar Grow in Antalya Region. In Proceedings of the Production, Processing and Marketing of Pomegranate in the Mediterranean Region: Advances in Research and Technology, Orihuela, Spain, 15–17 October 1998; Melgarejo, P., Martínez-Nicolás, J.J., Martínez-Tomé, J., Eds.; CIHEAM Options Méditerranéennes: Série A. Séminaires Méditerranéens. CIHEAM: Zaragoza, Spain, 2000; Volume 42, pp. 79–85.
44. Rodríguez, P.; Mellisho, C.D.; Conejero, W.; Cruz, Z.N.; Ortuño, M.F.; Galindo, A.; Torrecillas, A. Plant Water Relations of Leaves of Pomegranate Trees under Different Irrigation Conditions. *Environ. Exp. Bot.* **2012**, *77*, 19–24. [CrossRef]
45. Fawole, O.A.; Opara, U.L. Changes in Physical Properties, Chemical and Elemental Composition and Antioxidant Capacity of Pomegranate (Cv. Ruby) Fruit at Five Maturity Stages. *Sci. Hortic.* **2013**, *150*, 37–46. [CrossRef]
46. Mphahlele, R.R.; Fawole, O.A.; Stander, M.A.; Opara, U.L. Preharvest and Postharvest Factors Influencing Bioactive Compounds in Pomegranate (*Punica granatum* L.)—A Review. *Sci. Hortic.* **2014**, *178*, 114–123. [CrossRef]
47. Radunić, M.; Jukić Špika, M.; Goreta Ban, S.; Gadže, J.; Díaz-Pérez, J.C.; Maclean, D. Physical and Chemical Properties of Pomegranate Fruit Accessions from Croatia. *Food Chem.* **2015**, *177*, 53–60. [CrossRef] [PubMed]
48. Mir, M.M.; Umar, I.; Mir, S.A.; Rehman, M.U.; Rather, G.H.; Banday, S.A. Quality Evaluation of Pomegranate Crop—A Review. *Int. J. Agric. Biol.* **2012**, *14*, 658–667.

49. Kulkarni, A.P.; Aradhya, S.M. Chemical Changes and Antioxidant Activity in Pomegranate Arils during Fruit Development. *Food Chem.* **2005**, *93*, 319–324. [CrossRef]
50. Jacobo, C.M. *Cactus Pear Domestication and Breeding*; John Wiley and Sons, Inc.: Hoboken, NJ, USA, 2001; Volume 20, pp. 135–166.
51. de Wit, M.; Nel, P.; Osthoff, G.; Labuschagne, M.T. The Effect of Variety and Location on Cactus Pear (Opuntia Ficus-Indica) Fruit Quality. *Plant Foods Hum. Nutr.* **2010**, *65*, 136–145. [CrossRef]
52. Shulman, Y.; Fainberstein, L.; Lavee, S. Pomegranate Fruit Development and Maturation. *J. Hortic. Sci.* **1984**, *59*, 265–274. [CrossRef]
53. Zarei, M.; Azizi, M.; Bashir-Sadr, Z. Evaluation of Physicochemical Characteristics of Pomegranate (*Punica granatum* L.) Fruit during Ripening. *Fruits* **2011**, *66*, 121–129. [CrossRef]
54. Ozgen, M.; Durgaç, C.; Serçe, S.; Kaya, C. Chemical and Antioxidant Properties of Pomegranate Cultivars Grown in the Mediterranean Region of Turkey. *Food Chem.* **2008**, *111*, 703–706. [CrossRef]
55. Poyrazoğlu, E.; Gökmen, V.; Artik, N. Organic Acids and Phenolic Compounds in Pomegranates (*Punica granatum* L.) Grown in Turkey. *J. Food Compos. Anal.* **2002**, *15*, 567–575. [CrossRef]
56. Wang, S.Y. Effect of Pre-Harvest Conditions on Antioxidant Capacity in Fruits. *Acta Hortic.* **2006**, *712*, 299–306. [CrossRef]
57. Babbar, N.; Oberoi, H.S.; Uppal, D.S.; Patil, R.T. Total Phenolic Content and Antioxidant Capacity of Extracts Obtained from Six Important Fruit Residues. *Food Res. Int.* **2011**, *44*, 391–396. [CrossRef]
58. Ziogas, V.; Tanou, G.; Molassiotis, A.; Diamantidis, G.; Vasilakakis, M. Antioxidant and Free Radical-Scavenging Activities of Phenolic Extracts of Olive Fruits. *Food Chem.* **2010**, *120*, 1097–1103. [CrossRef]
59. Jukić Špika, M.; Liber, Z.; Montemurro, C.; Miazzi, M.M.; Ljubenkov, I.; Soldo, B.; Žanetić, M.; Vitanović, E.; Politeo, O.; Škevin, D. Quantitatively Unraveling Hierarchy of Factors Impacting Virgin Olive Oil Phenolic Profile and Oxidative Stability. *Antioxidants* **2022**, *11*, 594. [CrossRef]
60. Branchereau, C.; Hardner, C.; Dirlewanger, E.; Wenden, B.; Le Dantec, L.; Alletru, D.; Parmentier, J.; Ivančič, A.; Giovannini, D.; Brandi, F.; et al. Genotype-by-Environment and QTL-by-Environment Interactions in Sweet Cherry (*Prunus avium* L.) for Flowering Date. *Front. Plant Sci.* **2023**, *14*, 1142974. [CrossRef]
61. Deloire, A.; Rogiers, S.; Šuklje, K.; Antalick, G.; Zeyu, X.; Pellegrino, A. Grapevine Berry Shrivelling, Water Loss and Cell Death: An Increasing Challenge for Growers in the Context of Climate Change. *IVES Tech. Rev. Vine Wine* **2021**. [CrossRef]
62. Ubi, B.E. External Stimulation of Anthocyanin Biosynthesis in Apple Fruit. *Food Agric. Environ.* **2004**, *2*, 65–70.
63. Borochov-Neori, H.; Judeinstein, S.; Harari, M.; Bar-Ya'akov, I.; Patil, B.S.; Lurie, S.; Holland, D. Climate Effects on Anthocyanin Accumulation and Composition in the Pomegranate (*Punica granatum* L.) Fruit Arils. *J. Agric. Food Chem.* **2011**, *59*, 5325–5334. [CrossRef] [PubMed]
64. Zaouay, F.; Mars, M. Phenotypic Variation and Estimation of Genetic Parameters to Improve Fruit Quality in Tunisian Pomegranate (*Punica granatum* L.) Accessions. *J. Hortic. Sci. Biotechnol.* **2014**, *89*, 221–228. [CrossRef]
65. Jaakola, L.; Hohtola, A. Effect of Latitude on Flavonoid Biosynthesis in Plants. *Plant Cell Environ.* **2010**, *33*, 1239–1247. [CrossRef] [PubMed]
66. Welch, C.R.; Wu, Q.; Simon, J.E. Recent Advances in Anthocyanin Analysis and Characterization. *Curr. Anal. Chem.* **2008**, *4*, 75–101. [CrossRef]

agriculture

MDPI

Article

Organic and Mineral Fertilization on the Photosynthetic, Nutritional, and Productive Efficiency of (*Ficus carica* L.) Subjected to Conduction Systems in a Semi-Arid Region of Brazil

Agda Malany Forte de Oliveira [1], Vander Mendonça [1], Patrycia Elen Costa Amorim [1], Raires Irlenizia da Silva Freire [1], Lucas Rodrigues Bezerra da Silva [1], David Emanoel Gomes da Silva [1], Fagner Nogueira Ferreira [1], Semako Ibrahim Bonou [2], Luderlândio de Andrade Silva [3], Pedro Dantas Fernandes [2], Alberto Soares de Melo [4] and Francisco Vanies da Silva Sá [5,*]

[1] Department of Agronomic and Forestry Sciences, Federal Rural University of Semi-Arid, Mossoró 59625-900, RN, Brazil; agdamalany@hotmail.com (A.M.F.d.O.); vander@ufersa.edu.br (V.M.); patrycia_amorim@hotmail.com (P.E.C.A.); rairessilva10@hotmail.com (R.I.d.S.F.); lucasrodrigues86dix@gmail.com (L.R.B.d.S.); davidemanoel2110@gmail.com (D.E.G.d.S.); fagnernf@gmail.com (F.N.F.)

[2] Department of Agricultural Engineering, Federal University of Campina Grande—UFCG, Campina Grande 58428-830, PB, Brazil; bonouibrahim@gmail.com (S.I.B.); pedrodantasfernandes@gmail.com (P.D.F.)

[3] Center for Agrifood Science and Technology, Federal University of Campina Grande—UFCG, Pombal 58840-000, PB, Brazil; luderlandioandrade@gmail.com

[4] Department of Biology, State University of Paraiba—UEPB, Campina Grande 58429-500, PB, Brazil; alberto.melo@servidor.uepb.edu.br

[5] Department of Agrarian and Exact Sciences, State University of Paraíba—UEPB, Catolé do Rocha 54888-000, PB, Brazil

* Correspondence: vanies_agronomia@hotmail.com; Tel.: +55-(83)-9.9861-9267

Abstract: Fig tree growth and development are highly susceptible to variations influenced by abiotic factors and management practices, including fertilization and training systems. This study aimed to evaluate the effect of organic and mineral fertilization on the photosynthetic, nutritional, and productive efficiency of fig trees subjected to different training systems in semi-arid regions. The experimental design was randomized blocks in a 5 × 4 factorial scheme, with three blocks and three plants per plot. The treatments consisted of five fertilizer sources (mineral fertilizer (NPK) applied at a dose of 126 g N, 90 g P, and 90 g K per plant (M); and four organic sources—cattle manure (CM), organic compost (OC), chicken litter (CL), and sheep manure (SM), all applied at a dose of 10 kg per plant); and four types of training systems (plants with two branches (2B), three branches (3B), four branches (4B), and espalier). Our results demonstrated that the mineral fertilizer (M) and chicken litter (CL) treatments yielded the highest results, particularly in photosynthetic performance. Fig trees fertilized with mineral fertilizer and subjected to the 3B system showed enhanced net photosynthesis (36.96 μmol m^{-2} s^{-1}) and, consequently, higher productivity of 21.28 t ha^{-1}. Similarly, plants fertilized with chicken litter (CL) under the 4B system produced comparable results. These findings demonstrate that the use of mineral and organic fertilizers, combined with an appropriate training system, is a viable strategy for optimizing fig productivity and profitability in semi-arid conditions.

Keywords: fig tree; chlorophyll content; yield; gas exchange; pruning systems

https://doi.org/10.3390/agriculture15202128

1. Introduction

Ficus (Moraceae) is one of the largest genera of angiosperms and is classified as pantropical, comprising more than 800 species distributed across many continents. These species include deciduous and evergreen terrestrial trees, shrubs, hemi-epiphytes, holo-epiphytes, rheophytes, vines, and climbers that adapt well to tropical and subtropical climates [1]. *Ficus* species play a multifunctional role in ecosystems, resulting in considerable ecological importance. Their presence is vital for the survival of a multitude of species, including birds, mammals, and some reptiles and fish worldwide [2]. In Brazil, the genus is notably diverse. The morphological variation, complexity, and enormous diversification of *Ficus* in the Neotropics make this genus an essential element for understanding diversification in species-rich biomes, such as the two main blocks of neotropical rainforests. The Amazon (AM), with about 50 species, and the Atlantic Forest (AF), with about 35 species, play a fundamental role in tropical forest ecosystems as a keystone resource for fauna [3].

Among these, *Ficus carica* L., commonly known as the fig tree, is one of the earliest cultivated fruit trees in the world. It is the most commercially important species of the genus due to its value for food and medicinal purposes. Native to Southwest Asia and the Eastern Mediterranean, it has been cultivated for over 11,000 years and has spread globally due to its exceptional adaptability to diverse soil and climatic conditions [4]. Unpruned, it is generally a tree 15 to 20 feet tall, with numerous spreading branches and a trunk rarely more than 7 feet in diameter [5]. The largest producers are Türkiye, Egypt, Morocco, and Algeria, accounting for more than 50% of global production [6]. The fruit has substantial commercial value and is used in various forms, including fresh, dried, and canned, and it is an essential ingredient in many food products, such as jellies and jams [7].

This adaptability makes it a promising candidate for regions seeking to diversify their agricultural output. In this context, introducing the fig tree emerges as a potential option for expanding fruit production in the Brazilian Nordeste. Although it is considered to be a temperate climate species, the fig tree stands out for its exceptional adaptability to diverse edaphic and climatic environments [8]. However, Brazilian fig productivity is still considered to be low, and its growth is subject to variations influenced by management practices, such as fertilization and training systems [9–11]. Therefore, it is necessary to study and implement new cultivation systems to enhance the productivity of the fig tree.

Among these systems, pruning is a crucial technique to ensure optimal fig production as, without this management, plants exhibit disordered growth, self-shading, and low fruit production [12]. This relevance is attributed to the commercial value of the fruits, which not only represent an essential complement to human nutrition, providing energy compounds such as starches, sugars, and fibers, but are also significant sources of minerals essential to metabolism, including calcium and phosphorus, as well as vitamins, amino acids, and antioxidants [13].

Additionally, it has been demonstrated that figs have high commercialization potential, both in their natural form and for industrial use [14]. In Brazil, commercial fig cultivation is primarily based on a single cultivar, 'Roxo de Valinhos,' due to its high vigor, productivity, robustness, and ease of adaptation to the Brazilian intensive pruning management system [15,16].

This intensive pruning management is a hallmark of the current commercial fig production system. In the current commercial fig production system in the central Brazilian producing regions, it is common to perform drastic winter pruning, eliminating the branches that produced fruit in the previous cycle (fruit-bearing stems), resulting in four or five main structural branches on the tree, known as the open vessel system (goblet). Under these conditions, Brazilian fig productivity is still considered to be low [17–19]. Therefore, it is necessary to study and implement new cultivation systems to enhance the productivity of the fig tree.

Beyond pruning, another critical aspect of fig cultivation that requires attention is fertilization. The study of various fertilization methods is a common practice in research involving multiple species as it plays a crucial role in the development and production of fruits [10,19–22]. The growing demand for food, driven by population growth, underscores the need to optimize agricultural production [23].

This gap is particularly relevant given that chemical fertilizers are recognized for boosting plant productivity. However, their prolonged and uncontrolled use can have adverse effects on soil, such as acidification and a reduction in microbial biomass, compromising soil fertility and, consequently, the sustainable capacity to support the healthy growth and productivity of species [20].

In contrast to synthetic fertilizers, organic fertilization provides a viable and sustainable alternative for cultivating a wide range of plant species. This approach plays a crucial role in enhancing the long-term health of agricultural ecosystems. Improving soil organic matter and enriching microbial biomass and biodiversity create a more resilient and productive soil environment [24]. The use of organic fertilizers also helps to reduce reliance on external synthetic inputs, thereby mitigating adverse environmental impacts such as soil acidification and nutrient runoff, which are common in conventional systems. Despite the demonstrated benefits of organic fertilization, a critical gap remains in research investigating its specific effects on the physiological responses of fig trees, particularly in semi-arid environments [9,25].

In this context, despite the growing importance of fig trees in the northeast region, a critical gap remains in understanding how management practices, such as different pruning systems and fertilization methods, affect the plant's ecophysiological responses in semi-arid conditions. Understanding these responses, such as photosynthetic efficiency and gas exchange, is crucial for optimizing productivity and ensuring the sustainability of cultivation under these conditions. Specifically, there is a lack of studies investigating the physiological responses of fig trees to organic fertilization.

This gap is particularly relevant because, as Średnicka-Tober et al. [21] report, the use of organic fertilizers can influence plant metabolism, leading to an increase in the profile of secondary metabolites in plant tissues. They also explain that organic crops can function as a stress condition for the plant as nutrients are released slowly depending on the type of fertilizer used, differing entirely from mineral fertilization, which is characterized by the rapid and instantaneous availability of nutrients.

In this context, finding viable sources of organic nutrients is essential. This way, animal manures represent one of the most common, accessible, and economical sources of organic fertilizers for agricultural production [22]. Several animal sources, including cows, goats, sheep, horses, poultry, and pigs, provide manure that can be used as soil amendments to improve organic carbon content, enhance nutrient absorption, and promote healthy plant growth [26]. Animal manures differ in nutrient content, and their nutritional composition varies according to the type of animal, its diet, and the method of manure preservation [22]. These differences in nutrient content play an essential role regarding the impact of manure on soil properties.

Consequently, it is important to consider the long-term effects of using these agricultural inputs since repeated application of animal manure can lead to beneficial changes in soil properties [26]. Therefore, monitoring and sustainable management practices are necessary to ensure that the continued use of these organic fertilizers is beneficial without compromising soil health and productivity in the long term.

In this context, we hypothesized that combining a specific training system with organic fertilization can optimize nutrient availability and plant structure, promoting superior physiological responses and resulting in greater photosynthetic efficiency and fig productivity in a semi-arid environment.

We hope that the results of this study demonstrate that the adoption of training systems and organic fertilization are viable and beneficial practices for fig cultivation, providing support for optimizing management and increasing profitability for producers in the Brazilian semi-arid region and offering valuable insights for the management of the crop in other arid and semi-arid areas globally.

Therefore, this study aimed to evaluate the effect of organic and mineral fertilization on the photosynthetic, nutritional, and productive efficiency of fig trees subjected to different training systems in semi-arid regions.

2. Materials and Methods

2.1. Characterization of the Experimental Area

The study was carried out at the Rafael Fernandes Experimental Farm, belonging to the Federal Rural University of the Semi-Arid (UFERSA), Mossoró, western region of the state of Rio Grande do Norte, Brazil, whose geographic coordinates are 5°03′37″ S latitude and 37°23′50″ W longitude, with an altitude of 78 m and flat relief. According to Köppen, the region's climate is classified as 'BSh', a hot tropical semi-arid type, characterized by well-defined periods: dry (prolonged) and wet (short and irregular). The region's vegetation type is the open arboreal Caatinga, and the soil of the area is classified as Typical Dystrophic Red Argisol [27,28].

The experiment was conducted from August 2021 to April 2022 (9 months), covering the entire first production cycle of the 'Roxo de Valinhos' fig tree under the annual management system adopted in the region. The period was determined to capture the plant's immediate physiological and productive response to the different fertilization and training systems studied. During the experimental period (August 2021 to April 2022), the following meteorological data were collected from the ASA space meteorological station (Meliponário Imperatriz) and Rafael Fernandes experimental station. The fig (*Ficus carica* L.) production cycle lasted from August to April, beginning with drastic pruning in August. A complete absence of precipitation (0 mm) in August and October and high average temperatures (27.0 °C to 28.0 °C) characterized the climatic conditions during this initial period. This combination of high temperatures and drought (August to October) coincided with the most sensitive phases of the cycle: budbreak and initial vegetative growth. Beginning in December, when the crop reached the fruit expansion (syconium) stage and the beginning of harvest, there was a change in the water regime with the first rainfall (approximately 50 mm). The average temperature peaked in this month (28.5 °C). The final period, from February to April, corresponded to the peak of the harvest and was the most challenging. Average temperatures remained high (28.0 °C in February), and precipitation increased significantly, contributing to a cumulative total of 185 mm at the end of the cycle (April) (Figure 1).

The organic sources, cattle manure (CM), organic compost (OC), chicken litter (CL), and sheep manure (SM), used in this study were analyzed for their chemical characteristics according to the Embrapa protocol [29] (Table 1).

Before implementing the experiment, soil samples were collected from the 0–20 cm depth layer of the entire experimental area to characterize the initial conditions. A composite sample was prepared by homogenizing 10 random subsamples collected across the plot in a zigzag pattern. These samples were then dried in the shade, crushed, and passed through a 2 mm mesh sieve (TFSA) to facilitate chemical analyses [29] (Table 2).

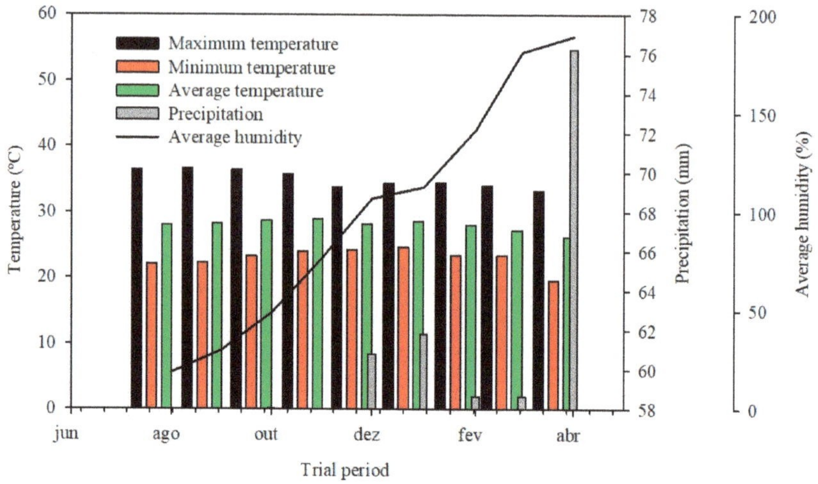

Figure 1. Air temperature (°C), relative humidity (%), and precipitation (mm) during the experiment (August 2021 to April 2022) in Mossoró, Rio Grande do Norte—Brazil.

Table 1. Chemical characteristics of the organic sources, cattle manure (CM), organic compost (OC), chicken litter (CL), and sheep manure (SM), used in the study. Mossoró, Rio Grande do Norte—Brazil.

Samples	pH	EC	OM	P	K^+	Na^+
	H_2O	dS m^{-1}	g kg^{-1}		mg dm^3	
CM	6.61	1.92	44.10	199.10	824.80	214.90
OC	7.30	5.43	41.61	513.60	521.50	683.30
CL	7.61	5.89	46.60	513.60	502.70	186.61
SM	8.02	7.33	39.12	234.00	240.20	866.10

Samples	Ca^{2+}	Mg^{2+}	Al^{3+}	(H + Al)	SB	CEC	V	PES
		cmol dm^3				%		
CM	10.48	0.77	0	0	14.877	14.877	100	14.39
OC	15.60	6.21	0	0	39.061	39.061	100	34.02
CL	5.88	6.99	0	0	30.287	30.287	100	29.24
SM	15.68	1.24	0	0	30.562	30.562	100	18.78

P, K^+, Na^+: Mehlich extractant 1; Al^{3+}; (H + Al) = 0; Ca^{2+}, Mg^{2+}: KCl extractant 1.0 mol L^{-1}; pH: hydrogen potential; EC: electrical conductivity; SB: sum of base; CEC: cation exchange capacity; OM: Walkley–Black wet digestion; PES: percentage of exchangeable sodium.

Table 2. Chemical characteristics of soil with mineral fertilization (SM), soil with cattle manure (SCM), soil with organic compost (SOC), soil with chicken litter (SCL), and soil with sheep manure (SSM) at the beginning of the experiment. Mossoró, Rio Grande do Norte—Brazil.

Samples	pH	EC	OM	P	K^+	Na^+
	H_2O	dS m^{-1}	g kg^{-1}		mg dm^3	
SM	7.36	0.07	22.77	73.50	62.50	0.06
SCM	7.93	0.16	23.96	74.30	62.50	0.13
SOC	7.50	0.08	23.30	74.31	66.90	0.07
SCL	7.42	0.08	23.84	58.80	49.90	0.07
SSM	7.95	0.15	23.71	53.70	62.50	0.11

Table 2. *Cont.*

Samples	Ca^{2+}	Mg^{2+}	Al^{3+}	(H+Al)	SB	CEC	V	PES
	cmolc dm^3						%	
SM	1.65	1.35	0	0	3.29	3.29	100	6.66
SCM	3.00	1.40	0	0	4.74	4.73	100	4.21
SOC	2.32	1.40	0	0	4.02	4.02	100	5.74
SCL	2.05	1.42	0	0	3.71	3.71	100	4.44
SSM	2.78	2.07	0	0	5.14	5.14	100	3.22

P, K^+, Na^+: Mehlich extractant 1; Al^{3+}; (H + Al) = 0; Ca^{2+}, Mg^{2+}: KCl extractant 1.0 mol L^{-1}; pH: hydrogen potential; EC: electrical conductivity; SB: sum of base; CEC: cation exchange capacity; OM: Walkley–Black wet digestion; PES: percentage of exchangeable sodium.

After the experimental period ended, soil samples were collected from the 0–20 cm depth layer of the entire experimental area to characterize the initial conditions. A composite sample was prepared by homogenizing 10 random subsamples collected across the plot in a zigzag pattern. These samples were then dried in the shade, crushed, and passed through a 2 mm mesh sieve (TFSA) to facilitate chemical analyses [29] (Table 3).

Table 3. Chemical characteristics of soil with mineral fertilization (SM), soil with cattle manure (SCM), soil with organic compost (SOC), soil with chicken litter (SCL), and soil with sheep manure (SSM) at the end of the experiment. Mossoró, Rio Grande do Norte—Brazil.

Samples	pH	EC	OM	P	K^+	Na^+
	H_2O	dS m^{-1}	g kg^{-1}		mg dm^3	
SM	7.91	0.09	6.00	1903.67	0.16	0.11
SCM	8.08	0.07	6.37	169.77	0.13	0.11
SOC	8.4	0.08	6.12	222.09	0.18	0.13
SCL	8.22	0.11	6.70	1438.88	0.21	0.14
SSM	8.17	0.10	6.25	213.19	0.16	0.15

Samples	Ca^{2+}	Mg^{2+}	Al^{3+}	(H+Al)	SB	CEC	V	PES
		cmolc dm^3					%	
SM	3.22	0.65	0	0	4.15	4.15	100	3.97
SCM	2.15	0.53	0	0	2.92	2.92	100	4.63
SOC	3.63	0.38	0	0	4.32	4.32	100	4.08
SCL	2.66	0.67	0	0	3.68	3.68	100	5.74
SSM	2.42	0.99	0	0	3.72	3.72	100	4.42

P, K^+, Na^+: Mehlich extractant 1; Al^{3+}; (H + Al) = 0; Ca^{2+}, Mg^{2+}: KCl extractant 1.0 mol L^{-1}; pH: hydrogen potential; EC: electrical conductivity; SB: sum of base; CEC: cation exchange capacity; OM: Walkley–Black wet digestion; PES: percentage of exchangeable sodium.

2.2. Materials and Growing Conditions

The plants in this study were of the *Ficus carica* L. species, 'Roxo of Valinhos'. They were planted in the experimental area in March 2017, with a spacing of 2 m by 3 m. Four months after planting the orchard, the first formative pruning was completed at a height of 40 cm. Four months after the first formative pruning, the second formative pruning was carried out to form the branches of each plant (primary vegetative branches).

The plants in the espalier system were pruned at a height of 40 cm from the ground, with their primary branches, measuring 75 cm on each side, tied horizontally to steel cables. The espalier structure was assembled with four posts 10 m apart. The steel cables were stretched 40, 80, and 1.80 cm above the ground. The first production pruning of the experiment was carried out on 27 August 2021. The plants were pruned, resulting in 5 cm long branches so that vegetative buds could be produced on each branch. Thirty days after production pruning, the excess productive branches were thinned, allowing each plant to remain with two productive branches (two branches—four productive branches; three branches—six productive branches; four branches—eight productive branches; espalier—six productive branches) (Figure 2).

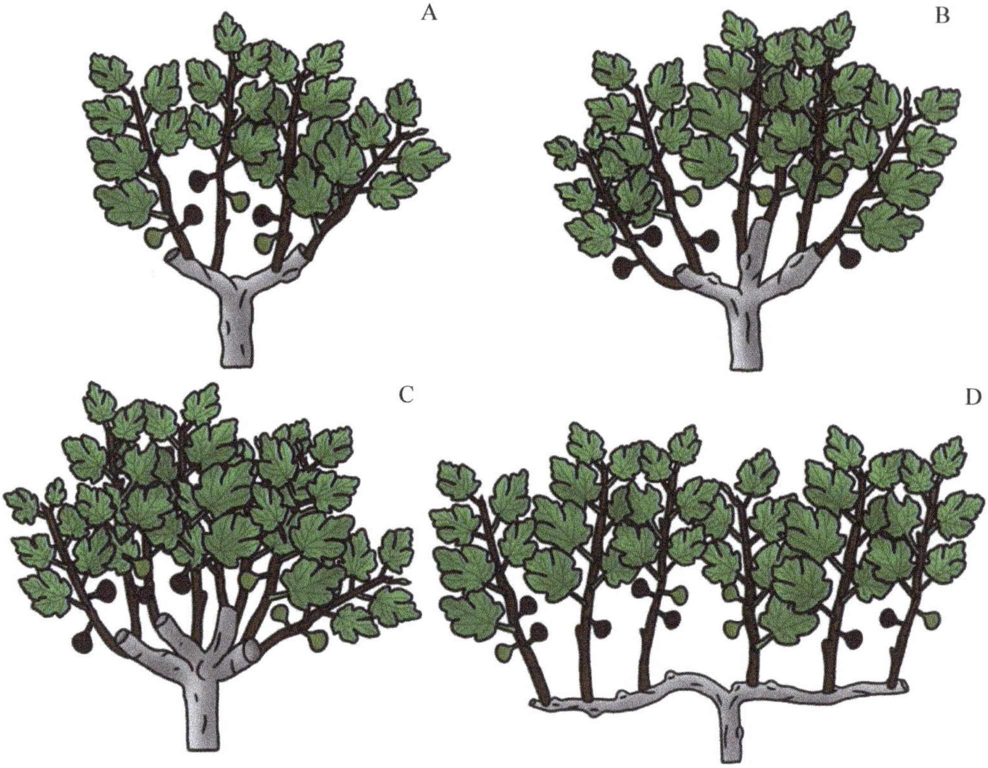

Figure 2. Training systems used in fig trees: two branches (**A**), three branches (**B**), four branches (**C**), and espalier (**D**).

The fig tree was fertilized using two systems: conventional (with mineral fertilizers) and organic. Plants under the organic system received an application of 10 kg per plant (fresh weight) of organic sources. This dosage was established based on the practices recommended by [10] for this crop in the semi-arid region. The organic material was applied superficially, distributed in a half-moon shape around the trunk, corresponding to the canopy projection area of the plant (ou: within the wetted zone of the micro-sprinkler). The sources used were cattle manure (CM), organic compost (OC), chicken litter (CL), and sheep manure (SM). The chicken litter was applied fresh, while the other organic sources were cured, as described in [10].

Mineral fertilizer was applied superficially via the soil in a half-moon shape, according to the essential fertilization recommendation for fig trees, receiving 126 g per plant^{-1} of nitrogen, 90 g per plant^{-1} of potassium, and 90 g per plant^{-1} of phosphorus per cycle [10,30]. The mineral sources used were nitrogen–ammonium sulfate (22% N), potassium–potassium chloride (60% K), and phosphate–simple superphosphate (52% P_2O_5). P fertilization was carried out once after pruning, and the other two, N and K, were applied at 15-day intervals, divided into three applications.

The irrigation system consisted of a micro-sprinkler system with a 75 L h^{-1} (Rivulis Rondo micro-sprinkler, Israel) on a 60 cm stake. It lasted 1 h daily, with one sprinkler per plant and a break on rainy days.

To control weeds, regular cleaning between the rows was carried out using a brush cutter and manual weeding. Whenever necessary, the thieving branches and new shoots were removed with pruning shears.

2.3. Experimental Design

The experimental design employed was a randomized block design in a 5 × 4 factorial scheme, with three blocks and three plants per plot, totaling 180 plants. Each plot measured 4 m (length) by 3 m (width), resulting in a total area of 12 m^2. In each plot, the central plant was considered the useful plant for data collection, and the two surrounding plants served as lateral borders. A minimum distance of 3.0 m was maintained between plots and between blocks to ensure the complete isolation of the treatments. The treatments consisted of five fertilizer sources: one conventional fertilizer, composed of mineral fertilizer (M) (N, P, and K), and four organic sources: cattle manure (CM), organic compost (OC), chicken litter (CL), and sheep manure (SM), and four types of training systems (plants with two branches, three branches, four branches, and espalier).

2.4. Variables Analyzed

The following variables were determined:

Gas exchange: The plants were evaluated only once, at the beginning of the production period (150 days after pruning), regarding gas exchange, between 7:00 and 9:00 a.m. The evaluations were performed on fully expanded healthy leaf per useful plant (for a total of three leaves per plot), located in the upper third of each plant, with a portable infrared carbon dioxide analyzer (IRGA) model LCPro+ Portable Photosynthesis System® (ADC Bio Scientific Limited, UK) LCPro+ with temperature control at 25 °C, irradiation of 1200 µmol photons m^{-2} s^{-1}, and airflow of 200 mL min^{-1}. Obtain net photosynthesis (A_N) in µmol (CO_2) m^{-2} s^{-1}, transpiration (E) in mmol (H_2O) m^{-2} s^{-1}, stomatal conductance (gs) in mol (H_2O) m^{-2} s^{-1}, internal CO_2 concentration (Ci) in mol (CO_2) m^{-2} s^{-1}, and leaf temperature (Lt) in °C. Using this data, the instantaneous water use efficiency (WUE) was quantified as the ratio of net photosynthesis to transpiration (A_N/E) in µmol m^{-2} s^{-1}/mmol (H_2O) m^{-2} s^{-1}. The instantaneous carboxylation efficiency (A_N/Ci) was quantified as the ratio of net photosynthesis to internal CO_2 concentration (A_N/Ci) [31].

Chlorophyll fluorescence was evaluated using a pulse-modulated fluorometer model OS5p from Opti Science. This protocol utilized an actinic illumination source with a saturating multi-flash pulse, coupled to a photosynthetically active radiation determination clip (PAR-Clip), after the leaves had been adapted to darkness for 30 min. Under these conditions, the maximum quantum efficiency of PSII (Fv/Fm) was estimated [31].

The chlorophyll content index was determined using an electronic chlorophyll content meter (ClorofiLOG®, CFL 1030) (Falker, Porto Alegre, RS, Brazil) only once at the beginning of the production period (150 days after pruning) in all plants comprising the experimental unit. In each plant, young and fully expanded leaves were selected from the upper middle third, without signs of pest or disease attack, for measurement at three equidistant points on the adaxial surface of each leaf. The values of chlorophyll a, chlorophyll b, and total were obtained.

The nutritional content of the leaves was determined by selecting young fully expanded leaves exposed to the sun from the upper middle third of each plant in the experimental unit without any signs of pest or disease attack [32]. Then, the contents of nitrogen (N), phosphorus (P), potassium (K), calcium (Ca), and magnesium (Mg) were determined after sulfuric digestion, and iron (Fe), zinc (Zn), manganese (Mn), and copper (Cu) after dry digestion [29].

Fruit yield: The productivity was determined through weekly harvesting and weighing the fresh fruits from the central useful plant of each experimental unit throughout the entire production cycle (August 2021 to April 2022). To determine productivity, the total commercial fresh weight per plot was calculated, and the ratio of commercial production to area was then calculated based on the number of plants and spacing. The results were expressed in tons per hectare (t ha^{-1}).

2.5. Statistical Analysis

The results were subjected to the Shapiro–Wilk normality test and the Bartlett homogeneity test. Subsequently, they were subjected to an analysis of variance using the F probability test ($p < 0.5$). The occurrence of a significant interaction was analyzed using Tukey's test to examine the unfolding of the interaction at the 5% probability level using the statistical program System for Analysis of Variance—SISVAR v. 5.6 (Lavras, MG, Brazil) [33]. The main effects were analyzed separately only when the interaction was non-significant (isolated factors).

3. Results

3.1. Gas Exchange

The studied factors, including fertilization sources (mineral and organic) and conduction systems, interacted significantly (based on the *p*-value less than the chosen significance level) and influenced the physiology of the fig trees. There were significant interactions for water use efficiency (WUE), transpiration (*E*), stomatal conductance (*gs*), leaf temperature (*LT*) ($p < 0.01$), internal CO_2 concentration (*Ci*) ($p < 0.01$), instantaneous carboxylation efficiency (A_N/Ci) ($p < 0.01$), net photosynthesis (A_N) ($p < 0.01$), and maximum quantum efficiency of PSII (Fv/Fm) ($p < 0.05$). Due to these significant interaction effects, the combined means were analyzed using the Tukey test ($p < 0.05$) to compare treatment combinations and interpret the results.

Plants that received mineral fertilization showed higher water use efficiency (WUE) in the 2B, 3B, and 4B systems (7.14, 8.19, and 6.84 (μmol m^{-2} s^{-1}) (mol H_2O m^{-2} s^{-1})$^{-1}$), respectively. Only in the espalier system did fertilization (OC) have a WUE of 5.35 μmol m^{-2} s^{-1} (mol H_2O m^{-2} s^{-1})$^{-1}$, similar to mineral fertilization of 5.72 μmol m^{-2} s^{-1} (mo H_2O m^{-2} s^{-1})$^{-1}$. Notably, the highest WUE with fertilization (M) was obtained in the systems with fewer branches, 2B and 3B, emphasizing the 3B system of 8.19 μmol m^{-2} s^{-1} (mo H_2O m^{-2} s^{-1})$^{-1}$, which was statistically superior to the other systems (Table 4).

OC, CM, and CL presented median and similar values among the organic sources. Only SM resulted in lower *WUE* for the conduction systems 2B (2.12 (μmol m^{-2} s^{-1}) (mo H_2O m^{-2} s^{-1})$^{-1}$), 3B (3.17 (μmol m^{-2} s^{-1}) (mo H_2O m^{-2} s^{-1})$^{-1}$), 4B (3.91 (μmol m^{-2} s^{-1}) (mo H_2O m^{-2} s^{-1})$^{-1}$), and espalier (3.76 (μmol m^{-2} s^{-1}) (mo H_2O m^{-2} s^{-1})$^{-1}$. Reductions in *WUE* of around 70.31%, 61.39%, 42.84%, and 34.27% were recorded, respectively (Table 4).

The lowest *E* was obtained with mineral fertilization in the 2B, 3B, and 4B systems, presenting transpiration rates of 5.02, 4.56, and 4.97 mmol H_2O m^{-2} s^{-1}, respectively. The highest *E* value was observed in the 2B system, using SM and CL of 9.22 mmol H_2O m^{-2} s^{-1} and 8.32 mmol H_2O m^{-2} s^{-1}, respectively, and in the 3B system, using SM and CL of 8.28 mmol H_2O m^{-2} s^{-1} and 8.17 mmol H_2O m^{-2} s^{-1}, with no statistically significant difference. With 4B, the organic fertilizers did not differ statistically from each other and had higher *E* than the mineral fertilizers. In the espalier system, the lowest *E* was obtained with CM and M of 6.17 and 6.34 mmol H_2O m^{-2} s^{-1}, respectively, and the highest E was obtained using CL of 7.99 mmol H_2O m^{-2} s^{-1} (Table 4).

Table 4. Water use efficiency ($WUE = A_N/E$) [(μmol m^{-2} s^{-1}) (mo H_2O m^{-2} s^{-1})$^{-1}$], transpiration (E) (mmol H_2O m^{-2} s^{-1}), and stomatal conductance (gs) (mol de H_2O m^{-2} s^{-1}) in fig trees subjected to different mineral and organic fertilization and training system sources.

FA	2 Branches	3 Branches	4 Branches	Espalier
		WUE		
M	7.14 ± 0.46 aAB	8.19 ± 1.11 aA	6.84 ± 0.25 aBC	5.72 ± 0.20 aC
CM	4.82 ± 0.13 bA	4.70 ± 0.08 bA	4.71 ± 0.18 bA	4.91 ± 0.08 abA
OC	5.45 ± 0.38 bA	4.51 ± 0.22 bA	4.77 ± 0.24 bA	5.35 ± 0.04 aA
CL	4.64 ± 0.13 bA	4.24 ± 0.29 bcA	5.22 ± 0.27 bA	4.62 ± 0.37 abA
SM	2.12 ± 0.006 cB	3.17 ± 0.02 cAB	3.91 ± 0.18 bA	3.76 ± 0.02 bA
CV (%)		11.45		
		E		
M	5.02 ± 0.26 cB	4.56 ± 0.63 cB	4.97 ± 0.06 bB	6.34 ± 0.26 cA
CM	6.32 ± 0.11 bC	7.12 ± 0.16 bAB	7.71 ± 0.28 aA	6.17 ± 0.29 cC
OC	6.42 ± 0.34 bA	6.73 ± 0.13 bA	6.83 ± 0.23 aA	6.96 ± 0.34 bcA
CL	8.32 ± 0.27 aA	8.17 ± 0.26 aA	6.98 ± 0.19 aB	7.99 ± 0.35 aA
SM	9.22 ± 0.08 aA	8.28 ± 0.06 aB	7.55 ± 0.24 aB	7.59 ± 0.08 abB
CV (%)		5.91		
		gs		
M	0.41 ± 0.04 bB	0.47 ± 0.08 bAB	0.57 ± 0.02 aA	0.51 ± 0.03 abAB
CM	0.41 ± 0.02 bB	0.50 ± 0.03 bAB	0.59 ± 0.04 aA	0.39 ± 0.03 bB
OC	0.43 ± 0.05 bA	0.48 ± 0.02 bA	0.46 ± 0.03 aA	0.48 ± 0.04 abA
CL	0.76 ± 0.08 aA	0.65 ± 0.02 aAB	0.48 ± 0.02 aC	0.62 ± 0.06 aBC
SM	0.77 ± 0.02 aA	0.53 ± 0.02 abB	0.47 ± 0.04 aB	0.50 ± 0.02 abB
CV (%)		11.69		

Means followed by the same lowercase letter in the column and uppercase letter in the row do not differ by Tukey's test ($p > 0.05$). Mean ± standard deviation. M: Mineral; CM: cattle manure; OC: organic compost; CL: chicken litter; SM: sheep manure.

The stomatal conductance (gs) of plants subjected to the 2B and 3B systems was similar, with the lowest gs observed with the M (0.41 and 0.47 mol (H_2O) m^{-2} s^{-1}), CM (0.41 and 0.50 mol (H_2O) m^{-2} s^{-1}), and OC (0.43 and 0.48 mol (H_2O) m^{-2} s^{-1}) fertilizations, respectively, which did not differ statistically among them. As in E, the highest gs values were verified using SM (0.77 and 0.53 mol (H_2O) m^{-2} s^{-1}) and CL (0.76 and 0.65 mol (H_2O) m^{-2} s^{-1}) in the 2B and 3B systems. In the espalier system, the highest gs values were obtained using a CL of 0.62 mol (H_2O) m^{-2} s^{-1}. Only the 4B system did not differ statistically between the sources studied (Table 4).

Plants fertilized with SM presented the highest Ci in the 2B, 3B, and espalier systems of 280, 218, and 200.33 μmol mol^{-1}, respectively, which may indicate internal CO_2 accumulation. In the system with 4B, there was a more significant accumulation of Ci with mineral fertilization, with 205.33 μmol mol^{-1}. With 3B, the M and CM sources presented the lowest Ci of 155.67 and 169.00 μmol mol^{-1}. In turn, 4B had the lowest Ci with CL (146.00 μmol mol^{-1}), and, in the espalier system, lower values were verified using OC and CM of 140.33 and 158.33 μmol mol^{-1}, respectively (Table 5).

The instantaneous carboxylation efficiency (A_N/Ci) was similar to the results of internal CO_2 concentration. The lower-efficiency plants (A_N/Ci) were observed to be fertilized with SM in the 2B (0.07), 3B (0.12), and espalier systems (0.14). In the 4B system, lower A_N/Ci values were also observed with M fertilization (0.17), although they did not differ statistically from those with SM (0.16). It is worth noting that lower A_N/Ci values led to reductions in the photosynthetic rate (Table 5).

Table 5. Internal CO_2 concentration (*Ci*) (μmol mol^{-1}), instantaneous carboxylation efficiency (A_N/Ci), and rate of CO_2 assimilation or photosynthesis (A_N) (μmol m^{-2} s^{-1}) in fig trees subjected to different mineral and organic fertilization and training system sources.

FA	2 Branches	3 Branches	4 Branches	Espalier
		Ci		
M	139.67 ± 13.02 bB	155.67 ± 21.73 bB	205.33 ± 12.72 aA	162.33 ± 6.12 abB
CM	163.33 ± 7.06 aB	169.00 ± 4.04 bA	166.00 ± 7.00 abA	158.33 ± 2.93 bA
OC	138.67 ± 19.63 bB	184.00 ± 9.54 abA	165.33 ± 8.95 abAB	140.33 ± 2.03 bB
CL	177.67 ± 7.17 bAB	189.00 ± 15.10 abA	146.00 ± 10.54 bB	169.00 ± 12.34 abAB
SM	280.00 ± 0.58 aA	218.00 ± 2.08 aB	183.33 ± 11.57 abB	200.33 ± 2.19 aB
CV (%)		9.82		
		A_N/Ci		
M	0.26 ± 0.03 aA	0.24 ± 0.03 aA	0.17 ± 0.02 bB	0.22 ± 0.01 aAB
CM	0.19 ± 0.01 aB	0.20 ± 0.003 abA	0.22 ± 0.02 abA	0.19 ± 0.003 abA
OC	0.26 ± 0.04 abA	0.16 ± 0.01 bcB	0.20 ± 0.01 abAB	0.26 ± 0.007 aA
CL	0.22 ± 0.009 abA	0.19 ± 0.03 abcA	0.26 ± 0.02 aA	0.22 ± 0.03 aA
SM	0.07 ± 0.001 bC	0.12 ± 0.001 cAB	0.16 ± 0.01 bA	0.14 ± 0.001 bA
CV (%)		15.92		
		A_N		
M	35.65 ± 1.04 aA	35.94 ± 0.64 aA	33.96 ± 1.30 abA	36.26 ± 0.78 aA
CM	30.45 ± 0.28 bB	33.45 ± 0.80 abAB	36.29 ± 1.20 aA	30.28 ± 1.11 bB
OC	34.72 ± 0.89 abAB	30.30 ± 1.11 bcB	32.50 ± 0.79 abBC	37.15 ± 1.55 aA
CL	38.53 ± 0.47 aA	34.63 ± 2.32 abA	36.37 ± 1.09 aA	36.71 ± 1.94 aA
SM	19.56 ± 0.21 cB	26.28 ± 0.08 cA	29.42 ± 0.40 bA	28.56 ± 0.18 bA
CV (%)		5.95		

Means followed by the same lowercase letter in the column and uppercase letter in the row do not differ by Tukey's test ($p > 0.05$). Mean ± standard deviation. M: Mineral; CM: cattle manure; OC: organic compost; CL: chicken litter; SM: sheep manure.

The highest A_N/Ci values varied among the studied conduction systems. In the 2B system, the highest A_N/Ci was obtained with M (0.26), which did not differ statistically from OC (0.26) and CL (0.22). With 3B, the highest A_N/Ci values were obtained with mineral fertilizer (0.24) and CM (0.20), which were statistically equal. In the 4B system, the organic fertilizers CL, CM, and OC presented efficiencies of 0.26, 0.22, and 0.20, respectively, which were statistically equal. In the espalier system, the highest efficiencies were obtained with OC (0.26), followed by CL (0.22) and M (0.22), which were statistically equal (Table 5).

The 2B and espalier systems showed similar behavior regarding fertilizer sources. The highest A_N rates were obtained with the use of CL, with 38.53 and 36.71 μmol m^{-2} s^{-1}, respectively, followed by M, with 35.65 and 36.26 μmol m^{-2} s^{-1}, respectively, and OC, with 34.72 and 37.15 μmol m^{-2} s^{-1}, respectively, statistically equal. It is also noteworthy that using the organic sources CL and OC, which are statistically equal, increased the A_N of fig trees by 2B to 49.23% and 43.66%, respectively. In the espalier system, the use of CL and OC fertilizers also increased the A_N of fig trees by 22.20% and 23.12%, respectively, compared to SM (Table 5).

Plants submitted to the 3B system presented higher A_N, with means of 35.94 μmol m^{-2} s^{-1}, 34.63 μmol m^{-2} s^{-1}, and 33.45 μmol m^{-2} s^{-1}, respectively, statistically equal. In this condition, using CL increased A_N by 24.11% compared to SM. The highest A_N was obtained in the 4B system, with a CL of 36.37 μmol m^{-2} s^{-1} and a CM of 36.29 μmol m^{-2} s^{-1}, which were statistically equal. In this system, these organic sources represented increases of 19.11% and 18.93% in the A_N of fig trees compared to SM (Table 5).

As in *WUE*, *E*, *gs*, *Ci*, and A_N/Ci, it was observed that the A_N of plants fertilized with SM was impaired in all the conduction systems studied, causing a considerable reduction in the photosynthesis of the plants in each system. In 2B, this reduction was more pronounced than in the other systems used, representing a 49.23% decrease, which is among the highest photosynthetic rates obtained in this system. In the different systems, 3B, 4B, and espalier, reductions of 26.88%, 19.11%, and 22.20%, respectively, were observed (Table 5).

The maximum quantum efficiency of PSII (Fv/Fm) varied according to the fertilizer sources and conductive systems used. There was no difference in the Fv/Fm values obtained in the 2B system for the fertilizer sources. The plants in the 3B system had higher Fv/Fm values, with an OC of 0.77, which did not differ statistically from CM and CL. The lowest Fv/Fm values were obtained in plants with M and SM of 0.70, which were statistically equal. With 4B, the organic sources had higher values, with CL fertilization of 0.76 standing out, and M fertilization provided the lowest value, 0.68. In the espalier system, the highest Fv/Fm values were obtained with OC, followed by SM, and the lowest values were obtained in CM, with a value of 0.68 (Table 6). The conduction systems used differed statistically only in bovine fertilization, where plants with 3B had the highest Fv/Fm values, 0.76, and espalier promoted the lowest value, 0.68 (Table 6).

Table 6. Maximum quantum efficiency of PSII (Fv/Fm) in fig trees subjected to different mineral and organic fertilization and training system sources.

FA	2 Branches	3 Branches	4 Branches	Espalier
		Fv/Fm		
M	0.72 ± 0.02 aA	0.70 ± 0.01 bA	0.68 ± 0.003 bA	0.72 ± 0.007 abA
CM	0.73 ± 0.02 aAB	0.76 ± 0.02 abA	0.73 ± 0.02 abAB	0.68 ± 0.003 bB
OC	0.75 ± 0.01 aA	0.77 ± 0.02 aA	0.73 ± 0.03 abA	0.76 ± 0.003 aA
CL	0.73 ± 0.01 aA	0.75 ± 0.02 abA	0.76 ± 0.01 aA	0.74 ± 0.01 abA
SM	0.73 ± 0.01 aA	0.70 ± 0.02 bA	0.70 ± 0.009 abA	0.75 ± 0.003 aA
CV (%)		3.98		

Means followed by the same lowercase letter in the column and uppercase letter in the row do not differ by Tukey's test ($p > 0.05$). Mean ± standard deviation. M: Mineral; CM: cattle manure; OC: organic compost; CL: chicken litter; SM: sheep manure.

3.2. Chlorophyll Content Index

The chlorophyll content index, composed of chlorophyll *a*, *b*, and total contents, was significantly influenced by the fertilizer sources (mineral and organic) and conduction systems at the level of ($p < 0.01$) for the fertilizer sources and conduction systems studied. Due to these significant interaction effects, the combined means were analyzed using the Tukey test ($p < 0.05$) to compare treatment combinations and interpret the results.

The chlorophyll *a* content varied between the fertilization sources and the 2B, 3B, and 4B conduction systems. The highest chlorophyll *a* content in the 2B system was obtained with an M of 27.10, and the lowest value was reported with a CM of 21.37. The other sources did not differ from each other in this system. However, with 3B, the highest chlorophyll *a* content was obtained with CM of 24.83, followed by CL and M. The lowest content was obtained with an SM of 21.80. The plants with 4B presented the highest chlorophyll *a* content, with a CM of 24.73 and an SM of 23.83, and the lowest value was recorded when an OC of 21.23 was used. The plants submitted to the espalier system did not differ statistically between the sources (Table 7).

The chlorophyll *b* levels exhibited similar behavior to those of chlorophyll *a* in the fertilizer sources and training systems. In the 2B and espalier systems, the highest chlorophyll *b* levels were obtained with M of 12.60 and 11.27, respectively. The lowest chlorophyll *b* values were obtained with a CM of 7.53 in the 2B system, and with espalier at a CM of

9.06 and CL at 9.53, which were statistically equal. However, when there was an increase in the number of branches using the 3B and 4B systems, the highest chlorophyll b levels were obtained using CM in the order of 11.63 and 10.57, respectively. The lowest chlorophyll b levels for the 3P and 4P systems were verified using OC with 8.35 and SM with 8.90, respectively (Table 7).

Table 7. Chlorophyll content index—chlorophyll *a*, chlorophyll *b*, and chlorophyll total in fig trees subjected to different mineral and organic fertilization and training system sources.

FA	2 Branches	3 Branches	4 Branches	Espalier
		Chlorophyll *a*		
M	27.10 ± 0.44 aA	24.63 ± 0.13 abB	23.40 ± 0.15 abB	24.63 ± 0.64 aB
CM	21.37 ± 0.23 cB	24.83 ± 0.39 aA	24.73 ± 2.37 aA	22.83 ± 0.43 aB
OC	23.57 ± 0.19 bA	22.83 ± 0.43 bcAB	21.23 ± 1.69 cB	24.00 ± 0.47 aA
CL	23.77 ± 0.48 bA	24.23 ± 0.52 abA	21.73 ± 3.23 bcB	23.27 ± 0.13 aAB
SM	24.03 ± 0.37 bA	21.80 ± 0.44 cB	23.83 ± 0.24 aA	23.93 ± 0.35 aA
CV (%)		3.34		
		Chlorophyll *b*		
M	12.60 ± 0.21 aA	11.10 ± 0.41 aB	9.63 ± 0.44 abC	11.27 ± 0.38 aAB
CM	7.53 ± 0.18 cC	11.63 ± 0.32 aA	10.57 ± 0.24 aA	9.06 ± 0.56 bB
OC	8.87 ± 0.45 bcAB	8.35 ± 0.38 cB	10.13 ± 0.41 abA	10.03 ± 0.45 abA
CL	9.50 ± 0.55 bA	10.33 ± 0.52 abA	10.15 ± 0.52 abA	9.53 ± 0.067 bA
SM	9.25 ± 0.28 bA	9.20 ± 0.31 bcA	8.90 ± 0.17 bA	9.93 ± 0.42 abA
CV (%)		6.53		
		Total chlorophyll		
M	40.36 ± 0.34 aA	34.96 ± 0.47 abBC	32.43 ± 0.12 bcC	36.88 ± 0.27 aB
CM	28.80 ± 0.23 dB	37.13 ± 0.83 aA	35.67 ± 0.62 aA	30.70 ± 0.77 cB
OC	32.50 ± 0.35 cAB	31.07 ± 0.89 cB	34.03 ± 1.09 abA	34.77 ± 0.70 abA
CL	32.63 ± 0.41 cA	34.37 ± 0.87 abA	29.53 ± 1.77 cB	32.47 ± 0.29 bcA
SM	36.53 ± 1.03 bA	32.60 ± 0.63 bcB	33.40 ± 0.49 abB	34.77 ± 0.37 abAB
CV (%)		3.74		

Means followed by the same lowercase letter in the column and uppercase letter in the row do not differ by Tukey's test ($p > 0.05$). Mean ± standard deviation. M: Mineral; CM: cattle manure; OC: organic compost; CL: chicken litter; SM: sheep manure.

For total chlorophyll values, the 2B and espalier systems presented the highest levels with mineral fertilization of 40.36 and 36.88, respectively. The lowest levels were verified using CMs of 28.80 and 30.70, representing reductions of 28.64% and 16.76%, respectively (Table 7).

However, when the 3B and 4B systems were used, the plants fertilized with CM presented the highest total chlorophyll values, 37.13 and 35.67, respectively, equivalent to increases in total chlorophyll content of 16.32% and 17.21%, respectively, compared to the lowest levels reported by OC of 31.07 and CL of 29.53, respectively (Table 7).

3.3. Nutritional Content

The two factors studied also affected the nutritional content of the leaves. There was a significant interaction for the macronutrients N, P, K, Ca, and Mg ($p < 0.01$), as well as for the micronutrients Fe, Zn, and Mn ($p < 0.01$). The conduction systems significantly influenced Cu in isolation ($p < 0.01$). Due to these significant interaction effects, the combined means were analyzed using the Tukey test ($p < 0.05$) to compare treatment combinations and interpret the results.

The 2B system presented the highest nitrogen content using M (50.84 g kg^{-1}) and CL (49.91 g kg^{-1}), which were statistically equal (Table 8). In the espalier system, the highest content was obtained with mineral fertilization (56.90 g kg^{-1}), followed by the second-highest value (CL; 49.90 g kg^{-1}). The lowest N contents were obtained using SM (36.08 g kg^{-1}) and OC (43.01 g kg^{-1}) in the 2B and espalier systems, respectively. In the 3B system, the highest contents were obtained with M (49.70 g kg^{-1}) and OC (47.90 g kg^{-1}), while the lowest was obtained with SM (32.43 g kg^{-1}). In the 4B system, the highest content was obtained using OC (49.71 g kg^{-1}), which did not differ statistically from CL (48.30 g kg^{-1}) (Table 8).

Table 8. Nitrogen (N), phosphorus (P), potassium (K), calcium (Ca), and magnesium (Mg) in fig trees subjected to different mineral and organic fertilization and training system sources.

FA	2 Branches	3 Branches	4 Branches	Espalier
		Nitrogen (N) (g kg^{-1})		
M	50.84 ± 1.11 aB	49.70 ± 1.16 aB	43.27 ± 0.23 bC	56.90 ± 2.17 aA
CM	48.04 ± 1.31 abA	42.00 ± 1.29 bcB	47.91 ± 1.69 bA	48.25 ± 1.04 bcA
OC	43.58 ± 0.009 bB	47.90 ± 0.02 abAB	49.71 ± 0.85 aA	43.01 ± 0.27 cB
CL	49.91 ± 0.23 aA	41.76 ± 0.58 cB	48.30 ± 3.23 abA	49.90 ± 0.15 bA
SM	36.08 ± 1.78 cB	32.43 ± 3.59 dB	31.97 ± 0.01 bB	45.24 ± 0.45 bcA
CV (%)		5.62		
		Phosphorus (P) (g kg^{-1})		
M	0.79 ± 0.01 cAB	0.65 ± 0.03 bB	0.61 ± 0.05 aB	0.94 ± 0.03 aA
CM	1.11 ± 0.06 bA	0.56 ± 0.01 bB	0.69 ± 0.01 aB	0.67 ± 0.004 abB
OC	0.60 ± 0.05 cA	0.81 ± 0.006 abA	0.54 ± 0.001 aA	0.61 ± 0.02 bA
CL	1.15 ± 0.09 bA	0.77 ± 0.01 abB	0.69 ± 0.01 aB	0.82 ± 0.18 abB
SM	1.55 ± 0.06 aA	1.04 ± 0.18 aB	0.78 ± 0.008 aB	0.79 ± 0.13 abB
CV (%)		15.72		
		Potassium (K) (g kg^{-1})		
M	22.50 ± 0.23 aAB	20.22 ± 0.62 bcB	21.42 ± 1.16 aB	21.47 ± 0.73 abA
CM	21.02 ± 0.46 abA	21.73 ± 0.54 abB	21.29 ± 0.46 aB	22.33 ± 0.62 aB
OC	20.87 ± 0.01 abA	23.18 ± 0.01 Aa	20.50 ± 0.008 aA	19.25 ± 0.58 bcA
CL	19.29 ± 0.08 bA	18.34 ± 0.15 Cb	19.63 ± 0.06 abB	18.00 ± 0.85 cB
SM	20.85 ± 0.99 abA	20.04 ± 1.25 bcB	17.90 ± 0.04 bB	20.09 ± 0.39 abcB
CV (%)		5.13		
		Calcium (Ca) (g kg^{-1})		
M	9.72 ± 0.06 aA	8.76 ± 0.49 abA	8.56 ± 0.24 bA	8.53 ± 0.20 cA
CM	9.15 ± 0.06 abA	10.51 ± 0.59 aA	9.35 ± 0.04 bA	10.18 ± 0.52 bcA
OC	7.86 ± 0.01 bC	7.93 ± 0.03 cB	10.25 ± 0.20 abB	12.09 ± 0.93 aA
CL	10.23 ± 0.17 aB	10.17 ± 0.08 bA	11.94 ± 0.78 aA	11.65 ± 0.43 abAB
SM	10.54 ± 0.85 aA	10.18 ± 0.69 aAB	8.86 ± 0.09 bB	11.14 ± 0.06 abA
CV (%)		7.64		
		Magnesium (Mg) (g kg^{-1})		
M	3.13 ± 0.16 aBC	3.64 ± 0.08 abA	2.96 ± 0.02 abC	3.67 ± 0.05 aA
CM	2.93 ± 0.10 aA	2.84 ± 0.20 aB	2.69 ± 0.03 bA	3.10 ± 0.17 bA
OC	3.21 ± 0.24 aA	2.97 ± 0.01 aB	3.49 ± 0.01 aA	3.38 ± 0.03 abA
CL	3.41 ± 0.13 aA	3.29 ± 0.21 abA	3.00 ± 0.03 abA	3.32 ± 0.01 abA
SM	3.24 ± 0.21 aAB	2.89 ± 0.10 bB	1.89 ± 0.01 cC	3.74 ± 0.18 aA
CV (%)		7.60		

Means followed by the same lowercase letter in the column and uppercase letter in the row do not differ by Tukey's test ($p > 0.05$). Mean ± standard deviation. M: Mineral; CM: cattle manure; OC: organic compost; CL: chicken litter; SM: sheep manure.

The espalier system's highest phosphorus (P) content was obtained using an M fertilizer of 0.94 g kg^{-1}, which did not differ statistically from the CL, SM, and CM sources. The lowest value was obtained with an OC of 0.61 g kg^{-1}. The highest content in the 2B system was obtained with an SM of 1.55 g kg^{-1}, while the lowest values were obtained with M

(0.79 g kg^{-1}) and OC (0.60 g kg^{-1}). The highest content in the 3B system was obtained with SM, with 1.04 g kg^{-1}, not differing statistically from CL and OC. The lowest contents were verified with M with 0.60 g kg^{-1} and CM with 0.56 g kg^{-1}. The plants with 4B did not differ statistically among the fertilizer sources (Table 8).

The highest potassium (K) content was verified using M fertilization at 22.50 g kg^{-1} in the 2B system, which did not differ statistically from the CM, OC, and SM sources. In the 3B system, the highest K content was obtained with OC, which was 23.18 g kg^{-1}. In the 4B system, the highest values were obtained using the M (21.42g kg^{-1}), CM (21.29 g kg^{-1}), and OC (20.50 g kg^{-1}) sources, which were statistically equal. The highest K content was obtained with CM of 22.33 g kg^{-1} in the espalier system, not differing statistically from the M fertilization of 21.47 g kg^{-1}. Fertilization with CL corresponded to the lowest K values (19.29, 18.34, and 18.00 g kg^{-1}) in the 2B, 3B, and espalier systems. In the 4B system, the weakest content was obtained from the SM of 17.90 g kg^{-1} (Table 8).

The highest calcium (Ca) contents in plants subjected to the 2B system were obtained using SM with 10.54 g kg^{-1}, CL of 10.23 g kg^{-1}, and M with 9.72 g kg^{-1}. In the 3B system, the highest contents were obtained with CM (10.51 g kg^{-1}), SM (10.18 g kg^{-1}), and CL (10.17 g kg^{-1}), statistically equal. In the 2B and 3B systems, the lowest Ca contents were verified with OC, corresponding to 7.86 and 7.93 g kg^{-1}, respectively. However, in the 4B system, fertilization with CL provided the highest value (11.94 g kg^{-1}). The lowest contents were obtained using CM, SM, and M, statistically equal. In the espalier system, the use of OC provided the highest value of 12.09 g kg^{-1}, followed by CL and SM, and the lowest Ca content was obtained with M fertilization of 8.53 g kg^{-1} (Table 8).

The plants submitted to the 2B system did not exhibit significant differences in magnesium (Mg) content among the different fertilizer sources used. In the 3B system, the highest magnesium content was obtained with M (3.64 g kg^{-1}), which was statistically equal to that of CL (3.64 g kg^{-1}); the other sources did not differ significantly. The plants with 4B presented the highest Mg contents using OC (3.49 g kg^{-1}), followed by CL (3.00 g kg^{-1}) and M (2.96 g kg^{-1}), which were statistically equal, and the lowest content was verified with SM (1.89 g kg^{-1}). In the espalier system, the highest Mg contents were obtained with SM (3.74 g kg^{-1}) and M (3.67 g kg^{-1}), and the lowest Mg content was obtained with the use of CM (3.10 g kg^{-1}) (Table 8).

The highest iron (Fe) content in plants subjected to the 2B system was obtained using an OC equivalent to 388.10 mg kg^{-1}, and the other sources did not differ significantly from this value. In the 3B system, the highest Fe content was verified in fig trees fertilized with OC of 437.28 mg kg$^{-1,}$ and the lowest content was verified in plants fertilized with CM of 207.64 mg kg^{-1}. In the 4B system, the highest contents corresponded to CL and OC fertilizations of 346.26 and 305.72 mg kg^{-1}, respectively, and the lowest value was obtained with M fertilization with 117.07 mg kg^{-1}. In the espalier system, the highest Fe content was obtained with CL with 308.21 mg kg^{-1}, and the lowest Fe contents were obtained using the OC and SM sources (Table 9).

Plants submitted to the 2B system did not differ significantly in zinc (Zn) content for the sources used. Plants with 3B presented higher Zn contents, as measured by SM, CM, and OC, which were statistically equivalent. The lowest contents were obtained with CL and M, which did not differ from each other. In the 4B and espalier systems, the highest Zn contents were also verified using SM of 47.76 and 43.61 mg kg^{-1}, respectively. The lowest values were observed in plants fertilized with a CL of 29.84 mg kg ^{1} and a CM of 36.34 mg kg ^{1} (Table 9).

The highest manganese (Mn) content in the 2B system was obtained using CL equivalent to 41.03 mg kg^{-1}, which did not differ from the other organic sources. The lowest Mn content was obtained with M fertilization (29.39 mg kg^{-1}). The plants with 3B presented the highest content of CM (46.79 mg kg^{-1}), not differing statistically from SM (38.52 mg kg^{-1}), and the lowest contents were observed in M (20.95 mg kg^{-1}) and OC (16.98 mg kg^{-1}). The plants with 4B had the highest Mn content, using CL of 41.05 mg kg^{-1}, and the lowest was obtained with SM of 16.58 mg kg^{-1}. In the espalier system, the highest contents were obtained with M fertilization (38.94 mg kg^{-1}), followed by SM (35.56 mg kg^{-1}), which were statistically equal, and the lowest contents were obtained using CM (22.24 mg kg^{-1}) and CL (2.98 mg kg^{-1}) (Table 9).

Table 9. Iron (Fe) (mg kg^{-1}), zinc (Zn) (mg kg^{-1}), and manganese (Mn) (mg kg^{-1}) in fig trees subjected to different mineral and organic fertilization and training system sources.

FA	2 Branches	3 Branches	4 Branches	Espalier
		Iron (Fe) (mg kg^{-1})		
M	180.27 ± 27.66 bBC	314.28 ± 48.38 bA	117.07 ± 0.62 cC	229.81 ± 74.36 abAB
CM	171.24 ± 7.21 bA	207.64 ± 12.58 cA	248.91 ± 4.42 abA	244.47 ± 5.55 abA
OC	388.10 ± 0.62 aAB	437.28 ± 0.73 aA	305.72 ± 1.27 aB	184.70 ± 11.59 bC
CL	275.65 ± 25.17 bA	259.97 ± 1.25 bcA	346.26 ± 7.77 aA	308.21 ± 15.71 aA
SM	233.06 ± 59.71 bA	230.10 ± 23.64 bcA	176.71 ± 0.62 bcA	176.40 ± 19.98 bA
CV (%)		3.34		
		Zinc (Zn) (mg kg^{-1})		
M	40.10 ± 3.43 aA	29.04 ± 1.46 bB	36.61 ± 1.89 bcA	38.48 ± 0.55 abA
CM	39.43 ± 0.96 aAB	45.43 ± 3.33 aA	40.14 ± 0.85 bAB	36.34 ± 1.47 bB
OC	43.64 ± 0.34 aA	42.86 ± 0.03 aA	39.32 ± 0.14 bA	41.61 ± 4.70 abA
CL	39.62 ± 0.12 aA	34.49 ± 0.55 bAB	29.84 ± 1.16 cB	38.41 ± 1.26 abA
SM	42.65 ± 0.76 aA	45.93 ± 2.39 aA	47.76 ± 0.16 aA	43.61 ± 0.31 aA
CV (%)		6.53		
		Manganese (Mn) (mg kg^{-1})		
M	29.39 ± 2.84 bAB	20.95 ± 1.60 cB	37.92 ± 6.21 abA	38.94 ± 4.94 aA
CM	36.56 ± 0.97 abB	46.79 ± 3.78 aA	29.26 ± 2.66 bBC	22.24 ± 0.61 bC
OC	37.12 ± 0.01 abA	16.98 ± 0.04 cB	39.92 ± 2.60 abA	30.46 ± 2.21 abA
CL	41.03 ± 0.09 aA	33.34 ± 0.07 bA	41.05 ± 1.66 aA	20.98 ± 0.67 bB
SM	33.40 ± 5.18 abA	38.52 ± 3.32 abA	16.58 ± 0.17 cB	35.56 ± 0.28 aA
CV (%)		3.74		

Means followed by the same lowercase letter in the column and uppercase letter in the row do not differ by Tukey's test ($p > 0.05$). Mean ± standard deviation. M: Mineral; CM: cattle manure; OC: organic compost; CL: chicken litter; SM: sheep manure.

The highest Cu contents were verified in the espalier system, equivalent to 1.39 mg kg^{-1}, which did not differ statistically from the 2B (1.15 mg kg^{-1}) and 3B (1.17 mg kg^{-1}) systems. The 4B system resulted in a lower content of 1.03 mg kg^{-1} (Table 10).

Table 10. Copper (Cu) (mg kg^{-1}) in fig trees subjected to different conduction systems.

Conduction Systems	Copper (Cu) (mg kg^{-1})
2 Branches	1.15 ± 0.03 ab
3 Branches	1.17 ± 0.04 ab
4 Branches	1.03 ± 0.06 b
Espalier	1.39 ± 0.14 a
CV (%)	21.94

Means followed by the same lowercase letter in the column do not differ by Tukey's test ($p > 0.05$). Mean ± standard deviation.

3.4. Productivity

The studied factors, fertilization sources (mineral and organic), and conduction systems interact significantly and influenced the productivity at the level of ($p < 0.01$). Due to these significant interaction effects, the combined means were analyzed using the Tukey test ($p < 0.05$) to compare treatment combinations and interpret the results (Figure 3).

Figure 3. Training systems used in fig trees: two branches (**A**,**E**), three branches (**B**,**F**), four branches (**C**,**G**), and espalier (**D**,**H**).

Although the plants that received organic sources exhibited net photosynthesis similar to those that received mineral sources, this was not the case for the productivity of fig trees. Plants that received mineral fertilization presented higher productivity than the organic sources used, emphasizing the 2B, 3B, and espalier systems (12.79, 21.28, and 7.36 t ha^{-1}, respectively). Further, 3B was statistically superior to the other systems used with mineral fertilization with 21.28 t ha^{-1}. The other organic sources used in these systems did not differ significantly from one another (Table 11).

Table 11. Productivity (t ha^{-1}) in fig trees subjected to different mineral and organic fertilization and training system sources.

FA	2 Branches	3 Branches	4 Branches	Espalier
	Productivity (t ha^{-1})			
M	12.79 ± 2.45 aB	21.28 ± 1.26 aA	7.82 ± 0.84 aC	7.36 ± 1.29 aC
CM	2.19 ± 0.07 bA	3.28 ± 0.23 bA	4.10 ± 0.14 bcA	2.66 ± 0.19 bA
OC	4.41 ± 0.24 bA	2.72 ± 0.36 bA	3.01 ± 0.31 cA	2.52 ± 0.23 bA
CL	4.46 ± 0.77 bAB	5.33 ± 1.01 bAB	6.51 ± 0.34 abA	3.22 ± 0.12 bB
SM	1.41 ± 0.02 bA	2.77 ± 0.11 bA	1.37 ± 0.09 cA	2.69 ± 0.24 bA
CV (%)	26.91			

Means followed by the same lowercase letter in the column and uppercase letter in the row do not differ by Tukey's test ($p > 0.05$). Mean ± standard deviation. M: Mineral; CM: cattle manure; OC: organic compost; CL: chicken litter; SM: sheep manure.

It is noteworthy that, in the 4P system, there was equivalence regarding the organic and mineral sources in the productivity of fig trees, with an emphasis on the CL source, which, in addition to having presented considerable net photosynthesis, provided a productivity of 6.51 t ha^{-1}, similar to the mineral 7.82 t ha^{-1} (Table 11).

4. Discussion

Successful fig tree management involves several practices that promote plant development and increased fruit production. Crop management practices, such as fertilization and pruning, can significantly affect the crop's phenological patterns to maximize output. In this study, we observed that mineral fertilization yielded higher productivity compared to other organic sources. The average fig productivity observed in our study in 2022 was 18.22 t ha^{-1} [34]. It is emphasized that only combining mineral fertilization with the 3B system had productivity above the national average for fig trees (21.28 t ha^{-1}). The other combinations studied were below the national productivity. Moura et al. [10] reported productivity values of 10.18 t ha^{-1} for figs under conventional and organic cultivation and training systems in semi-arid conditions.

The total amount of light received during a crop's growth phases is related to its yield and is influenced by pruning, which consequently increases light input to the plant. Light is assimilated during photosynthesis. In this process, plants convert light energy into chemical energy, playing a central role in plant metabolism [35]. Therefore, the combination of mineral fertilization and the 3B system may have provided better architecture and improved light absorption by plants, resulting in greater assimilation and productivity (Figure 3).

Furthermore, this result is primarily due to the ability of mineral fertilization to provide an immediate supply of essential nutrients (N and P), efficiently synchronizing with the peak nutritional demand of the fig tree after pruning. In contrast, organic sources (CM, OC, and SM), originating from different animals, exhibit slow release that, in a single cycle, usually fails to meet the instantaneous nutritional needs of high-yielding fruit crops. Furthermore, due to the pandemic, the fig trees were left unattended and unfertilized for a year before the commencement of this study. Therefore, it is believed that the fig trees under organic management have not recovered their full production potential. Since organic fertilizers release nutrients slowly, they often fail to meet the crop's critical nutritional needs [36,37].

Due to their high nutrient content, chicken litter and cattle manure are widely used as organic fertilizers in agriculture. This benefits the physical, chemical, and biological properties of the soil by releasing nutrients to plants, making them easily accessible to producers. Chicken litter is considered to be richer in nutrients, such as nitrogen, calcium, and phosphorus, than other organic sources, including cattle manure and organic compost [10,38].

The CL (chicken litter) source achieved similar productivity to M in system 4B due to its composition and higher mineralization rate (Figure 3). As reported by Moura et al. [10] and corroborated by our composition data (Table 1), CL has exceptionally high levels of P and N, resulting in a mineralization rate three times higher than that of other manures [39], which enables it to approach the efficiency of mineral fertilizers. However, as can be seen in Table 3, phosphorus apparently became concentrated in the soil since it is a nutrient that is not very mobile, indicating low absorption by fig trees, as evidenced by the low levels in the leaves (Table 8). This result confirms that the quality and decomposition rate of organic material are more critical factors than simply adding biomass.

Leaf analysis revealed that treatments with chicken manure litter (CL) resulted in high nitrogen (N) levels, a crucial component of the RuBisCO enzyme and chlorophyll (Tables 7 and 8, respectively). This increased N input directly correlated with higher carboxylation efficiency (Table 5), indicating that the primary mechanism behind the increased net photosynthesis was the optimization of the biochemical capacity of the leaf mesophyll rather than increased CO_2 availability through stomata.

The contrast between the poultry manure and sheep manure treatments in the two-branch system highlights this mechanism since, although both presented high stomatal conductance (gs), the low N content in the SM treatment resulted in severe biochemical limitation (A_N/Ci of 0.07, Table 5), resulting in a low photosynthetic rate (A_N). Additionally, the maintenance of a high quantum efficiency of photosystem II (Fv/Fm) in the organic treatments (Table 6) indicates that the photosynthetic apparatus was more resilient to stress [40].

Variations in final productivity are intrinsically linked to adjustments in photosynthetic machinery and water use. To assess physiological integrity, the maximum quantum efficiency of PSII (Fv/Fm) was evaluated, which is an essential photochemical quenching parameter that determines the maximum quantum yield of PSII. It is also the most frequently used parameter in chlorophyll fluorescence analysis. Under normal conditions, Fv/Fm shows minimal variation and is not influenced by species or various conditions [41]. The mean values of (Fv/Fm) remained high (close to 0.80—Table 6), indicating that the plants were healthy, and the chlorophyll *a/b* ratio was stable at 3:1 (Table 7). These results demonstrate that, under all the management systems (mineral and organic), the fig trees did not suffer significant photooxidative stress, maintaining the integrity of their photosynthetic apparatus. The absence of photoinhibition suggests that the observed limitations in productivity do not result from structural damage but rather from functional and regulatory constraints [42].

In this context, we observed that the mineral treatment achieved better water use efficiency (WUE) and lower leaf temperatures, especially in systems 2B and 3B (Tables 4 and 5, and Figure 3). This behavior reflects more efficient stomatal regulation, with lower transpiration (E) and stomatal conductance (gs) rates. This optimization in WUE is agronomically advantageous as it allows the plant to maximize carbon fixation while minimizing water loss under semi-arid conditions.

In contrast to the efficiency of mineral fertilization, fig trees fertilized with organic sources (CM, OC, and CL) tended to have more significant stomatal aperture (gs) and greater transpiration (E). Although a high gs suggests a greater potential for gas exchange and CO_2 uptake [43], the water use efficiency (WUE) obtained was only intermediate (Table 4). A similar A_N/Ci to that of mineral fertilization suggests that the plant was attempting to maximize carbon fixation. Still, this high gs did not translate into a higher A_N, resulting in greater water loss per unit of CO_2 fixed. This reinforces the hypothesis that nutritional limitation prevented the full realization of photosynthetic potential [44].

The sheep manure (SM) treatment, however, presented an even more severe restriction, being a failure point in the experiment. SM not only resulted in low WUE but also led to significantly lower carboxylation efficiency (A_N/Ci) in all the systems (Table 5), as evidenced by the high Ci levels. This accumulation of CO_2 in the mesophyll, without proper utilization, indicates a biochemical limitation in photosynthesis. The substrate was available but not used due to the inefficiency of the Rubisco enzyme or the low regeneration of cofactors (NADPH and ATP) [40,44–46].

Chatzistathis et al. [22] reported high Na levels in the sheep manure used in a study with different organic sources, emphasizing that one of the adverse effects of applying animal manure may be salt accumulation. This risk may be increased by the continuous application of animal manure, which depends on both its quality (resulting from its chemical composition due to its animal origin) and quantity. Therefore, the constant application of SM could be hazardous and lead to soil salinization.

The mineralization rate of sheep manure is generally lower than that of other frequently used organic sources, which is related to the characteristics of the diet of the animals responsible for producing this compound when sheep are fed forage. There may be an accumulation of recalcitrant components, such as lignin and tannins, which directly affect the mineralization rate of the manure and, consequently, the supply of nutrients [20].

Organic sources provide a slow and gradual release of nutrients throughout the production cycle, matching the net photosynthesis of fig trees that received mineral fertilization.

However, the explanations of how organic fertilization benefits the photosynthetic process remain uncertain. It is believed that the beneficial effect of comprehensive nutrition on the physical and chemical properties of soil may lead to enhanced nutrition for plants, increasing the profile of secondary metabolites in plant tissues and benefiting the physiological processes of plants [47,48].

Some authors only report the effects of organic fertilization on photosynthetic variables, whether positive, such as Efthimiadou et al. [48] in sweet corn and Ye et al. [49] in Jujube Pear, which reported higher rates of A_N with organic fertilizer, or harmful, such as Bilalis et al. [50] in tobacco, or no effect in kiwi [49].

Furthermore, such measurements are directly or indirectly related to all the stages of light-dependent photosynthetic reactions, including electron transport, water splitting, establishment of the pH gradient across the thylakoid membrane, and ATP synthesis, which is essential for assessing the fitness and integrity of the internal photosynthesis apparatus [51].

Regarding chlorophyll content, it was observed that the source with CM provided higher levels of chlorophyll *a*, *b*, and total under 3B and 4B conditions. Cattle manure is commonly used in crops because it has a low carbon–nitrogen (C: N) ratio, low lignin content, and high levels of cellulose, contributing to its rapid mineralization, in addition to having considerable amounts of nutrients, especially nitrogen, which is essential for the chlorophyll molecule and chloroplast activity [52,53].

While our results clearly demonstrate the impact of treatments on physiological indicators and the first harvest, it is essential to note that this study spanned 9 months; for perennial crops such as fig trees, the long-term effects of fertilizer sources on soil fertility and cumulative productivity may differ significantly. Therefore, we suggest that future studies be conducted across multiple production cycles to validate whether the observed trends persist, especially under organic management, which exhibits cumulative effects on soil.

The results of this study clearly demonstrate that mineral fertilization, especially when combined with the 3B training system, is the most efficient strategy for enhancing first-harvest productivity due to its ability to optimize physiological processes rapidly.

5. Conclusions

Fig trees fertilized with minerals and subjected to the 3B system proved to be the most efficient strategy as it maximized productivity and photosynthesis in the first cycle due to the immediate availability of nutrients.

Fig trees fertilized with chicken manure and subjected to the 4B system stood out as a more viable alternative, presenting photosynthetic and productive performance similar to that of the mineral treatment, which is due to their more favorable mineralization rate.

The use of sheep manure proved to be unviable, resulting in lower photosynthetic rates and productivity, leading to a series of biochemical limitations and the risk of using low-quality organic sources in the semi-arid environment.

Based on this evidence, the essential future direction is to conduct multi-cycle studies to quantify the cumulative effects of organic sources on improving soil properties and confirm the long-term economic and productive sustainability of the fig tree.

Author Contributions: Conceptualization: A.M.F.d.O., V.M. and F.V.d.S.S.; methodology: A.M.F.d.O., V.M., P.E.C.A., F.V.d.S.S. and A.S.d.M.; formal analysis and investigation: A.M.F.d.O., V.M., P.E.C.A., R.I.d.S.F., L.R.B.d.S., D.E.G.d.S., F.N.F., S.I.B., L.d.A.S., P.D.F. and F.V.d.S.S.; writing—original draft preparation: A.M.F.d.O., V.M., P.E.C.A., R.I.d.S.F., F.N.F. and F.V.d.S.S.; writing—review and editing: S.I.B., P.D.F., A.S.d.M. and F.V.d.S.S.; funding acquisition: V.M. and F.V.d.S.S.; resources: V.M., L.d.A.S., P.D.F.,A.S.d.M. and F.V.d.S.S.; supervision: V.M., P.D.F., and F.V.d.S.S. All authors have read and agreed to the published version of the manuscript.

Funding: This research received no external funding.

Institutional Review Board Statement: Not applicable.

Data Availability Statement: The original contributions presented in this study are included in the article. Further inquiries can be directed to the corresponding author.

Acknowledgments: The authors would like to extend their sincere appreciation to the Coordenação de Aperfeiçoamento de Pessoal de Nível Superior (CAPES, grant number 001) and the Conselho Nacional de Desenvolvimento Científico e Tecnológico (CNPq) for awarding grants to the researchers.

Conflicts of Interest: The authors declare that they have no conflicts of interest.

References

1. Elish, S.E.A.A.; Temraz, A.; Hassan Baky, M. Phytochemical diversity of genus *Ficus*: A mini review. *ERU Res. J.* **2023**, *2*, 502–524. [CrossRef]
2. Sawadogo, Y.; Belem, M.; Sabo, P.; Kabré, B.; Tiétiambou, F.R.S.; Ouédraogo, A. A critical review on the *Ficus* genus in Africa: Current knowledge and perspectives for its sustainable management. *Biodivers. Conserv.* **2024**, *33*, 2739–2756. [CrossRef]
3. Machado, A.F.P.; Rønsted, N.; Bruun-Lund, S.; Pereira, R.A.S.; de Queiroz, L.P. Atlantic forests to the all Americas: Biogeographical history and divergence times of Neotropical *Ficus* (Moraceae). *Mol. Phylogenet. Evol.* **2018**, *122*, 46–58. [CrossRef]
4. Gusella, G.; Gugliuzzo, A.; Guarnaccia, V.; Martino, I.; Aiello, D.; Costanzo, M.B.; Russo, A.; Groenewald, J.Z.; Crous, P.W.; Polizzi, G. Fungal species causing canker and wilt of *Ficus carica* and evidence of their association by bark beetles in Italy. *Plant Dis.* **2024**, *108*, 2136–2147. [CrossRef] [PubMed]
5. Isa, M.M.; Jaafar, M.N.; Kasim, K.F.; Mutalib, M.F.A. Cultivation of fig (*Ficus carica* L.) as an alternative high value crop in Malaysia: A brief review. In *IOP Conference Series: Materials Science and Engineering*; IOP Publishing: Bristol, UK, 2020; Volume 864, p. 012134. [CrossRef]
6. FAO. Food and Agriculture Organization of the United Nations. Faostat: Figures 2022. Available online: http://faostat.fao.org (accessed on 29 February 2024).
7. Rasool, I.F.U.; Aziz, A.; Khalid, W.; Koraqi, H.; Siddiqui, S.A.; Al-Farga, A.; Wing-Fu, L.; Ali, A. Industrial application and health prospective of fig (*Ficus carica*) by-products. *Molecules* **2023**, *28*, 960. [CrossRef] [PubMed]
8. Ecker, S.L.; Lugaresi, A.; Girardi, G.C.; Louis, B.; Galon, L.; Giacobbo, C.L. Conduction systems in fig tree with two cultivars, evaluating fruits quality and yield. *Rev. Sci. Agrár. Parana.* **2020**, *19*, 77–82. [CrossRef]
9. Ammar, A.; Aissa, I.B.; Mars, M.; Gouiaa, M. Comparative physiological behavior of fig (*Ficus carica* L.) cultivars in response to water stress and recovery. *Sci. Hortic.* **2020**, *260*, 108881. [CrossRef]
10. Moura, E.A.; Mendonça, V.; Souza Ferreira, E.; Oliveira, L.M.; Melo, B.E.F. Management systems on production and physicochemical traits of fruits of conventionally and organically grown fig. *Rev. Caatinga* **2020**, *34*, 867–878. [CrossRef]
11. Pramanick, B.; Mahapatra, B.S.; Datta, D.; Dey, P.; Singh, S.P.; Kumar, A.; Paramanike, B.; Awasthi, N. An innovative approach to improve oil production and quality of mustard (*Brassica juncea* L.) with multi-nutrient-rich polyhalite. *Heliyon* **2023**, *9*, e13997. [CrossRef]
12. Mendoza-Castillo, V.M.; Pineda-Pineda, J.; Vargas-Canales, J.M.; Hernández-Arguello, E. Nutrition of fig (*Ficus carica* L.) under hydroponics and greenhouse conditions. *J. Plant. Nutr.* **2019**, *42*, 1350–1365. [CrossRef]
13. Pereira, C.; Martín, A.; López-Corrales, M.; Córdoba, M.D.G.; Galván, A.I.; Serradilla, M.J. Evaluation of the physicochemical and sensory characteristics of different fig cultivars for the fresh fruit market. *Foods* **2020**, *9*, 619. [CrossRef] [PubMed]
14. Ferraz, R.A.; Leonel, S.; Souza, J.M.A.; Ferreira, R.B.; Modesto, J.H.; Arruda, L.L. Phenology, vegetative growth, and yield performance of fig in Southeastern Brazil. *Pesqui. Agropecu. Bras.* **2020**, *55*, e01192. [CrossRef]
15. Almeida, P.D.; Pavan, B.E.; Rodrigues, M.G.F.; Gaspar, G.D.F.; Giro, L.D.P.; Silva, R.F.D. Genetic parameters and gains with the selection of fig tree genotypes. *Acta Sci. Agron.* **2022**, *44*, e55796. [CrossRef]
16. Özkul, M.; Nikpeyma, Y.; Özen, M. Effects of different pruning intensities on growth and fruit quality in Sarılop fig varieties. *Ege. Üniv. Ziraat. Fak. Derg.* **2022**, *59*, 249–263. [CrossRef]

17. Souza, J.M.A.; Silva, M.D.S.; Ferraz, R.A.; Modesto, J.H.; Ferreira, R.B.; Bolfarini, A.C.B.; Tecchio, M.A.; Leonel, S. The use of hydrogen cyanamide or nitrogen fertilizer increases vegetative and productive performance of fig cv. Roxo de Valinhos. *Acta Sci. Agron.* **2021**, *43*, e50519. [CrossRef]
18. Pio, R.; Sanchez, O.M.P.; Ribeiro, C.H.M.; Fazenda, L.H.V.; Silva, A.D.D.; Peche, P.M. Decreased spacing with different numbers of branches to enhance green fig production. *Pesqui. Agropecu. Bras.* **2023**, *58*, e03135. [CrossRef]
19. Soliman, S.S.; Alebidi, A.I.; Al-Obeed, R.S.; Al-Saif, A.M. Effect of potassium fertilizer on fruit quality and mineral composition of fig (*Ficus carica* L. cv. Brown Turky). *Pak. J. Bot.* **2018**, *50*, 1753–1758.
20. Iqbal, A.; He, L.; Khan, A.; Wei, S.; Akhtar, K.; Ali, I.; Ullah, S.; Munsif, F.; Zhao, Q.; Jiang, L. Organic manure coupled with inorganic fertilizer: An approach for the sustainable production of rice by improving soil properties and nitrogen use efficiency. *Agronomy* **2019**, *9*, 651. [CrossRef]
21. Średnicka-Tober, D.; Barański, M.; Kazimierczak, R.; Ponder, A.; Kopczyńska, K.; Hallmann, E. Selected antioxidants in organic vs. conventionally grown apple fruits. *Appl. Sci.* **2020**, *10*, 2997. [CrossRef]
22. Chatzistathis, T.; Papadakis, I.E.; Papaioannou, A.; Chatzissavvidis, C.; Giannakoula, A. Comparative study effects between manure application and a controlled-release fertilizer on the growth, nutrient uptake, photosystem II activity and photosynthetic rate of *Olea europaea* L.(cv.'Koroneiki'). *Sci. Hortic.* **2020**, *264*, 109176. [CrossRef]
23. Bitew, Y.; Alemayehu, M. Impact of crop production inputs on soil health: A review. *Asian J. Plant. Sci.* **2017**, *16*, 109–131. [CrossRef]
24. Elouattassi, Y.; Ferioun, M.; El Ghachtouli, N.; Derraz, K.; Rachidi, F. Enhancing onion growth and yield through agroecological practices: Organic fertilization and intercropping. *Ecol. Front.* **2024**, *44*, 547–557. [CrossRef]
25. de Mesquita Alves, J.; de Lima, A.S.; de Mesquita, E.F.; Junior, S.D.O.M.; de Sousa Ferreira, R.; da Silva, F.L.; da Mota Santos, J. Gas exchange and chlorophyll content in tomato grown under different organic fertilizers and biofertilizer doses. *Afr. J. Agric. Res.* **2018**, *13*, 2256–2262. [CrossRef]
26. Chatzistathis, T.; Chatzissavvidis, C.; Papaioannou, A.; Papadakis, I.E. Independent or combinational application of sheep manure and litter from indigenous field vegetation of *Quercus* sp. influences nutrient uptake, photosynthesis, intrinsic water use efficiency, and foliar sugar concentrations in olive plants (*Olea europaea* L., cv."Koroneiki"). *Appl. Sci.* **2023**, *13*, 1127. [CrossRef]
27. Alvares, C.A.; Stape, J.L.; Sentelhas, P.C.; Gonçalves, J.D.M.; Sparovek, G. Köppen's climate classification map for Brazil. *Meteorol. Z.* **2013**, *22*, 711–728. [CrossRef]
28. Rêgo, L.G.D.S.; Martins, C.M.; Silva, E.F.D.; Silva, J.J.A.D.; Lima, R.N.D.S. Pedogenesis and soil classification of an experimental farm in Mossoró, state of Rio Grande do Norte, Brazil. *Rev. Caatinga* **2016**, *29*, 1036–1042. [CrossRef]
29. Embrapa–Empresa Brasileira de Pesquisa Agropecuária. *Manual de Análises Químicas de Solos, Plantas e Fertilizantes*; Embrapa Solos: Rio de Janeiro, Brazil, 2009.
30. Caetano, L.C.S.; Guarçoni, M.A.; de Melo Lima, I.; Ventura, J.A. *Recomendações Técnicas Para a Cultura da Figueira*; Incaper: Vitória, Brazil, 2013.
31. Sá, F.V.S.; Gheyi, H.R.; de Lima, G.S.; Pinheiro, F.W.A.; de Paiva, E.P.; Moreira, R.C.L.; Silva, L.A.; Fernandes, P.D. The right combination of N-P-K fertilization may mitigate salt stress in custard apple (*Annona squamosa* L.). *Acta Physiol. Plant.* **2021**, *43*, 59. [CrossRef]
32. Quaggio, J.Á.; Raij, B.; Piza, J.R.C.T. Frutíferas. In *Recomendações de Adubação e Calagem Para o Estado de São Paulo*, 2nd ed.; Raij, B., Ed.; Fundação Instituto Agronômico de Campinas: Campinas, Brazil, 1996; pp. 121–153.
33. Ferreira, D.F. SISVAR: A computer analysis system to fixed effects split plot type designs. *Braz. J. Biom.* **2019**, *37*, 529–535. [CrossRef]
34. FAO–Food and Agriculture Organization of the United Nations. Faostat: Figures 2023. Available online: https://www.fao.org/faostat/en/#data/QCL/visualize (accessed on 27 February 2024).
35. Santos, A.S.D.; Pereira, G.A.; Fonseca, W.L.; Zuffo, A.M.; da Cunha, J.G.; Soares, N.P.; Morais, E.M.; Nascimento, A.A.S.; Santos, D.P.; Almeida, M.S.; et al. Photosynthetic assimilation of the guava (*Psidium guajava*) cv. Paluma under different pruning and fruit thinning intensities. *Agronomy* **2023**, *13*, 1610. [CrossRef]
36. Chatzistathis, T.; Kavvadias, V.; Sotiropoulos, T.; Papadakis, I.E. Organic fertilization and tree orchards. *Agriculture* **2021**, *11*, 692. [CrossRef]
37. Leal, Y.H.; Dias, T.J.; Souza, A.D.G.; Bezerra, A.C.; Rodrigues, L.S.; Albuquerque, M.D.; Leal, M.P.S.; Silva, A.J.; Lucena, M.F. Gas exchanges and chlorophyll content in green pepper plants under bio-fertilization and times of application. *Biosc. J.* **2021**, *37*, 1981–3163. [CrossRef]
38. Silva, V.B.D.; Silva, A.P.D.; Dias, B.D.O.; Araujo, J.L.; Santos, D.; Franco, R.P. Decomposição e liberação de N, P e K de esterco bovino e de cama de frango isolados ou misturados. *Rev. Bras. Ciênc. Solo* **2014**, *38*, 1537–1546. [CrossRef]
39. Pitta, C.S.R.; Adami, P.F.; Pelissari, A.; Assmann, T.S.; Franchin, M.F.; Cassol, L.C.; Sartor, L.R. Year-round poultry litter decomposition and N, P, K and Ca release. *Rev. Bras. Ciênc. Solo* **2012**, *36*, 1043–1053. [CrossRef]
40. Taiz, L.; Zeiger, E.; Moller, I.A.; Murphy, A. *Fisiologia e Desenvolvimento Vegetal*, 6th ed.; Artmed: Porto Alegre, Brazil, 2017; p. 858.

41. Peng, Y.; Tong, H.; Yin, W.; Yuan, Y.; Yuan, Z. Effects of summer pruning on the growth and photosynthetic characteristics of pepper (*Capsicum annuum* L.). *Acta Sci. Pol. Hortorum Cultus* **2024**, *23*, 41–50. [CrossRef]
42. Hossain, A.B.M.S.; Salleh, A.; Mekhled, M.A.; Al-Saif, A.M. Chlorophyll fluorescence intensity, photosynthetic yield and flowering in fig fruit trees as affected by phloem stress. *Bulg. J. Agric. Sci.* **2010**, *16*, 547–552.
43. Rodríguez, N.C.; Melgarejo, L.M.; Blair, M.W. Purple passion fruit, *Passiflora edulis* Sims f. edulis, variability for photosynthetic and physiological adaptation in contrasting environments. *Agronomy* **2019**, *9*, 231. [CrossRef]
44. Bloukounon-Goubalan, A.Y.; Saïdou, A.; Obognon, N.; Amadji, G.L.; Igué, A.M.; Clottey, V.A.; Kenis, M. Decomposition and nutrient release pattern of animal manures biodegraded by fly larvae in Acrisols. *Can. J. Soil Sci.* **2018**, *99*, 60–69. [CrossRef]
45. Silva, J.R.; Souza, E.; Leite, M.L.D.M.; Barros, G.; Sales, A.T.; Antonino, A.C. Gas exchange and millet phytomass under organic fertilization and graywater irrigation. *Rev. Bras. Eng. Agric. Ambient.* **2022**, *26*, 111–118. [CrossRef]
46. Comas, L.H.; Trout, T.J.; DeJonge, K.C.; Zhang, H.; Gleason, S.M. Water productivity under strategic growth stage-based deficit irrigation in maize. *Agric. Water Manag.* **2019**, *212*, 433–440. [CrossRef]
47. Zhang, M.; Sun, D.; Niu, Z.; Yan, J.; Zhou, X.; Kang, X. Effects of combined organic/inorganic fertilizer application on growth, photosynthetic characteristics, yield and fruit quality of *Actinidia chinesis* cv 'Hongyang'. *Glob. Ecol. Conserv.* **2020**, *22*, e00997. [CrossRef]
48. Efthimiadou, A.; Bilalis, D.; Karkanis, A.; Froud-Williams, B. Combined organic/inorganic fertilization enhance soil quality and increased yield, photosynthesis and sustainability of sweet maize crop. *Aust. J. Crop Sci.* **2010**, *4*, 722–729. Available online: https://search.informit.org/doi/10.3316/informit.859484046375351 (accessed on 27 February 2024).
49. Ye, S.; Liu, T.; Niu, Y. Effects of organic fertilizer on water use, photosynthetic characteristics, and fruit quality of pear jujube in northern Shaanxi. *Open Chem.* **2020**, *18*, 537–545. [CrossRef]
50. Bilalis, D.; Karkanis, A.; Triantafyllidis, V.; Ladavos, A.; Bizos, D.; Patsiali, S.; Efthimiadou, A.; Papatheohari, Y. Effects of organic and inorganic fertilization on growth, yield and nicotine content of flue-cured and oriental tobacco (*Nicotiana tabacum* L.) seedlings grown in organic and conventional float system. *J. Food Agric. Environ.* **2010**, *8*, 585–589.
51. Verma, K.K.; Liu, X.H.; Wu, K.C.; Singh, R.K.; Song, Q.Q.; Malviya, M.K.; Xiu-Peng, X.; Singh, P.; Verma, C.L.; Li, Y.R. The impact of silicon on photosynthetic and biochemical responses of sugarcane under different soil moisture levels. *Silicon* **2020**, *12*, 1355–1367. [CrossRef]
52. Marchezan, C.; Ferreira, P.A.; Silva, L.S.; Bacca, A.; Krug, A.V.; Nicoloso, F.T.; Tarouco, C.P.; Tiecher, T.L.; Brunetto, G.; Ceretta, C.A. Nitrogen availability and physiological response of corn after 12 years with organic and mineral fertilization. *J. Soil Sci. Plant Nutr.* **2020**, *20*, 979–989. [CrossRef]
53. Oliveira, Z.V.S.R.; Mesquita, A.C.; Simões, W.L.; Salviano, A.M.; da Silva, J.S.; Felix, A.T.R. Trocas gasosas e análises bioquímicas em variedades de melancia sob cultivo orgânico no semiárido nordestino. *Sci. Plena.* **2022**, *18*, 1–14. [CrossRef]

agriculture

MDPI

Article

Effects of Mechanical Pruning on Tree Growth, Yield, and Fruit Quality of 'Arisoo' Apple Trees

Nay Myo Win [1,†], Juhyeon Park [1,†], Seonae Kim [1], Youngsuk Lee [1], Van Giap Do [1], Jung-Geun Kwon [1], Soon-Il Kwon [1], Jingi Yoo [2], In-Kyu Kang [3] and Hun-Joong Kweon [1,*]

1 Apple Research Center, National Institute of Horticultural and Herbal Science, RDA, Daegu 43100, Republic of Korea; naymyowin@korea.kr (N.M.W.); wngus1113@korea.kr (J.P.); seonaekim@korea.kr (S.K.); kongfo@korea.kr (Y.L.); giapbio@korea.kr (V.G.D.); dsms0@korea.kr (J.-G.K.); topapple@korea.kr (S.-I.K.)

2 Department of Horticulture & Life Science, Yeungnam University, Gyeongsan 38541, Republic of Korea; jingiyoo@yu.ac.kr

3 Department of Horticultural Science, Kyungpook National University, Daegu 41566, Republic of Korea; kangik@knu.ac.kr

* Correspondence: kweonhj@korea.kr

† These authors contributed equally to this work.

Abstract: Pruning is labor-intensive and increases production costs, while mechanical pruning offers a promising alternative. However, research on its effectiveness remains limited. To address this gap, we evaluated the effects of mechanical pruning over two consecutive years (2023 and 2024) on tree growth, yield, and fruit quality of 'Arisoo' apple trees. The treatment included hand (manual) pruning (HP), mechanical pruning (MP), and combined mechanical and hand pruning (MP + HP) applied during winter pruning in a super-spindle-slender-shaped apple orchard. MP significantly reduced pruning time; however, the amount of plant biomass removed was lower in the MP treatment than in the HP and MP + HP treatments. Canopy volume was higher in the HP treatment than in MP and MP + HP treatments; however, the pruning treatments did not affect trunk cross-sectional area or tree yield. Leaf chlorophyll and nitrogen contents were slightly lower in the MP treatment than in the HP treatment in 2023 but were not affected in 2024. The MP treatment also noticeably reduced light penetration within the canopy and produced smaller fruits with lower soluble solids content and poorer coloration at harvest compared to the HP and MP + HP treatments. In contrast, the HP and MP + HP treatments showed similar effects on light penetration, yield, fruit size, and fruit quality; however, the MP + HP treatment significantly reduced the pruning time compared with the HP treatment. Overall, this study found that MP reduced light penetration and produced smaller and poorly colored fruits, whereas a follow-up combination of HP after MP improved pruning efficiency, light penetration, fruit size, and fruit quality.

Keywords: pruning; tree canopy; light penetration; tree growth; yield; fruit quality

1. Introduction

Apples are a major crop worldwide, ranking third in global fruit production [1]. In the Republic of Korea, apples are the second most widely cultivated fruit crop. In 2022, the total cultivation area of apple crops was approximately 34,000 ha, and the annual average production was 566,000 metric tons. Domestic consumption of apples was recorded at approximately 10 kg per capita per year [2,3]. According to Korean statistics, the cultivation area of apple crops decreased by approximately 5% in 2023 due to a shortage of skilled workers, labor shortages, and higher labor costs [3]. This labor shortage and lack of skilled laborers could become a major challenge for apple production in the future, potentially leading to reduced apple cultivation areas and decreased fruit yield and economic value. These are pressing issues that require urgent attention.

In apple orchards, pruning is one of the most intensive tasks, requiring a large number of laborers and significant time, accounting for over 22% of total fieldwork in apple production [4]. Pruning is a fundamental and essential practice for maximizing fruit production [5]. In recent decades, traditional hand (manual) pruning (HP) has become the most widely used method in apple orchards. Mechanical pruning (MP) has been developed as an alternative to HP [6,7]. MP, also referred to as mechanical hedging, is a non-selective pruning method in which a cutting tool is attached to a tractor to cut branches at a predetermined distance along the sides of the trees [8]. This pruning method has shown positive effects on fruit yield and quality, resulting in lower pruning costs for citrus and olive trees [9,10]. However, it has been shown to limit shoot growth, yield, and quality in peach and apple trees [11]. Therefore, the efficiency of MP can vary depending on the training system, canopy structure, and crop [12,13]. In contrast, the MP method can be beneficial when apple trees are trained into tall or spindle leaders with narrow, vertical canopy structures in high-density apple orchards [14,15]. However, some studies have reported that HP is required as a follow-up practice after the use of MP [4]. Particularly, there is limited information about the combined effects of MP and HP on pruning efficiency and physiological changes in the apple trees.

In addition, pruning often alters the structure, canopy, and light distribution of trees [16]. Consequently, pruning can modify the microclimate around trees, directly influencing productivity and fruit quality [17]. For example, inadequate pruning can lead to unbalanced tree growth and reduced light penetration, whereas excessive pruning can lower fruit yield and productivity [8]. In apple production, sufficient light intensity is required not only for maintaining tree health, photosynthesis, and vegetative and reproductive growth but also for ensuring the desired shape and size, and reducing susceptibility to pathogens [18–20]. Additionally, the production of high-quality fruits, considering external (size, color, and visual appearance) and internal (texture, sugars, and acidity) qualities, is considerably influenced by pruning intensity and light availability [21,22]. Despite the positive effects of MP in reducing labor costs and fieldwork, its impact on the physiological responses, growth, yield, and fruit quality has been rarely reported. Moreover, HP is widely used in many apple orchards, possibly due to limited availability or information on MP techniques.

To address this gap in the literature, this study aimed to evaluate the effects of MP on vegetative growth, yield, and fruit quality of apple trees. A combined MP + HP treatment was also used to investigate its effects in comparison to MP or HP alone.

2. Materials and Methods

2.1. Experimental Design and Treatments

This study was conducted at an experimental station at the Apple Research Center, Gunwi-gun, Daegu, Republic of Korea (36.28 N, 128.47 E), over two consecutive years: 2023 and 2024. Apple trees planted under the same soil conditions (clay loam soil and pH 6.8) and an experimental block were selected. The detailed nutrient conditions of the soil are presented in Supplementary Table S1. Additionally, the average annual air temperatures and precipitation were 13.6 °C and 14.3 °C, and 1359 mm and 1265 mm for 2023 and 2024, respectively. The daily temperature and precipitation of the experimental orchard are presented in Supplementary Figure S1.

'Arisoo' (*Malus domestica* Borkh.), a mid-season apple, is a hybrid population of a cross between 'Yoko' and 'Senshu' [23]. 'Arisoo' is a low-vigor tree type with adequate numbers of tree branches, and its bearing type is on spurs and long shoots. 'Arisoo' is a relatively high-premium cultivar in the Republic of Korea due to its high-quality attributes (shape, taste, and color). Therefore, this apple cultivar was chosen in this study. The selected apple trees were grafted onto M.9 rootstocks and trained in a super-slender-spindle shape. All selected apple trees were four years old and planted at a spacing of 3.0 m × 1.0 m. The selected trees were uniform in canopy structure and height. The orchards were managed using an integrated pest management system and irrigated using a drip irrigation system.

Three pruning treatments were performed in this study: HP, MP, and a combination of mechanical and hand pruning (MP + HP). For the MP + HP treatment, MP was performed first, followed by HP. A completely randomized block design was used. Each block consisted of 20 apple trees per treatment, with three replicates (a total of 180 apple trees across all treatments). For HP, pruning shears (Hwashin Metal Co., Ltd., Daegu, Republic of Korea) were used following the orchard pruning techniques [24]. These techniques involved removing both internal and external parts of the trees, including dead and damaged branches, suckers, downward growing shoots, and water sprouts to open the canopy structure and light penetration of the trees by professional, skilled laborers with more than 10 years of experience in apple tree pruning. MP was performed using a CRF-340 mechanical pruning device (Rinieri, Forlí, Italy). This device consists of a trimmer-type fixed blade attached to a tractor that cuts branches vertically. The device and its characteristics are shown in Supplementary Figure S2. The branches were cut at a predetermined distance of approximately 40 cm from the trunk of the tree (Figure 1).

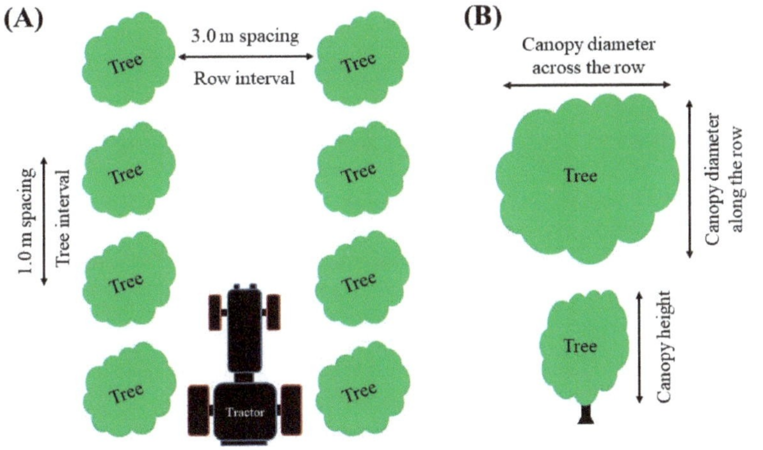

Figure 1. Tree canopy and spacing along (**A**) tree and row intervals, and (**B**) canopy height and canopy diameter across and along the row of apple trees.

During MP, the tractor was operated at 1 km/h by a skilled driver. In the combined pruning treatment, HP was performed in the internal parts of apple trees following the orchard pruning techniques after MP. All pruning treatments were performed during late winter, in the first week of March in both years (Figures 1 and 2). The flowers fully bloomed in the experimental orchard on 22–24 April 2023 and 2024, respectively.

Figure 2. Mechanical pruning of apple trees using (**A**,**B**) a CRF-340 mechanical pruning device during winter pruning, (**C**) apple trees after mechanical pruning in March, and (**D**,**E**) in May of 2023.

2.2. Measurement of Pruning Capacity and Plant Biomass Removal

The pruning capacity of HP was calculated based on the time required to prune an entire tree (minutes of pruning per tree per laborer) using both a handsaw and manual pruning shears, and was calculated as hours per hectare (h/ha) [13]. The pruning time for MP was calculated in a similar manner, determined by the tractor's forward speed and the time needed to prune both sides of a single tree, and expressed as h/ha. For the determination of plant biomass removal, the cut branches of each apple tree by each pruning method were separately collected, and the collected branches were measured as the fresh plant biomass removal by pruning treatments, and the results are expressed as kg per tree (kg/tree). For the combined MP + HP treatment, the cut branches after MP and HP were combined and weighed.

2.3. Measurement of Flower Set and Tree Growth Characteristics

The number of flower sets was recorded for each apple tree during the full bloom stage in 2024 and 2025 to analyze the effects of pruning treatments on flower density. Shoot length and canopy width (canopy diameter measured along and across rows) were measured using measuring tape (Digikey Co., Ltd., Suwon, Republic of Korea), and canopy height was measured using a tower ruler (Levenhuk Co., Ltd., Prague, Czech Republic) (Figure 2). Canopy volume was determined using the equation by Lodolini et al. [25]: canopy volume = canopy height × canopy width along the row × canopy width across the row. The trunk cross-sectional area (TCSA) was measured approximately 10 cm above the graft using digital calipers (CD-15APX, Kawasaki, Tokyo, Japan).

2.4. Measurement of Sunlight Penetration, Leaf Chlorophyll and Nitrogen Contents, and Photosynthetic Parameters

Sunlight penetration inside the tree canopy was assessed as photosynthetically active radiation (PAR) using a light sensor meter (3415FX; Spectrum Technologies Inc., Aurora, IL, USA). PAR was measured approximately 1 m above the ground on both sides of each apple tree on a sunny day in July, following the method described by Win et al. [26]. Leaf chlorophyll content was measured on 10 randomly selected mature apple leaves per tree, located in different positions within the middle canopy, using a chlorophyll meter (SPAD-502 Plus, Konica Minolta, Tokyo, Japan). Leaf nitrogen content was measured after SPAD assessment using an elemental analyzer (Kjeltec 8400, Fisher Scientific, Hoganas, Sweden), following the method outlined by Lee et al. [27].

Photosynthetic parameters, including assimilation rate (P_n) and stomatal conductance (G_s), were determined on fully expanded and mature leaves located on the 6th or 7th leaf from the tip of a growing shoot in the outer canopy [28]. Measurements were conducted using an open gas exchange system (LI-COR 6400, Li-COR Inc., Lincoln, NE, USA) equipped with a leaf chamber and a red–blue LED light source, set to a light intensity of 1000 μmol m^{-2} s^{-1} and a CO_2 concentration of 400 μmol m^{-2} s^{-1} during the measurement. Analyses were conducted from approximately 8:00 to 9:00 a.m. on a clear day in August 2024. Ten leaves (five per side of the canopy) per tree were used to measure photosynthetic parameters. For all measurements of chlorophyll, nitrogen content, and photosynthetic parameters, leaves located 100 cm above the ground were selected, following the method outlined by Bhusal et al. [28].

2.5. Measurement of Tree Production, Yield, and Fruit Quality Characteristics

All fruits were harvested on 4–5 September 2023 and 2024, respectively. Fruits were harvested by estimating the starch pattern index (SPI) of apples between 7.5 and 8.0, following Cornell's 8-point SPI scale method [29]. At harvest, the tree production (fruits/tree) and the tree yield were recorded for each apple tree. The individual weight of all harvested fruits was also measured. Fruit size (length and diameter) was assessed using a digital caliper, and fruit shape was calculated as the length-to-diameter ratio. Fruit color was measured using a chroma meter (CR-400; Konica Minolta, Tokyo, Japan). The fruit color index was calculated using the following equation [30]: fruit color index = $1000 \times a^*/(L^* \times b^*)$. Flesh firmness was measured using a firmness tester (TR-327; Forlí, Italy). Soluble solid content was measured using a refractometer (PR-201, Atago, Tokyo, Japan). Titratable acidity was measured using the malic acid reduction method described by Win et al. [26]. For all pruning treatments, five individual fruits per apple tree (300 fruits/treatment) were randomly taken and used for determination of fruit color, firmness, SSC, and TA values.

2.6. Statistical Analysis

The data were subjected to analysis of variance (one-way ANOVA), and the means of the pruning treatments for each experimental year were compared using Tukey's HSD test at $p < 0.05$. Data are presented as mean ± standard error. All analyses were performed using SPSS software (Version 26, IBM Corp. Inc., Armonk, NY, USA).

3. Results

3.1. Pruning Time and Plant Biomass Removal

In 2023, the pruning time for HP and MP + HP was approximately 88.9 h and 42.9 h, respectively, whereas MP required only 5.9 h per ha, reducing pruning time by 82.9 h (14.8 times) and 36.9 h (7.2 times) compared to HP and MP + HP, respectively (Table 1). The MP + HP method saved 46.0 h of pruning time compared with HP. The non-selective MP method removed approximately 1.5 kg/tree of plant biomass, whereas the HP method removed 2.8 kg/tree, and the MP + HP method removed 2.5 kg/tree.

Table 1. Pruning time and plant biomass removal of 'Arisoo' apple trees subjected to hand (HP), mechanical (MP), and mechanical + hand (MP + HP) pruning treatments during winter pruning.

Pruning Treatments	Experiment Year	Pruning Time (h/ha)	Plant Biomass (kg/tree)
HP	2023	88.9 ± 3.1 [z] a [y]	2.8 ± 0.6 a
MP		5.9 ± 0.1 c	1.5 ± 0.4 b
MP + HP		42.9 ± 1.6 b	2.5 ± 0.3 a
HP	2024	92.6 ± 4.9 a	3.0 ± 0.4 a
MP		6.0 ± 0.1 c	1.7 ± 0.3 b
MP + HP		46.8 ± 6.9 b	3.2 ± 0.5 a

[z] Data represent mean ± standard error. [y] Different letters among treatments for each experimental year indicate significant differences according to Tukey's test at $p < 0.05$.

In 2024, the pruning time for HP and MP + HP was approximately 92.6 h and 46.8 h per ha, whereas MP required only 6.0 h per ha, saving 86.6 h (15.4 times) and 40.8 h (7.8 times) compared to HP and MP + HP, respectively. The MP + HP method saved 45.8 h of pruning time compared with HP. However, the MP method removed approximately 1.7 kg/tree of plant biomass, whereas the HP method removed 3.0 kg/tree, and the MP + HP method removed 3.2 kg/tree of plant biomass (Table 1).

3.2. Tree Growth Characteristics

In both 2023 and 2024, TCSA was not statistically affected by the pruning treatments (Table 2). In 2023, shoot length was longer in the MP treatment than in the HP treatment; however, it was not statistically different from that in the MP + HP treatment. In 2024, shoot length was longer in the MP treatment than in the MP + HP treatment; however, it was not statistically different from that in the HP treatment. In both years, shoot lengths did not statistically differ between the HP and MP + HP treatments (Table 2).

Table 2. Tree growth characteristics, including trunk cross-sectional area (TCSA), shoot length, canopy height, canopy diameter, and canopy volume of 'Arisoo' apple trees subjected to hand (HP), mechanical (MP), and mechanical + hand (MP + HP) pruning treatments.

Pruning Treatments	Experiment Year	TCSA (cm^2)	Shoot Length (cm)	Canopy Height (m)	Canopy Diameter (m)		Canopy Volume (m^3)
					Along the Row	Across the Row	
HP	2023	3.9 ± 0.2 [z] a [y]	19.1 ± 0.6 b	3.5 ± 0.3 a	1.3 ± 0.1 b	1.5 ± 0.1 a	9.2 ± 1.1 a
MP	2023	3.7 ± 0.1 a	22.8 ± 2.0 a	3.3 ± 0.1 a	1.6 ± 0.0 a	1.2 ± 0.0 b	5.1 ± 0.1 b
MP + HP	2023	3.8 ± 0.1 a	20.6 ± 0.7 ab	3.6 ± 0.1 a	1.5 ± 0.1 ab	1.2 ± 0.1 b	6.5 ± 1.0 b
HP	2024	4.4 ± 0.2 a	18.8 ± 3.0 ab	3.7 ± 0.2 a	1.5 ± 0.2 a	1.7 ± 0.1 a	10.3 ± 1.4 a
MP	2024	4.2 ± 0.1 a	22.5 ± 1.8 a	3.4 ± 0.2 a	1.8 ± 0.2 a	1.3 ± 0.1 b	7.2 ± 1.1 b
MP + HP	2024	4.2 ± 0.1 a	17.9 ± 1.3 b	3.7 ± 0.2 a	1.6 ± 0.2 a	1.3 ± 0.1 b	7.3 ± 1.2 b

[z] Data represent mean ± standard error. [y] Different letters among treatments for each experimental year indicate significant differences according to Tukey's test at $p < 0.05$.

Canopy height was slightly shorter in the MP treatment in both years; however, the results were not statistically different from those of the other treatments (Table 2). The canopy diameter (along the row) was longer in MP than in HP in 2023; however, no differences were observed among treatments in 2024. However, the canopy diameter (across the rows) was greater in the HP treatment than in the MP and MP + HP treatments in both years. Additionally, the canopy volume was larger in the HP treatment than in the MP and MP + HP treatments in both years (Table 2).

3.3. Light Penetration, Leaf Chlorophyll and Nitrogen Content, and Photosynthetic Parameters

PAR was used as an indicator of sunlight penetration inside the tree canopies (Table 3). PAR was mainly affected by pruning treatment and was significantly lower in MP than in HP and MP + HP in both years. Leaf chlorophyll content (measured by SPAD) and leaf nitrogen content were lower in the MP treatment than in the HP treatment in 2023; however, these results were not statistically different among pruning treatments in 2024. No significant differences in SPAD or leaf chlorophyll content were observed between the MP + HP and MP or HP treatments in either year (Table 3).

Table 3. Photosynthetically active radiation (PAR) inside tree canopies, SPAD values, and leaf nitrogen content in mature leaves of 'Arisoo' apple trees subjected to hand (HP), mechanical (MP), and mechanical + hand (MP + HP) pruning treatments.

Pruning Treatments	Experiment Year	PAR (μmol m^{-2} s^{-1})	SPAD (μmol m^{-2})	Leaf Nitrogen (%)
HP		475.0 ± 37.7 [z] a [y]	53.1 ± 0.6 a	2.2 ± 0.1 a
MP	2023	227.7 ± 13.5 b	50.9 ± 0.7 b	2.0 ± 0.0 b
MP + HP		483.3 ± 23.7 a	51.7 ± 1.0 ab	2.0 ± 0.1 ab
HP		417. 8 ± 29.3 a	51.2 ± 1.5 a	2.0 ± 0.1 a
MP	2024	210. 8 ± 11.2 b	50.6 ± 0.9 a	1.9 ± 0.2 a
MP + HP		409.0 ± 17.2 a	52.5 ± 0.8 a	2.2 ± 0.1 a

[z] Data represent mean ± standard error. [y] Different letters among treatments for each experimental year indicate significant differences according to Tukey's test at $p < 0.05$.

Photosynthetic parameters (P_n and G_s) were measured on mature leaves of apple trees in 2024 (Figure 3). Although both P_n and G_s values were slightly lower in the MP, the results were not statistically different from those of the other pruning treatments (Figure 3).

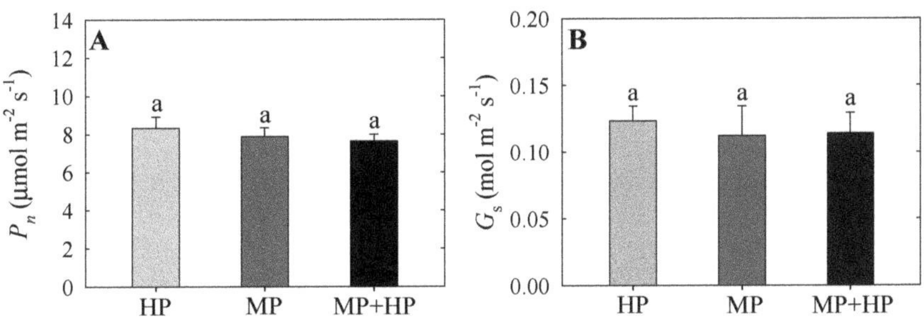

Figure 3. Photosynthesis rate (P_n) (**A**) and stomatal conductance (G_s) (**B**) of mature apple leaves from trees subjected to hand (HP), mechanical (MP), and mechanical + hand (MP + HP) pruning in 2024. Same lowercase letters indicate non-significant differences between treatments according to Tukey's HSD test at $p < 0.05$.

3.4. Flower Set, Tree Production, Yield, Fruit Weight, and Size

The number of flower set per tree were counted during the full bloom stage, but the results were not statistically different among pruning treatments in both years (Table 4). Production (fruit per tree) and yield per tree were assessed at harvest. In 2023, apple trees produced approximately 14.6 kg/tree (71.0 fruits/tree) under HP, 14.3 kg/tree (74.7 fruits/tree) under MP, and 13.7 kg/tree (65.0 fruits/tree) under MP + HP. Except for higher production in MP than in MP + HP treatments, yield was not statistically different among pruning treatments. In 2024, apple trees produced approximately 16.0 kg/tree (68.3 fruits/tree) under HP, 15.4 kg/tree (78.3 fruits/tree) under MP, and 15.7 kg/tree (63.4 fruits/tree) under MP + HP.

In both years, the HP and MP + HP treatments produced heavier fruits than the MP treatment did (Table 4). In 2023, fruit diameter was smaller in the MP treatment than in the MP + HP treatment. However, fruit length and the L/D ratio were not statistically different among treatments. In 2024, both fruit length and diameter were smaller in the MP treatment, particularly compared to the MP + HP treatment. Additionally, fruit length was smaller in the MP treatment than in the HP treatment (Table 4).

Table 4. Flower set per tree, production (fruits/tree), yield, fruit weight, and fruit size of 'Arisoo' apples trees subjected to hand (HP), mechanical (MP), and mechanical + hand (MP + HP) pruning treatments.

Pruning Treatments	Experiment Year	Flower Set/Tree (no.)	Fruits/Tree (no.)	Yield (kg/tree)	Fruit Weight (g)	Fruit Size (mm)		L/D Ratio
						Length (L)	Diameter (D)	
HP	2023	132.5 ± 14.3 [z] a [y]	71.0 ± 5.6 ab	14.6 ± 1.2 a	228.3 ± 3.0 a	73.9 ± 0.9 a	79.4 ± 0.9 ab	0.9 ± 0.0 a
MP		152.4 ± 18.7 a	74.7 ± 3.8 a	14.3 ± 0.7 a	218.4 ± 2.2 b	73.7 ± 1.4 a	77.2 ± 0.3 b	0.9 ± 0.0 a
MP + HP		144.2 ± 12.5 a	65.0 ± 2.1 b	13.7 ± 0.8 a	233.6 ± 8.1 a	72.1 ± 1.3 a	81.8 ± 0.9 a	0.9 ± 0.0 a
HP	2024	147.5 ± 12.5 a	68.5 ± 2.0 b	16.0 ± 0.5 a	225.3 ± 6.5 a	72.5 ± 0.8 a	76.0 ± 1.3 ab	0.9 ± 0.0 a
MP		165.5 ± 25.1 a	78.3 ± 4.1 a	15.4 ± 1.3 a	204.1 ± 5.8 b	70.1 ± 0.8 b	74.8 ± 0.8 b	0.9 ± 0.0 a
MP + HP		134.2 ± 9.6 a	63.4 ± 1.6 b	15.7 ± 0.6 a	235.1 ± 4.2 a	73.5 ± 0.8 a	78.7 ± 0.5 a	0.9 ± 0.0 a

[z] Data represent mean ± standard error. [y] Different letters among treatments for each experimental year indicate significant differences according to Tukey's test at $p < 0.05$.

3.5. Fruit Quality Characteristics

Fruit quality attributes were evaluated at harvest (Table 5). In 2023, SPI scores at harvest ranged from 7.4 to 7.8 scales; however, the results were not statistically different among treatments. L^* was not statistically affected by the pruning treatments; however, lower b^* values were observed in the HP treatment than in the MP and MP + HP treatments. The a^* value and fruit color index were significantly higher in the MP treatment than in the HP and MP + HP treatments.

Table 5. Fruit quality characteristics, including starch pattern index (SPI), flesh firmness, soluble solids content (SSC), titratable acidity (TA), fruit skin color index, and fruit skin color values (L^*, a^*, b^*) of 'Arisoo' apples trees subjected to hand (HP), mechanical (MP), and mechanical + hand (MP + HP) pruning treatments.

Pruning Treatments	Experiment Year	SPI Score (1–8)	Firmness (N)	SSC (°Brix)	TA (%)	Color Index	Fruit Skin Color Value		
							L^*	a^*	b^*
HP	2023	7.7 ± 0.2 [z] a [y]	70.2 ± 2.3 a	13.7 ± 0.2 a	0.5 ± 0.1 a	10.8 ± 0.6 a	59.1 ± 1.0 a	13.2 ± 0.6 a	20.1 ± 0.2 b
MP		7.8 ± 0.2 a	70.4 ± 0.5 a	13.3 ± 0.3 a	0.5 ± 0.1 a	8.7 ± 0.5 b	60.3 ± 0.6 a	11.2 ± 0.2 b	23.0 ± 0.8 a
MP + HP		7.4 ± 0.3 a	69.6 ± 1.6 a	13.8 ± 0.2 a	0.4 ± 0.0 a	10.3 ± 0.4 a	60.2 ± 0.6 a	13.4 ± 0.3 a	22.4 ± 0.4 a
HP	2024	8.0 ± 0.0 a	65.4 ± 0.5 ab	14.6 ± 0.2 a	0.3 ± 0.0 a	22.2 ± 1.2 a	43.6 ± 2.3 a	18.7 ± 0.6 a	15.7 ± 1.0 a
MP		8.0 ± 0.0 a	64.3 ± 1.2 b	13.8 ± 0.1 b	0.3 ± 0.0 a	18.3 ± 0.5 b	45.8 ± 0.8 a	16.8 ± 0.5 b	16.4 ± 0.4 a
MP + HP		7.9 ± 0.1 a	67.3 ± 0.8 a	14.8 ± 0.1 a	0.3 ± 0.0 a	21.5 ± 0.7 a	44.9 ± 1.6 a	18.9 ± 0.6 a	16.0 ± 0.8 a

[z] Data represent mean ± standard error. [y] Different letters among treatments for each experimental year indicate significant differences according to Tukey's test at $p < 0.05$.

In 2024, the SPI scores at harvest were nearly 8.0 across treatments. The SPI and titratable acidity (TA) values were not statistically different among treatments. However, fruit firmness in the MP treatment was lower than in the MP + HP treatment, but not statistically different from that in the HP treatment. Additionally, soluble solids content (SSC) was lower in the MP treatment than in the other treatments. L^* and b^* values were not statistically affected by pruning treatments. The color index and a^* values were significantly lower in the MP group than in the HP and MP + HP groups (Figure S3, Table 5).

4. Discussion

The application of automation in the agricultural sector is necessary not only for reducing fieldwork and timing but also for increasing crop quality and orchard profitability. In this study, MP significantly reduced pruning time, particularly when compared to HP. Additionally, this result was also consistent in MP treatment in both experimental years. A reduction in pruning time and fieldwork can help address labor storage and reduce pruning costs [31]. Lehnert [32] also reported that MP is a time-saving technique that can offer additional benefits in cause-and-effect development. However, the difference in pruning time between MP and HP can vary based on pruning intensity, orchard density, orchard type, tree age, canopy size, and the type of pruning device used. The removal of plant biomass with the MP method was considerably lower than that observed with other pruning treatments. This is likely because the MP device used in this study was non-selective, only trimming at a predetermined distance along the tree branches. Therefore, the extent of plant biomass removal from trees largely depends on the predetermined distance between the trees [8]. Additionally, the removal of plant biomass can also vary depending on the canopy size and the duration of MP application [7,13]. Moreover, the excessive removal of plant biomass by MP can negatively affect the vegetative and reproductive growth of trees, leading to reduced yield and productivity [33]. Therefore, MP should be carefully adjusted according to the canopy structure and training system of the trees, as not all crops are suitable for MP [4]. Additionally, the combined MP + HP treatment significantly reduced pruning time compared to HP, and the plant biomass removed under MP + HP was similar to that of HP. Because MP is not a complete replacement for HP, follow-up HP should be considered after MP to improve pruning efficiency [4].

Shoot length was greater under MP than HP in 2023 and greater under MP than MP + HP in 2024. The increase in shoot length observed under MP may have resulted from the removal of shoots and buds along the sides of the tree canopy, which then stimulated the production of longer shoots from the remaining buds. Similarly, Mika et al. [7] reported that MP trees produced longer shoots than HP trees. Canopy height did not differ among treatments in both years. However, canopy diameter along the row was greater under MP than under HP in 2023. This outcome may reflect the effects of cutting along the sides of trees under MP, which promoted the production of longer canopy diameters along the tree intervals. Conversely, a longer canopy diameter across rows was observed under HP than under MP or MP + HP. This may be attributable to the vertical cutting effect of MP along the sides of trees, resulting in a shorter canopy diameter across row intervals [32].

In fruit trees, TCSA is closely associated with fruit yield and productivity, as a larger TCSA can support a greater fruit load and higher yield [34]. Additionally, TCSA is responsible for transporting water and nutrients from the roots to the canopy [35]. In this study, TCSA was not considerably affected by the pruning treatments. Generally, pruning intensity can negatively affect TCSA, as severe pruning can reduce TCSA, leading to decreased fruit yield [8,22]. Therefore, the non-significant effect of MP on TCSA observed in this study may be linked to the lower pruning intensity and reduced plant biomass removal, resulting in TCSA values similar to those of the other pruning treatments. Moreover, MP treatment had no significant effect on TCSA and canopy height in both years, indicating that the differences in climatic conditions had little impact on vegetative growth under MP. The lower canopy volume observed in the MP and MP + HP treatments in both years may be attributed to the pruning effect of MP on the external parts of the tree, which limited canopy expansion.

PAR within the tree canopy was noticeably reduced under MP, whereas HP and MP + HP maintained higher PAR values across both years. The SPAD value was lower in MP trees than in MP + HP trees in 2023, but no difference was observed in 2024. Additionally, leaf nitrogen content was lower under MP than under HP in 2023. Light penetration and distribution inside the tree canopy are essential for improving photosynthesis [18,36]. Reduced light penetration under MP can limit photosynthesis and air circulation, resulting in reduced plant growth and health, and the production of low-quality fruits [21]. However, P_n and G_s did not differ significantly among treatments, consistent with the absence of significant differences in SPAD and leaf nitrogen content observed in 2024. Moreover, the reduction in PAR in this study may be linked to the solid-hedge structure created by MP. Follow-up pruning with HP after MP helped improve PAR. Thus, MP using a non-selective cutting bar creates a dense outer layer of the tree, which can become a barrier to sunlight penetration [37]. Consistent with our results, previous studies have reported that MP reduces light penetration into the interior parts of trees [7].

The tree production and yield were slightly increased in the MP treatment; however, fruit weight was significantly lower in the MP treatment than the HP and MP + HP treatments. In mandarins, MP produced a yield comparable to that of HP [33]. In olive trees, MP showed a non-significant reduction in yield over eight years, whereas combined MP + HP did not increase yield [38]. However, in citrus, follow-up HP after MP increased yield compared to HP alone [39]. Interestingly, the alternative annual use of MP and HP in mandarins resulted in a higher yield than the consecutive application of MP alone [31]. This indicates that the effect of MP may vary depending on the crop and the timing of application. The increase in fruit size is highly dependent on the relationship between the crop load and TCSA [40]. Unpruned trees typically produce smaller fruits than pruned trees, although fruit size can vary depending on pruning strategies and intensity [12,17]. Krueger et al. [41] reported that MP trees had greater fruit density and yield and produced more fruits than HP trees, but with smaller fruit size. Mika et al. [7] also reported that MP apple trees produce small or industrial fruit, as observed in this study. Additionally, the production of smaller fruits in the MP treatment was consistent across both experimental years, indicating that reduction in fruit weight and size of apples was mainly affected by pruning treatments and was not influenced by climatic conditions. Moreover, the absence of HP in the internal parts of trees in the MP treatment led to higher crop loads and the production of smaller fruits. Therefore, follow-up application of HP may be an effective approach to maintain crop load balance and ensure the production of marketable fruits [4].

In 2024, SSC was lower in the MP treatment than in the HP and HP + MP treatments, whereas the color index and a^* value in fruit skins (2023 and 2024 years) were significantly higher in the HP and MP + HP treatments than in the MP treatment. In this study, the production of poor-colored fruits was consistent across both experimental years, indicating that the improvement in red-blush color in fruit was primarily attributable to pruning treatments. Consequently, the improvement in fruit color in the MP + HP trees in this study may be attributed to the follow-up application of HP after MP treatment. Fruit firmness was higher in the MP + HP treatment than in the MP treatment in 2024, whereas TA and SPI were unaffected by the treatments. Both SSC and red blush color are important indicators of marketability, and lower values of these indicators indicate reduced marketability [42]. Generally, the development of red pigments in fruit skin is closely associated with light availability to the fruits [43,44]. Additionally, natural light and solar radiation promote not only fruit color but also sugar accumulation in fruits [45]. Therefore, the lower light intensities in the interior parts of trees in this study may have contributed to the reduced development of red-colored pigments and sugars in the fruit, resulting in the production of poor-colored and low-quality fruits, similar to MP trees [7]. Madhumala et al. [46] also reported that pruning enhanced fruit color and sugar accumulation by improving light penetration into fruits, similar to that observed in HP trees. Additionally, some studies also reported that the variations in climatic conditions can influence the quality of apples [47]. Therefore, the lower flesh firmness and SSC observed in MP treatment in 2024 may be attributed to difference in average daily temperature and precipitation during the fruit development and maturation stages.

5. Conclusions

In conclusion, MP significantly reduced pruning time, and a combination of MP + HP reduced pruning time compared with the HP treatment in both 2023 and 2024. The plant biomass removal was lower in MP than in HP and MP + HP. MP noticeably reduced the canopy volume of apple trees, although it did not affect TCSA and canopy height compared with HP. Furthermore, MP did not reduce flower set and tree yield, while remarkably reducing light distribution inside the tree canopies. Additionally, MP negatively affected fruit weight and size and resulted in poorly colored (reduced fruit color index and a^* value) and low-quality fruits. However, combining MP with HP improved these effects, resulting in large fruits with high-quality attributes. Moreover, the MP + HP treatment showed similar effects on tree yield and fruit quality. Overall, this study suggests that MP alone reduces light penetration, fruit size, and quality of apples. A combined MP + HP can partially reduce pruning time and field workloads and improve fruit quality. Therefore, MP should be followed by HP to improve pruning efficiency and fruit quality. However, the results obtained in this study were focused on the 'Arisoo' apple, and continuous studies are required on other apple cultivars to observe the potential variability of the results. Moreover, the long-term effects of MP on plant health, hormone levels, and nutrient metabolism should be considered in future studies.

Supplementary Materials: The following supporting information can be downloaded at: https://www.mdpi.com/article/10.3390/agriculture15202118/s1. Table S1: The detailed nutrient conditions of the soil; Figure S1: Daily air temperature and precipitation of the experimental orchard; Figure S2: A Rinieri CRF-340 mechanical pruning device: Front view and lateral view; Figure S3: Fruit samples of 'Arisoo' apples harvested from trees subjected to hand pruning, mechanical pruning, and mechanical + hand pruning in 2024.

Author Contributions: Conceptualization, N.M.W., J.P. and H.-J.K.; methodology, N.M.W., J.P. and H.-J.K.; software, N.M.W.; validation, N.M.W.; formal analysis, N.M.W.; investigation, N.M.W. and J.P.; resources, H.-J.K.; data curation, N.M.W.; writing—original draft preparation, N.M.W.; writing—review and editing, N.M.W., J.P. and H.-J.K.; visualization, S.K., Y.L., V.G.D., J.-G.K., S.-I.K., J.Y. and I.-K.K.; supervision, H.-J.K.; project administration, J.P. and H.-J.K.; funding acquisition, H.-J.K. All authors have read and agreed to the published version of the manuscript.

Funding: This work was funded by the "Cooperative Research Program for Agriculture Science and Technology Development (RS-2021-RD009831)" of the Rural Development Administration of the Republic of Korea. This work was supported by the 2025 RDA Fellowship Program of the National Institute of Horticultural and Herbal Science, Rural Development Administration, Republic of Korea.

Data Availability Statement: Data are available on request from the corresponding author.

Conflicts of Interest: The authors declare no conflicts of interest.

References

1. Statista. Global Fruit Production in 2023, by Selected Variety (in Million Metric Tons). Available online: https://statista.com/statistics/264001/worldwide-production-of-fruit-by-variety/ (accessed on 1 June 2025).
2. USDA. Korea's Apple Update. Available online: https://www.fas.usda.gov/data/south-korea-koreas-apple-update (accessed on 21 December 2023).
3. Korea Statistical Information Service (KOSIS). Available online: https://kosis.kr/eng/ (accessed on 1 June 2025).
4. He, L.; Schupp, J. Sensing and automation in pruning of apple trees: A review. *Agronomy* **2018**, *8*, 211. [CrossRef]
5. Zhang, L.; Koc, A.B.; Wang, X.N.; Jiang, Y.X. A review of pruning fruit trees. *IOP Conf. Ser. Earth. Environ. Sci.* **2018**, *153*, 062029. [CrossRef]
6. Karkee, M.; Adhikari, B.; Amatya, S.; Zhang, Q. Identification of pruning branches in tall spindle apple trees for automated pruning. *Comput. Electron. Agric.* **2014**, *103*, 127–135. [CrossRef]
7. Mika, A.; Buler, Z.; Treder, W. Mechanical pruning of apple trees as an alternative to manual pruning. *Acta Sci. Pol. Hortorum Caltus* **2016**, *15*, 113–121.
8. Kendell, A.; Alexander, T.R.; LaHue, G.T.; Miles, C.A. Summer mechanical hedging to prune eight cider apple cultivars. *HortTechnology* **2022**, *32*, 313–320. [CrossRef]
9. Vivaldi, G.A.; Strippoli, G.; Pascuzzi, S.; Stellacci, A.M.; Camposeo, S. Olive genotypes cultivated in an adult high-density orchard respond differently to canopy restraining by mechanical and manual pruning. *Sci. Hortic.* **2015**, *192*, 391–399. [CrossRef]

10. Albarracín, V.; Hall, A.J.; Searles, P.S.; Rousseaux, M.C. Responses of vegetative growth and fruit yield to winter and summer pruning in olive trees. *Sci. Hortic.* **2017**, *225*, 185–194. [CrossRef]
11. Biddlecombe, C.T.; Dalton, A. To investigate the effect of four timings of mechanical pruning on yield and fruit quality compared to a hand pruned control in an intensive 'Gala' M9 orchard planted as a fruit wall. *Acta Hortic.* **2018**, *1228*, 97–104. [CrossRef]
12. Barcia, F.; Prieto, J.; Trentacoste, E.R. Effects of mechanical box pruning intensity on bud development, vegetative growth, and yield components on cv. Cabernet-Sauvignon in Mendoza, Argentina. *OENO One* **2023**, *57*, 153–163. [CrossRef]
13. Lodolini, E.M.; Polverigiani, S.; Giorgi, V.; Famiani, F.; Neri, D. Time and type of pruning affect tree growth and yield in high-density olive orchards. *Sci. Hortic.* **2023**, *311*, 111831. [CrossRef]
14. Sazo, M.M. New advances to narrower canopy systems: Transitioning from 3-D to 2-D canopies or fruiting walls—Part 3. *Fruit Q.* **2018**, *26*, 31–36.
15. Scalisi, A.; O'Connell, M.G.; Stefanelli, D.; Zhou, S.; Pitt, T.; Graetz, D.; Dodds, K.; Han, L.; De Bei, R.; Stanley, J.; et al. Narrow orchard systems for pome and stone fruit—A review. *Sci. Hortic.* **2024**, *338*, 113815. [CrossRef]
16. Badrulhisham, N.; Othman, N. Knowledge in tree pruning for sustainable practices in urban setting: Improving our quality of life. *Proc. Soc. Behav. Sci.* **2016**, *234*, 210–217. [CrossRef]
17. Morales, P.; Davies, F.S.; Littell, R.C. Pruning and skirting affect canopy microclimate, yields, and fruit quality of 'Orlando' Tangelo. *HortScience* **2000**, *35*, 30–35. [CrossRef]
18. Simkin, A.J.; Kapoor, L.; Hofmann, T.A.; Doss, C.G.P. The role of photosynthesis related pigments in light harvesting, photoprotection and enhancement of photosynthetic yield in planta. *Photosyn. Res.* **2022**, *152*, 23–42. [CrossRef]
19. Fonte, A.; Garcerá, C.; Chueca, P. Influence of mechanical and manual pruning on the incidence of pests in 'Clemenules' mandarins. *Pest Manag. Sci.* **2023**, *79*, 4390–4402. [CrossRef]
20. Wu, W.; Chen, L.; Liang, R.; Huang, S.; Li, X.; Huang, B.; Luo, H.; Zhang, M.; Wang, X.; Zhu, H. The role of light in regulating plant growth, development and sugar metabolism: A review. *Front. Plant Sci.* **2025**, *15*, 1507628. [CrossRef]
21. Lin, L.; Niu, Z.; Jiang, C.; Yu, L.; Wang, H.; Qiao, M. Influences of open-central canopy on photosynthetic parameters and fruit quality of apples (*Malus* × *domestica*) in the Loess Plateau of China. *Hortic. Plant J.* **2022**, *8*, 133–142. [CrossRef]
22. Vosnjak, M.; Mrzlic, D.; Usenik, V. Summer pruning of sweet cherry: A way to control sugar content in different organs. *J. Sci. Food Agric.* **2022**, *102*, 1216–1224. [CrossRef] [PubMed]
23. Kwon, Y.S.; Kwon, S.I.; Kim, J.H.; Park, M.Y.; Park, J.T.; Lee, J. 'Arisoo', a midseason apple. *HortScience* **2021**, *56*, 1139–1141. [CrossRef]
24. Westerfield, B.; Chavez, D.; Smith, E. Home Orchard Pruning Techniques. UGA Cooperative Extension Circular 1087. 2020. Available online: https://fieldreport.caes.uga.edu/publications/C1087/home-fruit-orchard-pruning-techniques/ (accessed on 1 July 2025).
25. Lodolini, E.M.; de Ludicibus, A.; Lucchese, P.G.; Casas, G.L.; Torrisi, B.; Nicolosi, E.; Giuffrida, A.; Ferlito, F. Comparison of canopy architecture of five olive cultivars in a high-density planting system in Sicily. *Agriculture* **2023**, *13*, 1612. [CrossRef]
26. Win, N.M.; Yoo, J.; Do, V.G.; Yang, S.; Kwon, S.I.; Kweon, H.J.; Kim, S.; Lee, Y.; Kang, I.K.; Park, J. Effects of pneumatic defoliation on fruit quality and skin coloration in 'Fuji' apples. *Agriculture* **2024**, *14*, 1582. [CrossRef]
27. Lee, Y.; Kweon, H.J.; Park, M.Y.; Lee, D. Field assessment of macronutrients and nitrogen in apple leaves using a chlorophyll meter. *HortTechnology* **2019**, *29*, 300–307. [CrossRef]
28. Bhusal, N.; Bhusal, S.J.; Yoo, T.M. Comparisons of physiological and anatomical characteristics between two cultivars in bi-leader apple trees (*Malus* × *domestica* Borkh.). *Sci. Hortic.* **2018**, *231*, 73–81. [CrossRef]
29. Blanpied, G.D.; Silsby, K.J. Predicting Harvest Windows for Apples. Cornell Cooperative Extension Publication, Information Bulletin 221. 1992. Available online: https://www.researchgate.net/publication/277061616_Predicting_Harvest_Date_Windows_for_Apples (accessed on 1 July 2025).
30. Xu, Y.; Uoy, C.; Xu, C.; Zhang, C.; Hu, X.; Li, X.; Ma, H.; Gong, J.; Sun, X. Red and blue light promote tomato fruit coloration through modulation of hormone homeostasis and pigment accumulation. *Postharvest Biol. Technol.* **2024**, *207*, 112588. [CrossRef]
31. Fonte, A.; Torregrosa, A.; Garcera, C.; Mateu, G.; Chueca, P. Mechanical pruning of 'Clemenules' mandarins in Spain: Yield effects and economic analysis. *Agronomy* **2022**, *12*, 761. [CrossRef]
32. Lehnert, R. *Italian Researcher Studies Mechanical Pruning That Also Reaches Inside the Canopy*; Good Fruit Grower: Yakima, WA, USA, 2013; pp. 26–27. Available online: https://goodfruit.com/fruiting-walls-with-windows/ (accessed on 1 June 2025).
33. Martin-Gorriz, B.; Porras, I.; Torregrosa, A. Effect of mechanical pruning on the yield and quality of 'Fortune' mandarins. *Span. J. Agric. Res.* **2014**, *12*, 952–959. [CrossRef]
34. Atucha, A.; Merwin, I.A.; Brown, M. Long-term effects of four groundcover management systems in an apple orchard. *HortScience* **2011**, *46*, 1176–1183. [CrossRef]
35. Dogbatse, J.A.; Arthur, A.; Awudzi, G.K.; Quaye, A.K.; Konlan, S.; Amaning, A.A. Effects of organic and inorganic fertilizers on growth and nutrient uptake by young cacao (*Theobroma cacao* L.). *Int. J. Agron.* **2021**, *7*, 5516928. [CrossRef]

36. Laužikė, K.; Sigedaitė-Šėžienė, V.; Uselis, N.; Samuoliene, G. The impacts of stress caused by light penetration and agrotechnological tools on photosynthesis behavior of apple trees. *Sci. Rep.* **2020**, *10*, 9177. [CrossRef]
37. Cherbiy-Hoffmann, S.U.; Searles, P.S.; Hall, A.J.; Rousseaux, M.C. Influence of light environment on yield determinants and components in large olive hedgerows following mechanical pruning in the subtropics of the Southern Hemisphere. *Sci. Hortic.* **2012**, *137*, 36–42. [CrossRef]
38. Dias, A.B.; Peça, J.O.; Pinheiro, A. Long-term evaluation of the influence of mechanical pruning on olive growing. *Agron. J.* **2012**, *104*, 22–25. [CrossRef]
39. Velázques, B.; Fernández, E. The influence of mechanical pruning in cost reduction, production of fruit, and biomass waste in citrus orchards. *Appl. Eng. Agric.* **2010**, *26*, 531–540. [CrossRef]
40. Serra, S.; Leisso, R.; Giordani, L.; Kalcsits, L.; Musacchi, S. Crop load influences fruit quality, nutritional balance, and return bloom in 'Honeycrisp' apple. *HortScience* **2016**, *51*, 236–244. [CrossRef]
41. Krueger, W.H.; Niederholzer, F.; Fichtner, E. Investigation of pruning strategies for dried plums including hand, mechanical and combinations. *Acta Hortic.* **2013**, *985*, 201–207. [CrossRef]
42. Musacchi, S.; Serra, S. Apple fruit quality: Overview on pre-harvest factors. *Sci. Hortic.* **2018**, *234*, 409–430. [CrossRef]
43. Do, V.G.; Lee, Y.; Kim, J.H.; Kwon, Y.S.; Park, J.T.; Yang, S.; Park, J.; Win, N.M.; Kim, S. The synergistic effects of environmental and genetic factors on the regulation of anthocyanin accumulation in plant tissues. *Int. J. Mol. Sci.* **2023**, *24*, 12946. [CrossRef]
44. Peavey, M.; Scalisi, A.; Islam, M.S.; Goodwin, I. Fruit position, light exposure and fruit surface temperature affect color expression in a dark-red apple cultivar. *Horticulturae* **2024**, *10*, 725. [CrossRef]
45. Mei, Z.; Li, Z.; Lu, X.; Zhang, S.; Liu, W.; Zou, Q.; Yu, L.; Fang, H.; Zhang, Z.; Mao, Z.; et al. Supplementation of natural light duration promotes accumulation of sugar and anthocyanins in apple (*Malus domestica* Borkh.) fruit. *Environ. Exp. Bot.* **2023**, *205*, 105133. [CrossRef]
46. Madhumala, K.; Kumar, V.; Kumar, K. Physiology and mechanism of pruning in fruit crops. *Int. J. Adv. Biochem. Res.* **2024**, *8*, 443–445. [CrossRef]
47. Bui, T.A.T.; Stridh, H.; Molin, M. Influence of weather conditions on the quality of 'Ingrid Marie' apples and their susceptibility to grey mould infection. *J. Agric. Food Res.* **2021**, *3*, 100104. [CrossRef]

agriculture

MDPI

Article

Multi-Objective Nitrogen Optimization in Tea Cultivation: A Pathway to Achieve Sustainability in Cash Crop Systems

Jinze Pei [1,2,†], Hongyu Yang [1,†], Menghan Huang [3], Xiaojun Yan [4], Xinran Zeng [5], Lijin Guo [1] and Liangquan Wu [1,*]

[1] International Magnesium Institute, College of Resources and Environment, Fujian Agriculture and Forestry University, Fuzhou 350002, China; peijinze2025@163.com (J.P.); 52408031069@fafu.edu.cn (H.Y.); guolijin2022@fafu.edu.cn (L.G.)
[2] Fujian Yinfeng Poverty Alleviation Service Center, Fuzhou 350100, China
[3] College of Rural Revitalization, Fujian Agriculture and Forestry University, Fuzhou 350002, China; hannahhuang0128@163.com
[4] State Key Laboratory of Nutrient Use and Management, College of Resources and Environmental Sciences, Ministry of Education, China Agricultural University, Beijing 100193, China; 16565710@cau.edu.cn
[5] Forestry College, Fujian Agriculture and Forestry University, Fuzhou 350002, China; 19264975203@163.com
* Correspondence: 000q814056@fafu.edu.cn
† These authors contributed equally to this work.

Abstract: Excessive nitrogen fertilizer application in tea plantations is a common problem that leads to soil acidification and reductions in both yield and economic returns. To evaluate the impacts of varying nitrogen input levels (0–600 kg N ha^{-1} yr^{-1}) on yield, as well as environmental and economic outcomes, a two-year field experiment was conducted. Results demonstrated that when nitrogen application exceeded 150 kg N ha^{-1} yr^{-1}, key efficiency indicators—agronomic efficiency (NAE), recovery efficiency (NRE), and partial factor productivity (NPFP) declined markedly. Among all treatments, the 150 kg N ha^{-1} yr^{-1} rate achieved the highest NRE at 28.01%. Moreover, environmental burdens including global warming potential, acidification, and eutrophication intensified with increasing nitrogen input. Benefit analysis revealed that yield-based economic returns peaked between 150 and 165 kg N ha^{-1} yr^{-1}, whereas the highest ecosystem economic benefit, accounting for both profit and environmental cost, occurred at 120 kg N ha^{-1} yr^{-1}. Thus, 120 kg N ha^{-1} yr^{-1} is identified as the optimal application rate for maximizing integrated economic and environmental returns while maintaining yield. These findings provide valuable guidance for promoting sustainable nitrogen management in tea cultivation worldwide.

Keywords: tea; nitrogen fertilizer application; economic benefits; sustainable tea production

1. Introduction

Nitrogen (N) plays a vital role in plant growth and has a substantial impact on both crop yield and quality [1]. In China, the consumption of synthetic nitrogen fertilizers increased from 9.43 million tons in 1980 to 25.7 million tons by 2020. Furthermore, the average nitrogen application rate per hectare of arable land has steadily risen over the past decades, reaching 154 kg N ha^{-1} in 2020. While appropriate nitrogen input supports soil fertility and promotes sustainable agricultural production [2], excessive application—commonly observed in China—often exceeds crop requirements, thereby posing considerable threats to the environment and human health [3,4]. According to national estimates, only 28% of nitrogen fertilizers applied is utilized by crops [5], and the pollution loads of TN increased from 3.64×10^6 t in 1978 to 1.05×10^7 t tons in 2017 [6]. The goal of sustainable agriculture is to minimize adverse impacts to the environment while providing a sustained level of agriculture production [7]. Therefore, investigating the sustainable management strategy for N fertilizers is essential for maintaining food production, improving nitrogen use efficiency, and minimizing environmental effects, all of which are vital for both crop yield and environmental conservation.

Previous research has demonstrated that nitrogen management strategies employing active canopy sensor-based precision techniques have the potential to enhance nitrogen use efficiency while reducing environmental impacts [8]. Similarly, Singh reported that applying nitrogen at fixed times using tools such as leaf meters and leaf color charts not only achieved the highest grain yields but also improved nitrogen utilization efficiency in wheat cultivation [9]. Kumar further observed that determining fertilizer application rates based on the soil nitrogen bank approach could effectively lower environmental risks without compromising crop productivity [10]. Despite the promise of these advanced technologies, their adoption by farmers remains limited primarily due to a lack of access to information and insufficient technical understanding [11]. Therefore, there is a pressing need to develop nitrogen management strategies that address these knowledge gaps and practical constraints.

An effective starting point for nitrogen (N) management is to evaluate the relationship between N application rates and both crop yield and economic returns (i.e., income minus production costs) [12,13]. In maize, recent evidence showed that optimal nitrogen accumulation and remobilization under low-N conditions can synergistically improve yield and nitrogen-use efficiency, highlighting the potential of refining N management strategies [14]. In rice, yield continued to rise with increasing N inputs, with the greatest economic return was observed at an application rate of 263 kg ha^{-1} [15]. While such practices are widely accepted in cereal crop production [16,17], economic crops still hold substantial potential for reducing N input based on yield and profit optimization. For example, synthetic N fertilizer use in intensive greenhouse vegetable systems could be reduced by approximately 40% without sacrificing productivity [18]. Similarly, recent studies on major cropping systems in China have revealed substantial mitigation potential, showing that net greenhouse gas emissions can be reduced through improved management practices and larger-scale land operations that enhance carbon emission productivity [19,20]. These findings highlight the value and applicability of N management strategies centered on yield and economic benefit optimization, particularly for high-value economic crops.

In recent years, growing attention has been directed toward the ecosystem-level economic benefits of nitrogen management, which incorporate production costs alongside environmental and human health considerations. This integrated approach has become central to advancing sustainable agricultural practices [21,22]. Life cycle assessment (LCA) has emerged as a valuable tool for identifying optimal nitrogen management strategies that minimize ecological impacts [23,24]. Many of those studies concentrated on cereal crops; for example, the experiments conducted by Ma showed that enhancing nitrogen management in wheat production can significantly boost economic benefits for the ecosystem [24], and Yao proposed that N deep placement enhanced net ecosystem economic benefit relative to N broadcasting in rice fields [22]. But observation of the cash crop system is locking or was based on organic substitution management [25]. Additionally, the primary focus was on the impacts of various types of fertilizers, including fast-release and slow-release fertilizers [26]. Tea (*Camellia sinensis* L.) has a long history as a beverage and holds significant economic value in China and many other countries [27]. Anxi County is a major region for tea cultivation in China, with over 40,000 hectares of land dedicated to tea farming, contributing more than 2.93% of the national total tea production [28]. Tea production has become the main agricultural sector in this region, with more than 70% of farmers involved in the tea industry. However, it has encountered issues such as adverse effects on the environment and health risks for people [29,30]. These problems stem from excessive fertilization practices, particularly the over-application of nitrogen fertilizer, which averages 553 kg per hectare per year, far exceeding what is necessary for optimal yields [31].

To explore the comprehensive effects of nitrogen fertilizer application on tea yield, yield benefit, and ecosystem economic benefits in acidic red soil, we set up different nitrogen application treatments and conducted a systematic evaluation in two growing seasons. We proposed two hypotheses: (1) There is an optimal threshold for nitrogen fertilizer application and tea yield and yield benefit: increasing the application amount will increase yield and benefit within a certain range, but the promotion effect will weaken or turn negative after exceeding the threshold. (2) Ecosystem economic benefits and nitrogen fertilizer application have an inverted "U" relationship: when the application amount is low, the benefit increases with the application amount, and after exceeding the optimal level, the environmental cost caused by excessive nitrogen fertilizer increases, and the benefit begins to decline.

2. Materials and Methods

2.1. Experiment Site

The field study was carried out at a representative oolong tea production site in Anxi County (23.16° N, 118.15° E), Quanzhou City, located in the southwestern region of Fujian Province, China (Figure 1). The selected tea plantations had been under continuous cultivation for over a decade. The experiment spanned two growing seasons (2019–2021) and formed part of a long-term research project initiated in 2017. The site is characterized as a southern subtropical monsoon climate, which corresponds to the Cfa type (humid subtropical climate) in the Köppen–Geiger classification system, with an average annual temperature of 20.3 °C. Total precipitation amounted to 1349 mm and 1313 mm during the 2019–2020 and 2020–2021 growing seasons, respectively (Figure 2). The soil was classified as an acidic red soil in the Chinese Soil Taxonomy, corresponding to Ferralsols in the WRB 2023 system, with a clay loam texture, and the baseline properties of the 0–20 cm topsoil layer included pH 4.02, organic matter content of 23.0 g kg^{-1}, ammonium nitrogen 18.59 mg kg^{-1}, nitrate nitrogen 17.25 mg kg^{-1}, available phosphorus 108.10 mg kg^{-1}, and available potassium 171.22 mg kg^{-1}.

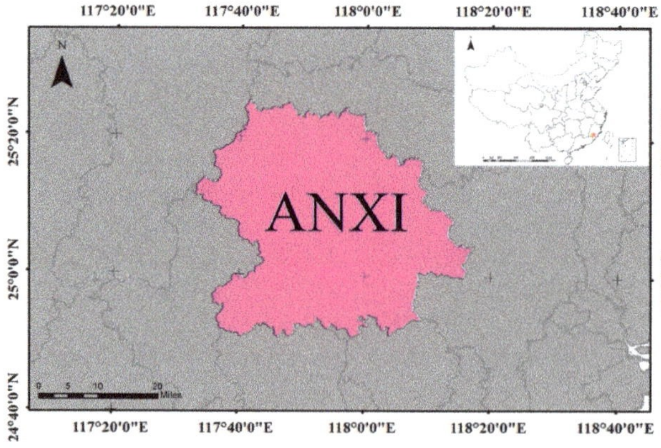

Figure 1. Location of Anxi County.

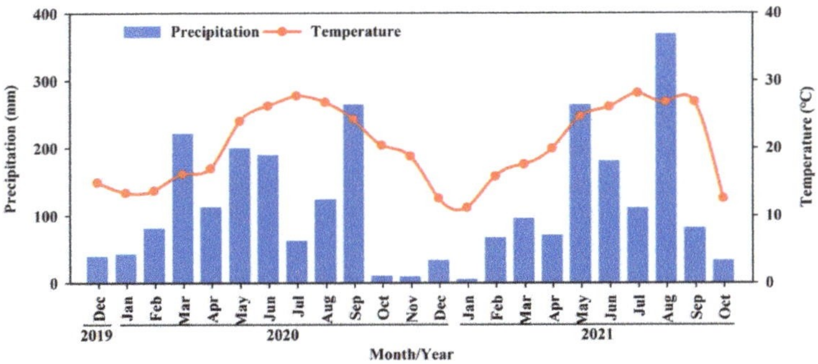

Figure 2. The precipitation and monthly mean air temperature over the tea growing seasons in Anxi County, Fujian Province, China.

2.2. Experimental Design

The experiment adopted a completely randomized block design with five nitrogen fertilizer treatment groups and three replicates for each treatment. The spacing between tea trees in the tea garden was 0.35 m × 0.85 m (intra-row and inter-row distances). Each plot measured 20 m^2. Nitrogen, phosphorus, and potassium were supplied using urea (46% N), superphosphate (20% P_2O_5), and potassium sulfate (60% K_2O), respectively, and applied during each tea growing season. Urea is applied as a solid and applied to the soil. The five nitrogen application levels were set at 0, 150, 300, 450, and 600 kg N ha^{-1} yr^{-1}, while phosphorus and potassium were consistently applied at 100 kg P_2O_5 ha^{-1} yr^{-1} and 125 kg K_2O ha^{-1} yr^{-1} across all treatments. These treatments were designated as N0, N150, N300, N450 (reflecting local farming practices), and N600.The selected N input levels (0–600 kg N ha^{-1} yr^{-1}) reflect both local farmer practices (≈450 kg N ha^{-1} yr^{-1}) and reduction scenarios recommended in regional guidelines and previous studies. Details regarding conventional tea cultivation management, including fertilization timing and rates, pruning, and harvesting schedules, are provided in Figure 3.

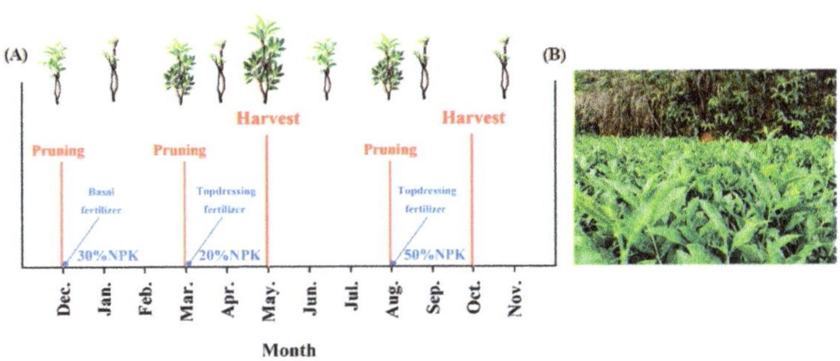

Figure 3. Conventional farming practices (**A**) and before harvest of tea shoots (**B**) in Anxi County, Quanzhou City, Fujian Province, China.

2.3. Measurements and Calculation

2.3.1. Tea Sampling and N Analysis

The bud density (tea bud number within 0.1 m^2), hundred-bud weight, and fresh weight yield during harvest in each plot were determined for all tea samples. For tea shoots N analysis, dry sample was first digested at 180~250 °C in H_2SO_4-H_2O_2 at 180~250 °C and then was analyzed using a continuous flow analyzer (AA3, Bran+Luebbe, Hamburg, Germany) [32]. Tea N uptake was calculated based on the sum of the dry matter and N concentration of plant parts.

2.3.2. Estimation of N Use Efficiency

Nitrogen use efficiency is typically assessed through three commonly applied indicators: agronomic efficiency (NAE), recovery efficiency (NRE), and partial factor productivity (NPFP). These indicators were computed using the following equations [33]:

$$\text{NAE}\ (\text{kg kg}^{-1}) = \frac{\text{Tea yield in N fertilized} - \text{Tea yield without N fertilized}}{\text{The amount of N fertilizer applied}}$$

$$\text{NRE}\ (\%) = \frac{\text{Tea N uptake in N fertilized} - \text{Tea N uptake without N fertilized}}{\text{The amount of N fertilizer applied}} \times 100\%$$

$$\text{NPFP}\ (\text{kg kg}^{-1}) = \frac{\text{Tea yield}}{\text{The amount of N fertilizer applied}}$$

2.3.3. System Boundary and Nitrogen Losses

The system boundary was defined around the tea production phase, encompassing the life cycle stages related to the manufacture and transportation of agricultural inputs such as chemical fertilizers, pesticides, and diesel. Nitrogen losses were categorized into three primary pathways: surface runoff and leaching, ammonia (NH_3) volatilization, and nitrous oxide (N_2O) emissions, with respective emission factors of 10.0%, 10.8%, and 1.2% [20]. The calculation of the carbon footprint of nitrogen input follows the methodology recommended by the IPCC Guidelines (2019), which is widely used in agricultural life cycle assessment studies. Specifically, published emission factors for N_2O emissions, NH_3 volatilization, and nitrate leaching are used [34].

2.3.4. Estimation of Environmental Impacts

Per hectare environmental impacts, including global warming potential (expressed as CO_2 equivalents), acidification potential (as SO_2 equivalents), and eutrophication potential (as PO_4 equivalents), were calculated by summing the equivalent contributions of individual inputs. These impact categories were quantified based on the following calculation formulas [20,35]:

$$\text{EI}_j = \sum_{i=1}^{n} \left(\text{UP}_{ij} + \text{PS}_{ij}\right) \times \text{Rate}_i$$

Here, EI_j denotes the environmental impact in category j, which includes global warming potential (CO_2 eq), acidification potential (SO_2 eq), and eutrophication potential (PO_4 eq) per hectare. UPij refers to the upstream emission factor of input i contributing to impact category j during the pre-planting stage of tea production. PSij indicates the emission factor associated with the application of input i in impact category j during the tea cultivation stage. Rate_i represents the quantity of each input utilized throughout the tea production process.

$$\text{SEI}_j = \frac{\text{EI}_j}{\text{Tea yield}}$$

where SEI_j represents the impact category which includes global warming (CO_2 eq), acidification (SO_2 eq), and eutrophication (PO_4 eq) potential per unit yield (in tonne, t).

2.3.5. Estimation of Benefits

The estimated of yield benefits (YB) only considered the costs of production. The benefits were estimated using the following equation:

$$\text{YB} = \text{Tea yield} \times \text{Price} - \text{Production}_{\text{cost}}$$

The local market prices for spring tea and autumn tea are 1.54 \$ kg^{-1} and 1.85 \$ kg^{-1}, respectively, and the price of nitrogen fertilizer (urea) is 0.53 \$ kg^{-1}. The production costs include agricultural material cost and the labor cost associated with N fertilizer application.

The estimation of ecosystem economic benefits (EEB) considered the costs of production, environmental, and human health. The benefits were estimated using the following equation:

$$EEB = \text{Tea yield} \times \text{Price} - \text{Production}_{cost} - \text{Environmental}_{cost} - \text{Human health}_{cost}$$

$$\text{Environmental}_{cost} = \text{Global warming}_{cost} + \text{Acidification}_{cost} + \text{Eutrophication}_{cost}$$
$$= (CO_2 \times 0.0204) + (1.87 \times NH_3 + 0.021 \times N) + (1.12 \times NO_3 + 0.24 \times NH_3 + 0.0018 \times N)$$

$$\text{Human health}_{cost} = 0.30 \times N_2O + 0.20 \times NO_3 + 3.30 \times NH_3$$

The market valuation of CO_2 was estimated at \$0.0204 per kilogram. The remediation costs associated with eutrophication caused by NO_3^- and NH_3 were \$1.12 and \$0.24 per kilogram, respectively. For acidification impacts, the cost to mitigate damage induced by NH_3 emissions was \$1.87 per kilogram. Additionally, the production of 1 kg of nitrogen fertilizer was associated with environmental damage costs of \$0.021 for soil acidification and \$0.0018 for eutrophication [15,36]. Regarding human health, the external costs per kilogram of N_2O, NO_3^-, and NH_3 emissions were \$0.30, \$0.20, and \$3.30, respectively [36,37].

2.4. Statistical Analysis

Data were processed using Microsoft Excel 2016, while statistical analyses were carried out in SPSS 23.0. One-way ANOVA followed by the least significant difference (LSD) test was employed to assess the effects of different nitrogen application rates on tea yield, bud density, hundred-bud weight, nitrogen uptake, and nitrogen use efficiency. SigmaPlot 14.0 was utilized to perform linear regression analyses exploring the relationships between tea yield and nitrogen uptake, as well as nitrogen application rate and associated environmental impacts. In addition, SAS 9.4 was applied to fit linear-plateau models describing the relationship between nitrogen input levels and both yield and economic benefits.

3. Results

3.1. Tea Yield and N Uptake

The results demonstrated that tea yield in Anxi County ranged from 7.83 to 11.22 t ha^{-1} in 2020, and from 13.16 to 16.19 t ha^{-1} in 2021. In both years, tea yield increased with nitrogen (N) application rates, with the first fertilized treatment at 150 kg N ha^{-1} yr^{-1} showing the most pronounced increase. Beyond this level (N300, N450, and N600), no further significant yield gains were observed. In 2020, the tea yield first rose and then fell as the nitrogen application rate increased, with the peak yield recorded under the N300 treatment. In contrast, in 2021, tea yield increased up to the N150 treatment and then remained relatively stable, with minimal variation observed from N150 to N600. These results suggest that moderate nitrogen input enhances tea yield, while excessive application does not confer additional benefits and may even be counterproductive under certain conditions (Figure 4).

Bud density and hundred-bud weight are key indicators for evaluating tea yield performance. Relative to the N0 treatment, bud density increased significantly under the N150, N300, N450, and N600 treatments by 53.6, 67.4, 94.4, and 58.9 buds per m^2, respectively (Figure 5A). In contrast, hundred-bud weight showed no significant variation among the five nitrogen treatments, with values ranging from 95.4 to 103.5 g (Figure 5B).

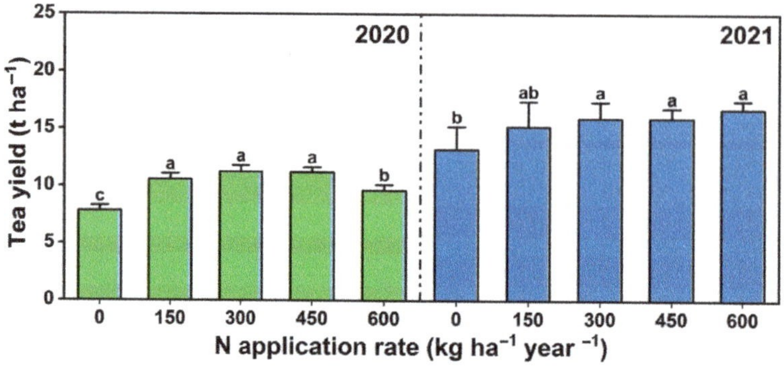

Figure 4. Tea yield under different N treatments in 2020 and 2021. Bars indicate mean ± S.D. (n = 4). Different letters on the bars indicate significant differences among different N rates at $p < 0.05$ according to LSD.

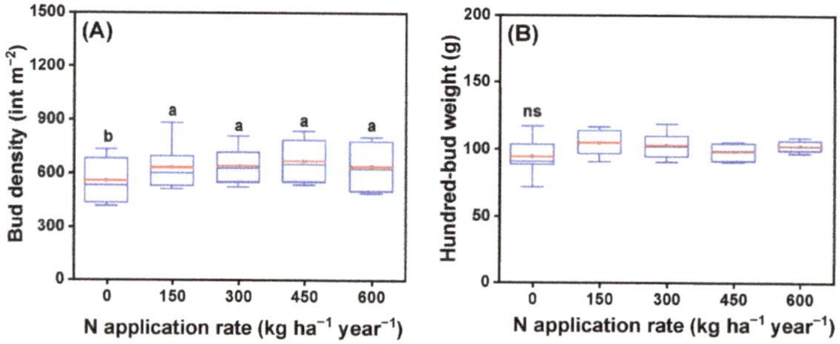

Figure 5. Bud density (**A**) and hundred-bud weight (**B**) under different N treatments. The letters at the top of the box plot indicate significant differences among different N rates at $p < 0.05$ according to LSD. The red line represents the mean.

N uptake was significantly positively correlated with tea yield and was increased by 4.65 kg ha^{-1} for every 1 t ha^{-1} increase in tea yield ($p < 0.05$) (Figure 6).

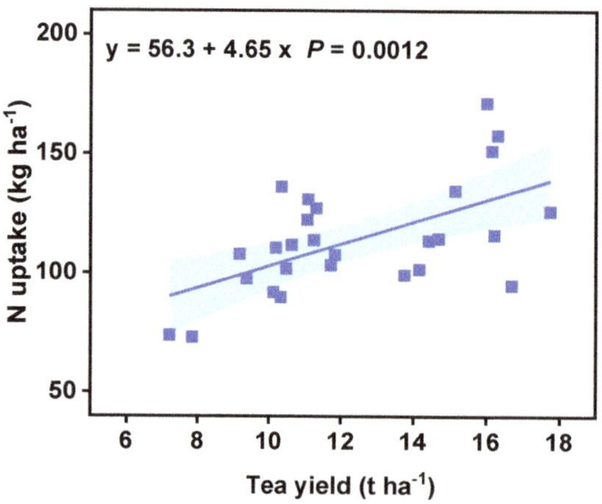

Figure 6. Correlations between tea yield with N uptake.

3.2. N Input, Output, and Surplus

In Fengshan Village, Anxi County, annual nitrogen (N) inputs for tea cultivation showed a marked increase, rising from 41 kg ha^{-1} under the N0 treatment to 641 kg ha^{-1} with the N600 treatment—representing a 15.63-fold escalation. This rise was predominantly driven by the use of synthetic nitrogen fertilizers, which increased from 0 to 600 kg ha^{-1} across these treatments. Corresponding to this input, crop nitrogen uptake varied between 89.17 and 130.63 kg ha^{-1}. A net nitrogen surplus was evident in most fertilized treatments (N150–N600), with excess N ranging from 34.86 to 397.66 kg ha^{-1}. Conversely, the N0 treatment displayed a negative nitrogen balance, as plant uptake exceeded input levels, suggesting a reliance on and depletion of existing soil nitrogen reserves (Figure 7).

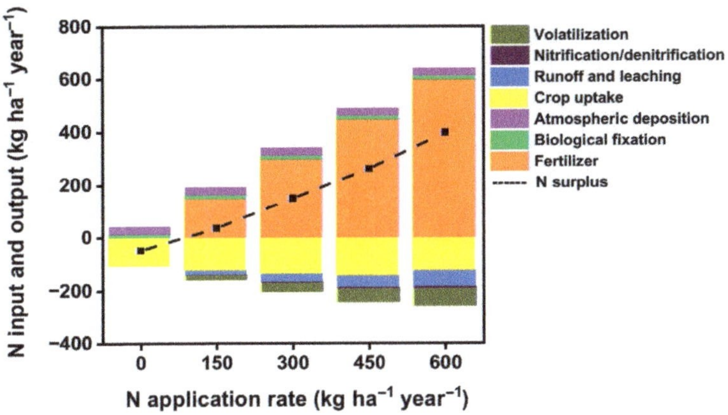

Figure 7. N input, output, and N surplus under different N treatments.

Nitrogen fertilizer use efficiency was significantly influenced by the rate of nitrogen application. As nitrogen input increased, all three key indicators of nitrogen use efficiency including NAE, NRE, and NPFP declined. Among the four nitrogen application treatments, the highest nitrogen use efficiency was achieved under the N150 treatment, with values significantly greater than those of other treatments ($p < 0.05$), indicating that a moderate nitrogen input level optimized the balance between crop uptake and fertilizer input. Although the average NAE did not differ significantly between the N150 and N300 treatments, a remarkable difference was observed in NRE, suggesting that the higher N300 application did not result in a proportionate increase in nitrogen uptake by the tea plants. This implies diminishing returns in nitrogen utilization efficiency with excessive fertilization. These findings highlight the importance of applying nitrogen at an optimal rate to maximize fertilizer use efficiency while diminishing potential environmental risks associated with nitrogen surplus (Table 1).

Table 1. Difference in N use efficiencies under different N treatments.

N Treatment	Agronomic Efficiency NAE (kg kg^{-1})	Recovery Efficiency NRE (%)	Partial Factor Productivity NPFP (kg kg^{-1})
N0	/	/	/
N150	26.89 ± 16.31 a	28.01 ± 6.53 a	92.72 ± 24.69 a
N300	10.43 ± 5.00 ab	9.80 ± 1.90 b	43.34 ± 7.26 b
N450	7.60 ± 2.57 b	8.94 ± 4.58 b	29.54 ± 5.55 b
N600	5.03 ± 3.95 b	3.18 ± 2.22 b	21.49 ± 6.91 b

Different letters indicate significant differences between different N rates at $p < 0.05$ according to LSD.

3.3. Environmental Impacts

Among all nitrogen (N) application treatments, the N0 treatment exhibited the lowest environmental impacts, while the N600 treatment showed the highest across all evaluated categories. When environmental burdens were assessed on a per hectare basis, the N150 treatment—representing a moderate nitrogen input—resulted in relatively low impacts, with a global warming potential (GWP) of 2.77 t CO_2 eq ha^{-1}, acidification potential (AP) of 43.23 kg SO_2 eq ha^{-1}, and eutrophication potential (EP) of 13.89 kg PO_4 eq ha^{-1}. In contrast, the N450 treatment, which reflects conventional nitrogen application rates commonly used by local farmers, led to substantially higher environmental pressures, with GWP, AP, and EP reaching 7.00 t CO_2 eq ha^{-1}, 124.75 kg SO_2 eq ha^{-1}, and 40.38 kg PO_4 eq ha^{-1}, respectively (Figure 8).

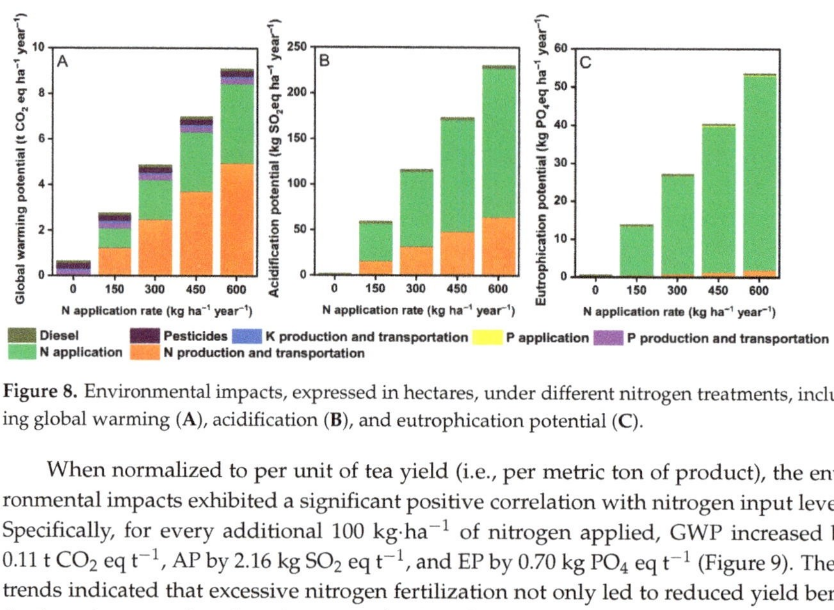

Figure 8. Environmental impacts, expressed in hectares, under different nitrogen treatments, including global warming (**A**), acidification (**B**), and eutrophication potential (**C**).

When normalized to per unit of tea yield (i.e., per metric ton of product), the environmental impacts exhibited a significant positive correlation with nitrogen input levels. Specifically, for every additional 100 kg·ha^{-1} of nitrogen applied, GWP increased by 0.11 t CO_2 eq t^{-1}, AP by 2.16 kg SO_2 eq t^{-1}, and EP by 0.70 kg PO_4 eq t^{-1} (Figure 9). These trends indicated that excessive nitrogen fertilization not only led to reduced yield benefits, but also exacerbated environmental risks related to greenhouse gas emissions, acid deposition, and nutrient runoff. Therefore, optimizing nitrogen application—particularly around the N150 level—can play a crucial role in balancing high-yield tea production with environmental sustainability.

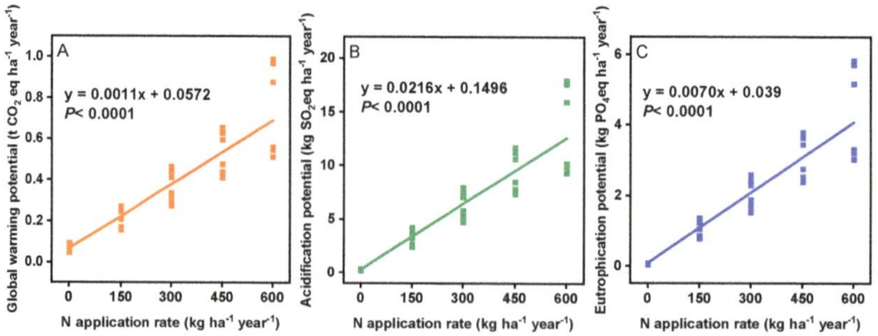

Figure 9. Correlations between nitrogen application rates per metric ton of tea production and environmental impacts, including global warming (**A**), acidification (**B**), and eutrophication potential (**C**).

3.4. Economic Assessment

Over the two years, the mean tea yield income initially slightly increased and then decreased; meanwhile, the planting, ecological, and health costs all increased with increasing N application rate (Figure 10A). The mean values of yield benefits in N0, N150, N300, N450, and N600 treatments were 12,419, 16,069, 16,577, 16,539, and 16,136 \$ ha^{-1} year^{-1}, respectively. In addition, the ecosystem economic benefits were also significantly improved by N up to the second rate (150 kg ha^{-1} year^{-1}) but did not significantly increase with higher dose spreading (Figure 10B).

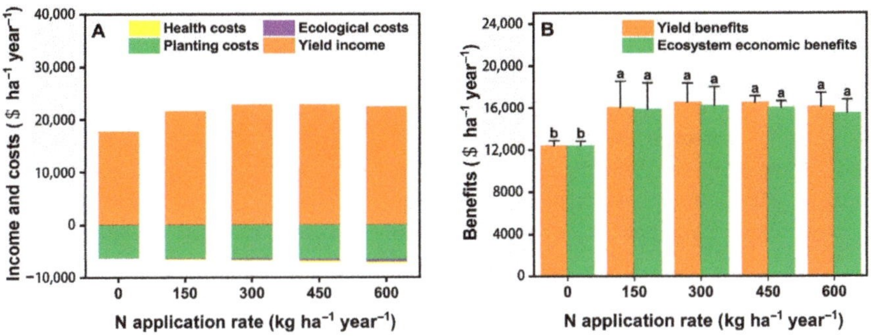

Figure 10. Income and costs (**A**) and benefits (**B**) of tea production under different N treatments. Different letters on the bars indicate significant differences among different N rates at $p < 0.05$ according to LSD.

3.5. Optimal N Application Rate

The linear-plateau model described the relationship between N application rate with tea yield and benefits well, and the optimal N application rate can be obtained. The optimal N application rate for tea yield, yield benefits, and ecosystem economic benefits was 204, 165, and 120 kg ha^{-1} year^{-1}, respectively (Figure 11).

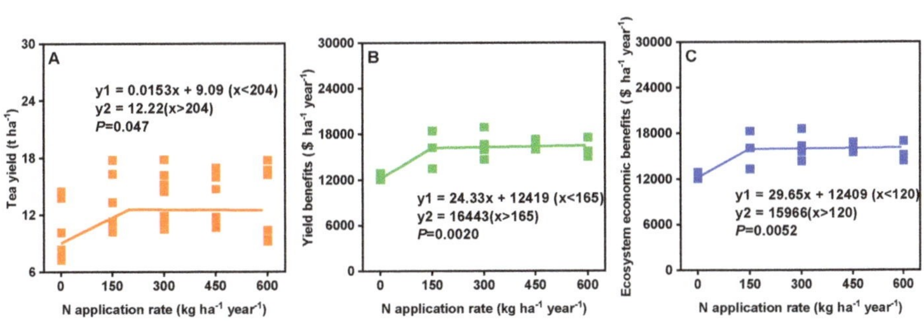

Figure 11. Tea yield (**A**), yield benefits (**B**), and ecosystem economic benefits (**C**) associated with different N application rates.

4. Discussion

4.1. Effects of N Rates on Tea Yield and N Balance

Rational application of N fertilizer is one of the most influencing factors for achieving a higher yield [38]. In this study, we observed notable increases in tea yield with nitrogen application rates between 0 and 150 kg per hectare per year; however, no significant yield improvements were seen beyond 150 kg per hectare per year (Figure 4). Based on the practices of local farmers in the study region, the annual N applied rate was over

450 kg ha^{-1}. Hence, they may have great potential for fertilizer saving under rational N fertilization in tea production in China. Further analysis indicated that the affected bud density by N applied rate was higher than that of hundred-bud weight (Figure 5). N application increased the expression of key genes for chlorophyll synthesis in tea leaves and thus improved the bud density that has been advanced to explain this phenomenon [39].

Our findings demonstrated a significant positive correlation between nitrogen uptake and tea yield ($p < 0.05$), with annual N uptake ranging from 89.17 to 130.63 kg ha^{-1} yr^{-1}. Meanwhile, nitrogen surplus increased substantially with higher N application rates, spanning from −48.10 to 397.66 kg ha^{-1} yr^{-1} (Figures 6 and 7). The magnitude of N surplus was strongly influenced by both fertilizer input and plant uptake, aligning with earlier research conducted in China [40]. In the N150 treatment, the annual surplus was approximately 35 kg ha^{-1}, closely matching the national average of 40 kg ha^{-1}. However, under the N450 treatment, which reflects local farmer practices, the surplus exceeded 260 kg ha^{-1}—far surpassing figures reported for other regions using the same estimation method, including the United States (27 kg ha^{-1} in 2017), the European Union (47 kg ha^{-1} in 2014), and India (134 kg ha^{-1} in 2018). Previous studies have indicated that when nitrogen surplus exceeds 20% of crop N uptake, the risk of nitrogen-related pollution increases markedly [41]. Therefore, based on the uptake values observed from N150 to N600 treatments, maintaining the surplus below 25 kg ha^{-1} yr^{-1} in this region would be advisable to reduce both environmental risks and fertilizer expenditure.

Nitrogen use efficiency (NUE) is a multifaceted process involving both nitrogen uptake and internal utilization mechanisms. In this study, the N150 treatment consistently exhibited significantly higher efficiency metrics across both years: agronomic efficiency (NAE), recovery efficiency (NRE), and partial factor productivity (NPFP) increased by 158–434%, 186–781%, and 114–331%, respectively, when compared to the N300, N450, and N600 treatments (Table 1). This enhanced efficiency under the N150 treatment may be attributed to the lower nitrogen input, which did not compromise tea yield or nitrogen uptake relative to the higher-rate treatments. These findings are in line with previous studies on cotton, wheat, and tomato, where nitrogen use efficiency varied across application rates and informed best management practices [42–44]. Moreover, our results indicate that nitrogen input had a more pronounced effect on tea yield than on nitrogen uptake (Table 1), likely due to the perennial nature of tea plants, where a portion of absorbed nitrogen is stored in woody tissues such as roots and stems rather than contributing immediately to harvestable biomass.

4.2. *Effects of N Rates on Environmental and Benefits Impacts*

Extensive research has demonstrated that excessive nitrogen fertilization in agriculture contributes to growing nitrogen surpluses, thereby accelerating environmental degradation [3,45]. It is therefore essential to consider both nitrogen uptake efficiency and associated environmental impacts when developing optimal nitrogen management strategies. In the present study, environmental consequences—including global warming, acidification, and eutrophication—were quantified using life cycle assessment (LCA). The results indicated a linear increase in environmental burdens with rising nitrogen application rates (Figures 8 and 9), a pattern also observed in maize production across different genotypes [46]. As a leaf-harvested economic crop, tea demands relatively high nitrogen inputs, and previous meta-analyses have evaluated the influence of nitrogen rates on soil characteristics, yield, and shoot quality [47]. However, limited research has assessed the environmental consequences of varying nitrogen levels in tea production systems, despite evidence that even recommended application rates may lead to significant ecological risks [48]. This study further contributes by examining the relationship between nitrogen input and ecosystem economic benefits in acidic tea garden soils.

When expressed per ha of tea production area, compared with local farmers' practices (N450 treatment), N150 treatment significantly reduced the global warming by 60%, acidification potential by 65%, and eutrophication potential by 66%, and such reductions were mainly attributed to the reduced N runoff and leaching, NH_3 volatilization, and N_2O emission during the tea production stage (Figure 8). The same situation was observed when environmental impacts were expressed per metric ton of tea yield production (Figure 9). Chen also highlights the importance of strengthening N management in pomelo production, based on agricultural practices that could achieve higher yields with lower environmental impacts and thereby avoid tradeoffs between productivity and ecosystem economic costs [20]. In addition, the precipitation amounts and intensity have a strong negative influence on N retention in soil [49], so it is necessary to effectively supply N to tea garden and minimizing N losses, especially in the subtropical monsoon climate zones (Figure 2).

Economic return is often a primary factor influencing farmers' choice of management practices [50]. In this study, yield benefits increased markedly as nitrogen application rose from 0 to 150 kg ha^{-1} yr^{-1}, but showed a slight decline when rates exceeded this threshold. When environmental and health costs were incorporated, ecosystem economic benefits further declined under higher N inputs (Figure 10). Excessive nitrogen use not only reduced yield-related economic returns but also significantly lowered ecosystem-level benefits. These findings are consistent with results reported by Xia for rice systems in the Taihu Lake region, where both yield and ecosystem economic benefits followed a quadratic relationship with nitrogen input [15]. Furthermore, previous research in Henan province (1987–2010) indicated a positive correlation between N fertilization and cereal aphid outbreaks, suggesting that improved N management may also reduce pest-related production costs [51]. However, our current study did not assess the effects of nitrogen application on pest dynamics, and further investigation is warranted to address this gap.

4.3. Options for Improving N Management

Nitrogen management strategies have primarily focused on maximizing yield and farmer profits, with limited attention paid to associated ecosystem risks [52,53]. In recent years, however, increasing concern has emerged regarding the environmental and human health consequences of nitrogen overuse in agriculture [4]. Optimizing nitrogen input is now recognized as a fundamental approach for achieving high yields while minimizing negative ecological and health impacts [54]. Given the prevalent overapplication of nitrogen in tea production, reducing fertilizer input based on crop yield response is a necessary first step toward enhancing nitrogen use efficiency, reducing nitrogen loss, and lowering input costs [55]. In this study, nitrogen recommendation rates based on yield response and yield economic return were determined to be 204 and 165 kg N ha^{-1} yr^{-1}, respectively (Figure 10A,B). These values are substantially lower than the average rate of 347 kg N ha^{-1} yr^{-1} suggested by the Nutrient Expert system [56], likely due to that system's reliance on nutrient uptake and efficiency models, which may underestimate the actual nitrogen use efficiency under field conditions [57]. Importantly, excess nitrogen (calculated as the difference between inputs and outputs) accumulates in the environment over time, contributing to long-term ecological damage [58]. Split nitrogen application has been widely recommended as an effective strategy to enhance nitrogen use efficiency and reduce fertilizer loss in crop production [59], a finding that is also supported by our results. Once nitrogen recommendations based on yield and economic return are established, the next step toward sustainable nitrogen management involves incorporating ecosystem economic benefits. Our analysis showed that, when considering environmental costs, the optimal nitrogen rate could be further reduced to 120 kg N ha^{-1} yr^{-1} (Figure 11C). Recently, growing interest has been directed toward integrating ecological and economic outcomes in fertilizer management. For instance, Yao demonstrated that urea deep placement significantly improved ecosystem economic benefits by up to 48%, compared to conventional broadcasting methods [22]. Similarly, Xu emphasized the importance of alternative fertilization strategies that balance environmental protection with profitability in intensive vegetable production systems in China [25]. Therefore, nitrogen management frameworks that incorporate ecosystem economic benefits offer promising pathways for promoting sustainable agricultural practices and reducing environmental burdens, with broad implications for global sustainable development [60]. The optimized N range (120–165 kg ha^{-1} yr^{-1}) provides a practical basis for regional fertilizer recommendations and subsidy programs, enabling farmers to reduce input costs and environmental risks while supporting sustainable tea production.

Despite these contributions, several limitations should be acknowledged. First, although our modeling approach was based on field experimental data, further multi-site and multi-year validations are required to confirm the robustness of the results, as the lack of broader calibration introduces uncertainties in extrapolating our findings to other tea-growing regions. Second, while N–yield response parameters were fitted using a linear-plateau model, more comprehensive parameterization and reporting, including uncertainty ranges, are needed to enhance transparency and reproducibility. Finally, soil type and climate variability, which strongly influence N dynamics and fertilizer use efficiency, were not fully incorporated into the present analysis. Future research should therefore integrate multi-regional soil characteristics and detailed weather variability to improve the reliability and general applicability of the optimization framework.

5. Conclusions

This study systematically evaluated the synergistic effects of nitrogen management on tea yield, environmental impact, and economic performance. The results identified 150 kg N ha^{-1} yr^{-1} as an optimal application rate that significantly improved tea yield and nitrogen use efficiency, while maintaining nitrogen surplus below 40 kg ha^{-1} yr^{-1}—only 15% of that under conventional local practices (450 kg N ha^{-1} yr^{-1}). Life cycle assessment showed that, compared to the conventional rate, this treatment reduced greenhouse gas emissions, acidification, and eutrophication potentials by 60%, 65%, and 66%, respectively. Significantly, the N rate optimized for ecological-economic benefits (EEB) was decreased to 120 kg ha^{-1} yr^{-1}, which is 27% to 41% lower than conventional guidelines. This adjustment still resulted in 95% of the maximum yield and improved EEB by more than 35%. Integrating EEB into decision-making was shown to reduce nitrogen loss risk by 52% and enhance resource use efficiency by 40%. The "yield–N uptake–EEB" triadic framework proposed in this study offers a practical approach for the sustainable transformation of tea production and broader green agricultural development.

Author Contributions: Conceptualization, J.P. and H.Y.; writing—original draft preparation J.P. and H.Y.; validation, M.H.; data curation, X.Y.; formal analysis, X.Z.; resources, L.G.; writing—review and editing, L.W. All authors have read and agreed to the published version of the manuscript.

Funding: This research was supported by the following projects: National Natural Science Foundation of China Project No. 42407365 and Carbon-neutral Tea Garden Construction Project (Project KH250054A).

Institutional Review Board Statement: Not applicable.

Data Availability Statement: The raw data supporting the conclusions of this article will be made available by the authors on request.

Conflicts of Interest: The authors declare no conflicts of interest.

References

1. Tataridas, A.; Folina, A.; Mavroeidis, A.; Kousta, A.; Chachalis, D. Evaluation of various nitrogen indices in N-fertilizers with inhibitors in field crops: A review. *Agronomy* **2021**, *11*, 418. [CrossRef]
2. Lassaletta, L.; Sanz-Cobena, A.; Aguilera, E.; Billen, G.; Garnier, J.; Bouwman, L.; Mueller, N.D.; Vitousek, P.M. Nitrogen dynamics in cropping systems under Mediterranean climate: A systemic analysis. *Environ. Res. Lett.* **2021**, *16*, 064050. [CrossRef]
3. Estavillo, J.M.; Huerfano, X.; Dunabeitia, M.K.; Aranguren, I.; Artetxe, R.; Campillo, C.; Lasa, B. Response of wheat storage proteins and breadmaking quality to dimethylpyrazole-based nitrification inhibitors under different nitrogen fertilization splitting. *Plants* **2021**, *10*, 703. [CrossRef] [PubMed]
4. Školníková, M.; Škarpa, P.; Ryant, P.; Kozáková, Z. Response of winter wheat (*Triticum aestivum* L.) to fertilizers with nitrogen-transformation inhibitors and timing of their application under field conditions. *Agronomy* **2022**, *12*, 223. [CrossRef]
5. Cong, W.F.; Ying, H.; Ying, F.; An, Z. Green technology for increasing grain crop production and efficiency: Innovation and application in China. *Front. Agric. Sci. Eng.* **2025**, *12*, 431–440. [CrossRef]
6. Fu, C.; Xiang, L.; Liu, Y.; Zhou, T. Response of nitrogen emissions to land use changes and driving forces analysis in Jiangxi province under multiple scenarios. *Sci. Rep.* **2025**, *15*, 29599. [CrossRef]
7. Wang, C.; Harrison, M.T.; Liu, D.L.; Yang, R.; Zhou, M. Dire need for quantification of environmental impacts associated with breeding climate-resilient crops. *Agric. Syst.* **2025**, *211*, 103861. [CrossRef]
8. Cao, Q.; Miao, Y.; Feng, G.; Gao, X.; Liu, B.; Liu, Y.; Li, F.; Khosla, R.; Mulla, D.J.; Zhang, F. Improving nitrogen use efficiency with minimal environmental risks using an active canopy sensor in a wheat-maize cropping system. *Field Crops Res.* **2017**, *214*, 365–372. [CrossRef]
9. Singh, G.; Rathore, M.; Yadav, H. Nitrogen-management strategy through leaf-colour chart and SPAD meter for optimizing the productivity in irrigated wheat (*Triticum aestivum*). *Indian J. Agron.* **2023**, *68*, 2. [CrossRef]
10. Kumar, S.; Patel, D.; Singh, N. Advancements in precision nitrogen management for sustainable agriculture. In *Sustainable Agroecosystems—Principles and Practices*; IntechOpen: London, UK, 2025. [CrossRef]
11. An, Z.; Yang, Y.; Yang, X.; Ma, W.; Jiang, W.; Li, Y.; Huang, J. Promoting sustainable smallholder farming via multistakeholder collaboration. *Proc. Natl. Acad. Sci. USA* **2024**, *121*, e2319519121. [CrossRef]
12. Bai, H.; Xiao, D.; Liu, D.L.; Tao, F. Impacts of future climate change and management practices to yield, eco-efficiency and global warming potential for rice–wheat rotation system. *J. Sci. Food Agric.* **2025**, *105*, 6138–6150. [CrossRef]
13. Zheng, W.; Liu, Z.; Zhang, M.; Shi, Y.; Zhu, Q.; Sun, Y.; Zhou, H.; Li, C.; Yang, Y.; Geng, J. Improving crop yields, nitrogen use efficiencies, and profits by using mixtures of coated controlled-released and uncoated urea in a wheat-maize system. *Field Crops Res.* **2017**, *205*, 106–115. [CrossRef]
14. Li, X.; Piao, L.; Duan, W.; Bai, Y.; Zhu, N.; Tang, Q.; He, F. Optimal nitrogen accumulation and remobilization can synergistically improve maize yield and nitrogen-use efficiency under low-nitrogen conditions. *Agronomy* **2025**, *15*, 1159. [CrossRef]
15. Xia, Y.; Yan, X. Ecologically optimal nitrogen application rates for rice cropping in the Taihu Lake region of China. *Sustain. Sci.* **2012**, *7*, 33–44. [CrossRef]
16. Hu, Q.; Gu, Y.; Lu, X.; Jiang, W.; Zhang, K.; Zhu, H.; Li, G.; Xu, F. Effects of one-time reduced basal application of controlled-release nitrogen fertilizer with increased planting density on yield and nitrogen utilization of rice. *Agronomy* **2024**, *14*, 3072. [CrossRef]
17. Li, T.; Xue, L.; Liu, P.; Zhang, S.; Huang, J.; Sun, J. The impact of cold storage facilities implemented at the production site on mitigating fruit and vegetable losses in China. *J. Clean. Prod.* **2025**, *463*, 146460. [CrossRef]
18. Sun, Y.; Ren, B.; Liu, C.; Yan, B.; Lin, L.; Zhao, Y.; Xu, H. Sustainability of different straw incorporation strategies in northeastern China: Impacts on rice yield formation, nitrogen use efficiency, and temporal soil properties. *Agronomy* **2025**, *15*, 729. [CrossRef]
19. Yin, L.; Tao, F.; Chen, Y.; Wang, Y.; Ciais, P. Spatiotemporal changes and drivers of net greenhouse gas emissions from major cropping systems in China during 1984–2018. *Resour. Conserv. Recycl.* **2026**, *224*, 108569. [CrossRef]
20. Chen, X.; Xu, X.; Lu, Z.; Zhang, W.; Yang, J.; Hou, Y.; Wang, X.; Zhou, S.; Li, Y.; Wu, L. Carbon footprint of a typical pomelo production region in China based on farm survey data. *J. Clean. Prod.* **2020**, *277*, 124041. [CrossRef]
21. Wang, X.; Miao, Y.; Dong, R.; Chen, Z.; Guan, Y.; Yue, X.; Fang, Z.; Mulla, D.J. Developing Active Canopy Sensor-Based Precision Nitrogen Management Strategies for Maize in Northeast China. *Sustainability* **2019**, *11*, 706. [CrossRef]
22. Yao, Y.; Zeng, K.; Deji, Z.; Zhao, Z.; Wang, H. The split injection of water-soluble fertilizers effectively reduces N_2O, CH_4 and NH_3 emissions while simultaneously improving rice yield and harvest index. *Field Crops Res.* **2024**, *305*, 109250. [CrossRef]
23. Lei, Y.; Zhou, X.; Cheng, H. Assessment of emergy, environmental and economic sustainability of the mango orchard production system in Hainan, China. *Sustainability* **2025**, *17*, 7030. [CrossRef]

24. Ma, J.; Zhang, K.; Fang, B.; Wang, X.; Wang, S. Optimization of nitrogen allocation and remobilization improves nitrogen use efficiency of winter wheat in the North China Plain. *Eur. J. Agron.* **2025**, *171*, 127782. [CrossRef]
25. Xu, W.; Yang, B.; Jia, Z.; Han, W. Effects of Different Water and Nitrogen on Nitrate Distribution and Leaching in Soil, Nitrogen Uptake, Utilization and Translocation by Winter Wheat. *SSRN* **2025**, *48*, 5292864. [CrossRef]
26. Zhao, Y.; Zheng, M.; Zhang, L.; Hao, L.; Zhang, Z.; Lv, L.; Zhang, J. Micro-Sprinkling Fertigation Enhances Wheat Grain Yield and Nitrogen Use Efficiency by Reducing C, N Redundancy and Increasing Root-Water-Nitrogen Spatiotemporal Coordination. *SSRN* **2025**, *36*, 5294535. [CrossRef]
27. Sarkar, L. Assessment of heavy metals in Gangaram tea garden with special reference to statistical construal and pollution indices. *J. Environ. Sci. Univ. Kashmir* **2024**, 1–12.
28. Xie, K.; Zhu, Y.; Ma, Y.; Chen, Y.; Chen, S. Willingness of tea farmers to adopt ecological agriculture techniques based on the UTAUT extended model. *Int. J. Environ. Res. Public Health* **2022**, *19*, 15351. [CrossRef]
29. Yan, Z.; Zhang, W.; Liu, X.; Li, Z.; Zhao, Y. Complex factor combinations driving tea growers to adopt ecological agricultural practices in tea gardens. *Front. Ecol. Evol.* **2024**, *12*, 1431779. [CrossRef]
30. Weng, C.; Li, Y.; Huang, W. Exploration of Factors Influencing Oolong Quality and Control Technology in Anxi County. *Agric. Sci. Technol.* **2014**, *15*, 1400.
31. Lv, X.; Kong, H.; Luo, Y.; Dong, D.; Liu, W.; Wu, D.; Ye, Z.; Ma, J. The impact of magnesium on the growth, physiology and quality of tea (*Camellia sinensis* L.) plants under acid stress. *Agronomy* **2024**, *14*, 767. [CrossRef]
32. Chen, H.; Jia, Y.; Xu, H.; Wang, Y.; Zhou, Y.; Huang, Z.; Yang, L.; Li, Y.; Chen, L.S.; Guo, J. Ammonium nutrition inhibits plant growth and nitrogen uptake in citrus seedlings. *Sci. Hortic.* **2020**, *272*, 109526. [CrossRef]
33. Srivastava, R.; Panda, R.; Chakraborty, A.; Halder, D. Enhancing grain yield, biomass and nitrogen use efficiency of maize by varying sowing dates and nitrogen rate under rainfed and irrigated conditions. *Field Crops Res.* **2018**, *221*, 339–349. [CrossRef]
34. IPCC. *2019 Refinement to the 2006 IPCC Guidelines for National Greenhouse Gas Inventories*; Intergovernmental Panel on Climate Change: Geneva, Switzerland, 2019; ISBN 978-92-9169-162-3. Available online: https://www.ipcc-nggip.iges.or.jp/public/2019rf/index.html (accessed on 15 June 2025).
35. Wang, X.; Zou, C.; Zhang, Y.; Shi, X.; Liu, J.; Fan, S.; Liu, Y.; Du, Y.; Zhao, Q.; Tan, Y. Environmental impacts of pepper (*Capsicum annuum* L.) production affected by nutrient management: A case study in southwest China. *J. Clean. Prod.* **2018**, *171*, 934–943. [CrossRef]
36. Sun, G.; Dong, Y.; Yin, W.; Wu, Y. A new core-inhibitor coated urea that increases wheat yield and nitrogen use efficiency by reducing nitrogen loss and regulating soil nitrogen supply. *Field Crops Res.* **2025**, *333*, 110080. [CrossRef]
37. Luo, L.; Luo, B.; Tai, A.P.K. Reactive nitrogen from agriculture: A review of emissions, air quality, and climate impacts. *Curr. Pollut. Rep.* **2025**, *11*, 41. [CrossRef]
38. Wang, F.; Guo, Y.; Li, P.; Wu, X.; Qiu, H.; Yin, W.; Zhao, L. Organic fertilizer substitution enhances maize yield and quality under reduced irrigation. *J. Integr. Agric.* **2025**, *24*, 1356–1368.
39. Wang, B.; Wang, S.; Li, G.; Fu, L.; Chen, H. Reducing nitrogen fertilizer usage coupled with organic substitution improves soil quality and boosts tea yield and quality in tea plantations. *J. Sci. Food Agric.* **2025**, *105*, 3301–3312. [CrossRef] [PubMed]
40. Chen, P.; Ma, J.; Yue, X.; Zeng, H.; Wang, C.; Huang, Q. Nitrate dynamics in deep soils of the Loess Plateau: Impact of different land use types. *Ecol. Indic.* **2025**, *175*, 113578. [CrossRef]
41. Zhao, Q.; Wang, Y.; Tang, Y.; Liu, B. *Handbook of Sustainable Fertilization and Crop Nutrient Management*, 3rd ed.; Science Press: Beijing, China, 2023; p. 572, ISBN 9787030739481.
42. Zhang, D.; Li, W.; Xin, C.; Tang, W.; Eneji, A.E.; Dong, H. Lint yield and nitrogen use efficiency of field-grown cotton vary with soil salinity and nitrogen application rate. *Field Crops Res.* **2012**, *138*, 63–70. [CrossRef]
43. Ma, J.; Shang, C.; Zhang, X.; Yin, B.; Zhen, W. Optimized agronomic management in North China Plain to maintain wheat (Triticum aestivum L.) yield while reducing water and fertilizer inputs. *Agronomy* **2025**, *15*, 1053. [CrossRef]
44. He, X.; Wu, P.; Wang, Z.; Zhang, L.; Wu, S.; Liu, X.; Yang, X. Subsurface irrigation system with ceramic emitters stabilizes soil nitrogen availability to enhance tomato yield and nitro-gen use efficiency in greenhouse systems. *Plant Soil* 2025, *in press*. [CrossRef]
45. Zhang, W.F.; Dou, Z.X.; He, P.; Ju, X.T.; Powlson, D.; Chadwick, D.; Norse, D.; Lu, Y.L.; Zhang, Y.; Wu, L. New technologies reduce greenhouse gas emissions from nitrogenous fertilizer in China. *Proc. Natl. Acad. Sci. USA* **2013**, *110*, 8375–8380. [CrossRef]
46. Liu, Z.; Zhou, W.; Wang, Z.; Zhao, Y.; Cheng, J.; Wang, X.; Ren, H. Quantifying contributions of efficiency-enhanced nitrogen fertilizers to maize production and environmental sustainability: Evidence for policy frameworks. *Agric. Syst.* **2025**, *230*, 104488. [CrossRef]
47. Yin, R.; Li, L.; Li, X.; Liu, H.; Yao, J.; Ma, C.; Pu, L. Positive effects of nitrogen fertilization on flavor ingredients of tea (Wuniuzao), soil physicochemical properties, and microbial communities. *Curr. Res. Food Sci. Technol.* **2025**, *37*, 103911. [CrossRef]
48. Shi, Y.; Lu, W.; Lin, L.; Li, Z.; Chen, H. Does environmental regulation affect China's agricultural green total factor productivity? Considering the role of technological innovation. *Agriculture* **2025**, *15*, 649. [CrossRef]

49. Martínez-García, L.B.; De Deyn, G.B.; Pugnaire, F.I.; Kothamasi, D.; van der Heijden, M.G. Symbiotic soil fungi enhance ecosystem resilience to climate change. *Glob. Change Biol.* **2017**, *23*, 5228–5236. [CrossRef]
50. Yu, H.; Sun, Y.; Cui, Y.; Tan, X.; Hou, Y.; Kuang, Z. The impact of farmers' cognition and attitudes on the adoption of conservation tillage: An empirical study based on the Lishu Model. *Sustainability* **2025**, *17*, 5649. [CrossRef]
51. Wang, L.; Hui, C.; Sandhu, H.S.; Li, Z.; Zhao, Z. Population dynamics and associated factors of cereal aphids and armyworms under global change. *Sci. Rep.* **2015**, *5*, 18801. [CrossRef] [PubMed]
52. Fallah, S.; Abedini Dastgerdi, H.; Kaul, H.P.; Salehi, A.S. Subsurface banding of poultry manure enhances photosynthetic efficiency, yield, and nutrient uptake in buckwheat (Fagopyrum esculentum Moench). *Plants* **2025**, *14*, 2700. [CrossRef] [PubMed]
53. Haro Altamirano, J.P.; Carrillo Barahona, W.E.; Quinde, L.M.F.; Altamirano, M.R. Determination of the sustainability of soil quality and crop health of two varieties of coffee, Rio Blanco, Ecuador. *Trop. Subtrop. Agroecosystems* **2025**, *28*, 006. [CrossRef]
54. Tang, X.Q.; Jiang, C.M.; Li, H.J.; Tian, J.; Li, D.W.; Zhang, X.; Ge, X.L.; Liu, X.; Gao, W.J.; Liu, G.L.; et al. Combined application of biochar and phosphorus influenced maize production and soil properties in the Yellow River Delta: A comparison between contrasting weather conditions. *Plant Soil. Environ.* **2025**, *71*, 104488. [CrossRef]
55. Yang, Y.; Ni, X.; Zhou, Z.; Yu, L.; Liu, B.; Yang, Y.; Wu, Y. Performance of matrix-based slow-release urea in reducing nitrogen loss and improving maize yields and profits. *Field Crops Res.* **2017**, *212*, 73–81. [CrossRef]
56. Tang, S.; Zheng, N.; Ma, Q.; Zhou, J.; Sun, T.; Zhang, X.; Wu, L. Applying Nutrient Expert system for rational fertilisation to tea (*Camellia sinensis*) reduces environmental risks and increases economic benefits. *J. Clean. Prod.* **2021**, *305*, 127197. [CrossRef]
57. He, M.; Xin, X.; Meng, L.; Yan, X.; Zhao, C.; Cai, Z. Long-term appropriate N management can continuously enhance gross N mineralization rates and crop yields in a maize-wheat rotation system. *Biol. Fertil. Soils* **2023**, *59*, 763–778. [CrossRef]
58. Zhang, X.; Davidson, E.A.; Mauzerall, D.L.; Searchinger, T.D.; Dumas, P.; Shen, Y. Mitigation strategies for soil acidification based on optimal nitrogen management. *Front. Agric. Sci. Eng.* **2024**, *11*, 229–242. [CrossRef]
59. Li, T.; Zhang, X.; Gao, H.; Li, B.; Wang, H.; Yan, Q.; Ollenburger, M.; Zhang, W. Exploring optimal nitrogen management practices within site-specific ecological and socioeconomic conditions. *J. Clean. Prod.* **2019**, *241*, 118295. [CrossRef]
60. Liu, S.; Zhang, H.; Liu, Q.; Zong, H.; Zhang, X. Effect of long-term application of manure and nitrogen fertilizer on infiltration for a wheat–maize rotation system. *Land Degrad. Dev.* **2023**, *34*, 326–338. [CrossRef]

agriculture

MDPI

Article

Comparison of Ray Tracing Software Performance Based on Light Intensity for Spinach Growth

Chengyao Jiang [1,†], Kexin Zhang [1,†], Yue Ma [2,†], Yu Song [2,*], Mengyao Li [1], Yangxia Zheng [1], Tonghua Pan [3] and Wei Lu [1,*]

[1] College of Horticulture, Sichuan Agricultural University, Chengdu 611130, China; catherinejiang@126.com (C.J.); zhangkexin0722@126.com (K.Z.); limy@sicau.edu.cn (M.L.); zhengyangxia@163.com (Y.Z.)
[2] Crop Research Institute, Xinjiang Academy of Agricultural Sciences, Urumqi 830091, China; my0988221@163.com
[3] School of Agricultural Engineering, Jiangsu University, Zhenjiang 212013, China; panthujs12345@ujs.edu.cn
* Correspondence: songyu150@163.com (Y.S.); luwei-1213@163.com (W.L.)
† These authors contributed equally to this work.

Abstract: With the development of modern agricultural technology, plant factories have become an important way to achieve efficient and sustainable crop production. Accurate understanding of the light received by plants is the key to improving the light energy utilization efficiency of lamps and ensuring the benefits of plant factories. Ray tracing technology, as one of the key technologies in plant factories, is of great significance to analyze the growing light environment of vegetables. Spinach has high nutritional value and is loved by the public and is one of the main crops grown in plant factories. In this paper, LightTools, TracePro, and Ansys Lumerical FDTD Solution, which are currently mature light environment tracking software in the field of lighting, are selected as the research objects to investigate their performance in simulating the light environment of spinach leaf surfaces under different planting arrangements and different lamp source distances. The results show as follows: Under the rectangular planting arrangement, the leaves received more light, and the plants grew faster. Different planting arrangements of plants had little effect on the simulation effect of the same kind of software, but the simulation effect of the three kinds of software under the same planting arrangement was significantly different, and the difference between the simulated value and the measured value of TracePro was the least. Further, TracePro was used to trace and simulate the leaf surface light conditions of spinach under a rectangular planting arrangement at different lighting distances, and the simulation results showed that there was no significant difference between the software simulation value and the measured value, and the simulation accuracy was the highest when the distance from the light source was 30 cm. Therefore, TracePro software can accurately simulate the light intensity of spinach leaves during the growth process and is most suitable for monitoring the change of light environment of spinach growth in plant factories.

Keywords: plant factories; ray tracing software; spinach; software comparison

1. Introduction

Spinacia oleracea L., an annual or biennial herbaceous green leafy vegetable belonging to the genus Spinacia in the family Amaranthaceae, is an important vegetable crop [1]. In 2016, China became the world's largest producer and consumer of spinach [2]. Spinach has a sweet and tender taste and is rich in nutritional value, containing high levels of protein, various vitamins, inorganic salts, etc. Among them, the vitamin A content in spinach is comparable to that in carrots [3]. Spinach has shown a wide range of beneficial effects on human health, such as lowering blood sugar and blood lipids [3].

During the growth of spinach, its surrounding environment, such as light, air, water, temperature, etc., will affect its growth, among which light is one of the important factors affecting the growth and development of spinach. Previous studies have respectively shown that different light conditions have significant impacts on the germination rate and growth and development of spinach seeds [4–6]. Therefore, when cultivating spinach in plant factories, the growth cycle of spinach can be regulated by changing the light duration and light intensity irradiated on spinach [5–7], so as to realize the year-round supply or off-season marketing of spinach. In plant factories, light-emitting diodes (LEDs) are often used as artificial light sources. This material was invented in the 1960s [8], and with the development of technology, LEDs are now widely used in various fields. LED light sources have the advantages of high photoelectric conversion efficiency, small size, long service life, and low energy consumption [9–11], so they have become the most commonly used light source materials in modern agricultural production [12]. In particular, the use of LED light sources allows precise control of the light spectrum, and they can be automatically controlled, with energy-saving and high efficiency, making them the best choice for the light environment of plant growth in plant factories [13–15]. However, for plants, as the change in canopy morphology during the growth process accentuates the unevenness of the light distribution received by the leaf surface, the middle and lower leaves are often in a shaded state. Even when the upper leaves have reached light saturation, the middle and lower leaves still fail to meet the optimal light conditions. This structural light deficiency is particularly prominent in traditional array-type LED light source systems, leading to a difference in the growth rate of plants in the same batch of more than 30% and seriously restricting the yield per unit area [5–7]. Therefore, how to optimize the LED array form and adopt innovative cultivation layouts has become a research hotspot in the efficient production of plant factories in recent years.

With the development of modern agricultural production, digital plant technology [16–18], which digitizes plants through computer technology to realize the perception and understanding of the behaviors of plant life systems and agricultural production systems [18], has gradually been widely applied in the intelligent production monitoring of horticultural crops. These technologies are also driving continuous innovation in the field of LED plant lighting [19–26]. Through light tracing and tracking in digital plant technology, a spinach model can be created in optical simulation software, and then the light conditions on various plant leaves can be analyzed using the light tracing function in the corresponding software. Ray tracing technology, also known as light tracing, is a general technology derived from geometric optics that obtains the model of the light path by tracking the light rays interacting with optical surfaces. With the continuous development of computer graphics processing technology, ray tracing technology has gradually received hardware-level support. In this experiment, three simulation software, namely Light Tools, TracePro, and Ansys Lumerical FDTD Solution, were selected to track and analyze the light on spinach leaves. LightTools is an optical system modeling software developed by Optical Research Associates (ORA) in the United States in 1995, and it is equipped with an illumination template supporting the main program, which solves the problem of computer-aided design for lighting systems [24]. The software is characterized by high flexibility in lighting systems with various light sources and observation angles [25,26]. TracePro is a set of optical simulation software commonly used in lighting systems, optical analysis, radiometric analysis, and photometric analysis [27–29]. This software has the characteristics of an intuitive interface, simple operation, and low time consumption. Ansys Lumerical FDTD Solution is simulation software based on the solution of Maxwell's equations, with an intuitive interface, powerful simulation functions, and good visual performance, which is often used in scenarios requiring three-dimensional analysis of data [30]. However, so far, the application of these three ray-tracing software in the field of plant lighting, especially in the intelligent production of horticultural crops relying on plant factories, has rarely been involved.

The purpose of this study is to fully understand the distribution of light intensity on spinach leaves in plant factories, use three relatively advanced existing light simulation software programs, namely LightTools, TracePro, and Ansys Lumerical FDTD Solution, for modeling and analysis, and investigate their performance in simulating the light environment on spinach leaves under different planting arrangements and different light source distances from the aspects of operating environment, modeling process, and reliability comparison. It aims to screen reliable, intuitive, and effective light environment monitoring and analysis software for spinach growth so as to provide technical and theoretical references for improving the light energy utilization efficiency of plant factories and enhancing the benefits of spinach cultivation.

2. Materials and Methods

Mainly data cultivation and plant verification work of this study were conducted in Chengdu and Urumqi, China. And the whole experiment period lasted from 2024 autumn to 2025 spring.

2.1. Materials Preparation

2.1.1. Plant Materials

The spinach variety selected was 'Shengxia Xianfeng', which has a moderate growth rate and an upright plant type and is widely used in plant factory production. All plants were provided by the Vegetable Laboratory of Sichuan Agricultural University.

2.1.2. Light Condition Instruments

The LED lamp was designed by the Light Environment Innovation Group at Sichuan Agricultural University. The LED lamp for plant production was in a uniform size of 1.2 m in length, 1800 lm in luminous flux, and full spectrum. Each LED lamp bead was 2.8 mm in length, 3.5 mm in width, and 0.8 mm in thickness. It was with a power of 0.2 W and a luminous flux of 25 lm. So each LED unit contains 72 lamp beads (1800/25 = 72).

The illuminance sensor used in this study was RS485, produced by Jianda Renke, Hangzhou, with a measurement range of 0–130,000 lux and a measurement accuracy of $\pm$6%.

2.1.3. Light Tracing Software

Three kinds of professional and widely used light tracing software were used in this study: LightTools 8.6, TracePro 7.4.3, and Ansys Lumerical FDTD Solution 2024.

2.2. Test Methods

2.2.1. Arrangement of Spinach Cultivation and Determination of Leaf Illuminance

Spinach seeds were sown in seedling trays in a germination room with a temperature of 30 °C and relative humidity of 90% for 3 days [5]. Then, the plants were transported into a plant cultivation room with a light cycle of 16 h/8 h (day/night), a temperature of 28 °C/20 °C (day/night), and a relative humidity of 70 $\pm$ 3% [5]. When plants grew to a stage with 2–3 true leaves, 64 seedlings with uniform growth were selected and divided into groups of 16. Four planting patterns, according to the results of the preliminary experiment, were set: equal difference (T1), rectangle (T2), hexagon (T3), and diamond (T4), as shown in Figure 1. A light substrate was used for seedling cultivation, with the substrate ratio of peat: vermiculite: decomposed organic fertilizer = 2:1:1. The experiment was conducted in the plant factory of Sichuan Agricultural University.

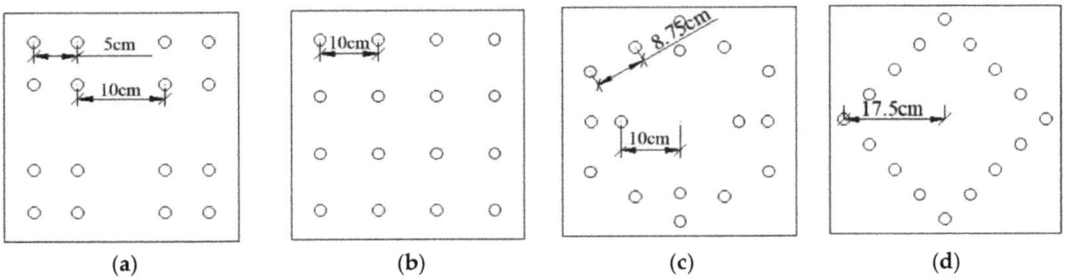

Figure 1. Spinach Cultivation Arrangement: (**a**) equal difference (T1), (**b**) rectangle (T2), (**c**) hexagon (T3), and (**d**) diamond (T4).

The illuminance on the spinach leaf surface was continuously measured using an illuminance sensor, and the average plant height of spinach was recorded. After fixed planting and grouping, plant height measurements were conducted on the 2nd, 9th, 16th, 23rd, 30th, and 37th days, respectively. The probes of the illuminance sensor were arranged on the top leaves, with 8 random measuring points set in each group, and data were recorded at 1 h intervals. Meanwhile, a movable cultivation frame (also designed by Light Environment Innovation Group) was used to record the leaf surface illumination data when the plant leaves were 10 cm, 20 cm, and 30 cm away from the LED lamps. All data were collected and recorded by a large-scale data logger (CR1000X; Campbell Scientific, Inc., Logan, UT, USA). The experiment was repeated 3 times.

2.2.2. Simulation Process of Leaf Surface Light Environment

Establishment of Spinach Model

First step: A real image of spinach was selected as a reference for modeling. A spinach leaf was chosen, and the dimensional data of the leaf's shape, as well as relevant data such as the length and thickness of the stem, were recorded. A reference plane was selected in SolidWorks (SOLIDWORKS 2018), and based on the recorded data, the outline of the spinach leaf was sketched within the reference plane. Then, the 3D model of the spinach leaf was generated using the extrusion function in SolidWorks, as shown in Figure 2a,b. Regarding the spinach stem, first sketch the cross-sectional contours of both the top and bottom ends of the spinach stem, add a guide path along the height direction of the stem, and use the lofting function in SolidWorks to generate the 3D model of the stem, as shown in Figure 2c,d.

Second step: Based on the actual images of spinach, compare the size and shape of two actual spinach leaves as well as the length and thickness of their stems. According to the differences between the two actual spinach leaves, on the basis of the first spinach leaf model, adjust the size and shape of the leaf as well as the length and thickness of the stem to draw the second spinach leaf model. Repeat this modeling step to create the remaining spinach leaf models.

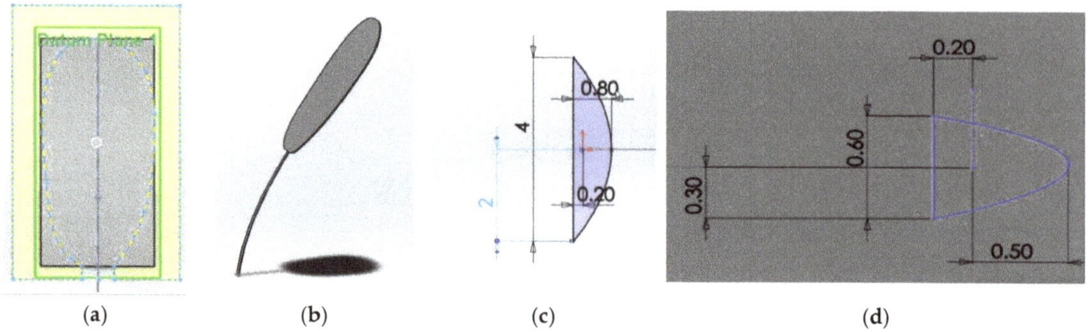

Figure 2. Sketch of Spinach Leaf and Stem: (**a**) contour of spinach leaf, (**b**) single leaf model, (**c**) contour of the stem base, and (**d**) contour of the stem top.

Third step: Assemble all the created spinach leaf models together according to the actual images of spinach to form an integral spinach model, and color it, as shown in Figure 3.

Figure 3. Overall model of spinach: (**a**) schematic diagram of spinach compound leaf assembly, (**b**) completed diagram of the model.

All models were constructed in batches and multiple times according to the growth stages of spinach leaves, in conjunction with plant height measurements, for subsequent use in illuminance simulation.

Establishment of the LED Lamp Model

First, based on the length, width, and height of the LED lamp and lamp beads, a cross-sectional contour was drawn, respectively, and then the 3D graphics were generated using the extrusion function, as shown in Figure 4a,b. Secondly, the LED lamp body, lampshade, and lamp bead models were built accordingly. Then, based on the side width of the LED lamp and lampshade, the side cover model was constructed, as shown in Figure 4c,d. Next, the assembly model of each component was established. During the assembly process, the LED lamp body model was first taken out, and then all lamp beads were installed on one side of the LED lamp body model at equal distances according to the required number of lamp beads. Then, the lampshade was installed on the same side as the lamp beads, and finally, the side covers were assembled, as shown in Figure 4e,f.

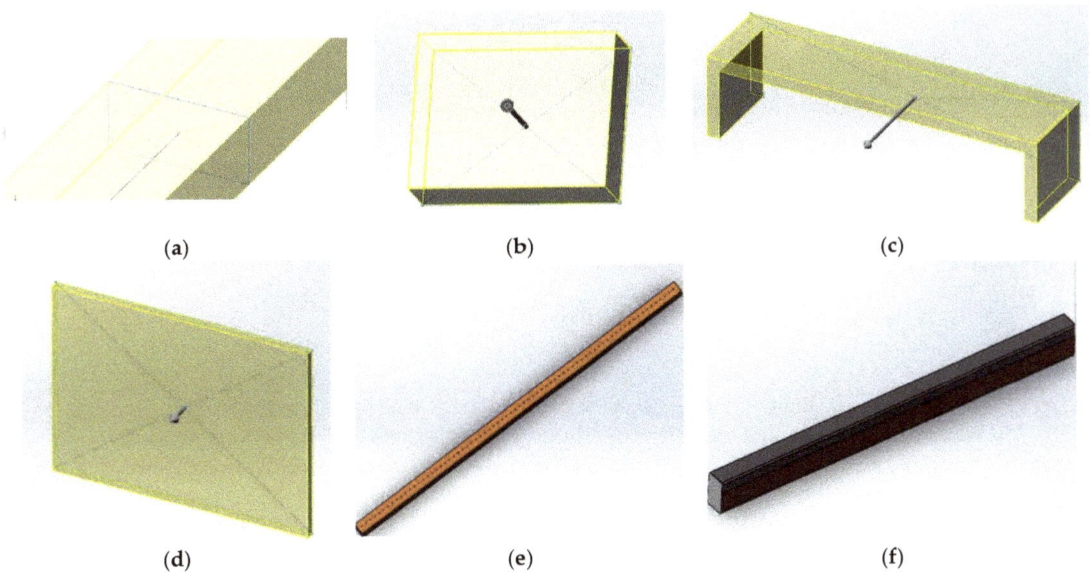

Figure 4. Composition diagram of the LED lamp model: (a) outline of the LED lamp, (b) outline of the lamp bead, (c) outline of the lampshade, (d) outline of the side cover, (e) schematic diagram of lamp bead assembly, (f) completed model of the LED lamp.

Setting of lampshade material, calculation of extinction coefficient, and determination of light propagation loss

The material of the lamp beads was defined and edited according to Wen et al. [27], as shown in Table 1 (a full-spectrum LED lamp was used with a transmittance of approximately 0.9). The temperature was set to 300 K, and the light wavelengths were set to four types: 0.4 μm, 0.5 μm, 0.6 μm, and 0.7 μm. The absorption value was set to 0.008 mm, and the extinction coefficients for each wavelength were obtained according to Beer–Lambert Law:

$$A = \varepsilon c L \quad (1)$$

Table 1. Properties of lamp bead material.

Wavelength (μm)	Temperature (K)	Absorption Value (mm)	Extinction Coefficient ($L \cdot \mu mol \cdot \mu m^{-1}$)
0.4	300	0.008	2.546
0.5	300	0.008	3.183
0.6	300	0.008	3.819
0.7	300	0.008	4.456

In the formula, A is the absorbed light intensity, ε is the molar extinction coefficient or molar absorption coefficient ($L \cdot mol \cdot cm^{-1}$), c is the content of the light-absorbing substance ($mol \cdot L^{-1}$), and L is the optical path length (cm).

The parameters of the overall LED lampshade material are shown in Table 2. Based on Table 1 and design standards, the wavelength of the LED lampshade was 0.5461, the refractive index was approximately 1.59, the absorption coefficient was 0.008, and the transmittance was around 0.9 (according to Beer–Lambert Law)

$$T = 10^{-A} \quad (2)$$

Table 2. Parameters of the lampshade material.

Wavelength (μm)	Refractive Index	Absorption Coefficient	Transmittance
0.5461	1.59	0.008	0.92

In the formula, T is the transmittance, and $1 - T$ is the proportion of absorbed light.

The index values calculated above meet the target values. To simplify the modeling process, the transmittance of the opaque parts of the spinach and LED lamp models was set to 0.

Setting of light source parameters on the surface of lamp beads.

According to Zheng et al. [24], first, set the emission form of 72 lamp beads as luminous flux, with the unit being photometry. The minimum number of light rays is set to 10, the total number of light rays is set to 5000, the field angle distribution is the Lambertian luminous field type, and the luminous flux is 25 lumens. Then, set different wavelengths, respectively: Divide the 72 lamp beads into 18 groups, with every 4 lamp beads numbered 1, 2, 3, and 4; then, group the lamp beads with the same number in the 18 groups into a large group, totaling 4 large groups. The wavelength of the lamp beads in the first large group is set to 0.45 μm (since the wavelength of 0.4 μm exceeds the effective wavelength range of the plastic material (0.4358 μm–1.052 μm), the wavelength of the first group of lamp beads is set to 0.45 μm), the second large group to 0.5 μm, the third large group to 0.6 μm, and the fourth large group to 0.7 μm. The luminous flux of each group is set to 25 lm. The minimum number of light rays of each group is set to 10 strips and the total number of light rays of is set to 5000 strips.

2.3. Data Statistics and Analysis

Data analysis was performed using SPSS 27.0 software and GraphPad Prism 9.5.0 software, and the experimental data (including plant morphology data, simulation, and actual measurement of light intensity) were analyzed by Tukey's HSD test. A paired two-sample t-test was conducted to compare the measured and simulated values with TracePro under different distances from the light source.

3. Results

3.1. Data on Changes in Spinach Plant Height and Measurements of Leaf Surface Illuminance

Four planned spinach cultivations, T1 (equal difference), T2 (rectangle), T3 (hexagon), and T4 (diamond), were investigated for leaf surface illuminance. After the spinach grew 2–3 true leaves, the average plant height of 4 groups of spinach was recorded every 8 days (Figure 5), and the illuminance on the leaf surfaces of the 4 groups of spinach was measured. The average illuminance of spinach with four planting arrangements under different plant heights is shown in Table 3. It can be seen from the data in the table that under the light-controlled conditions with the same output, the four planting arrangements resulted in different growth and development statuses of spinach, and the light received by the leaf surfaces also varied. In general, the plant height in the T2 group increased relatively rapidly, though some data in other groups showed no significant difference; for example, some plant height in T3, the absolute value of the T2 group is highest. Using the regression equations of plant height of each group shown in Figure 5 and combining them with the light integral formula, it is inferred that under the same height, the light received by the leaves in the T2 group and T3 group partially was significantly higher than that in other groups at the same height, leading to a faster increase in plant height (Table 3 and Figure 5).

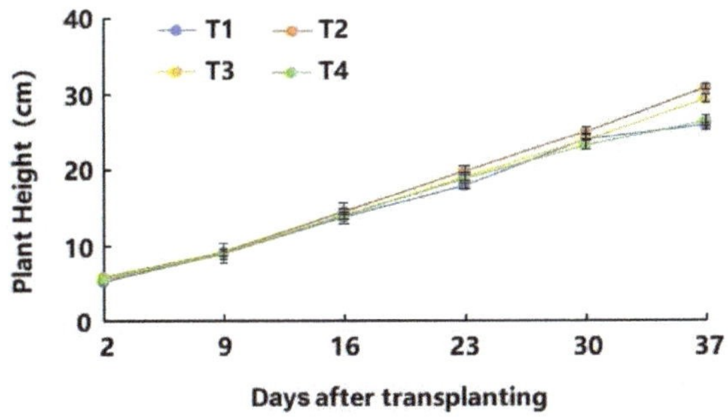

Figure 5. Changes in spinach plant height of T1 (equal difference), T2 (rectangle), T3 (hexagon), and T4 (diamond). The data of the four treatments were compared within the same number of days. Bars represent mean ± SD (n = 3).

Table 3. Average illuminance on spinach leaf surfaces under four cultivation arrangements at different plant heights (Klux).

Treatment	10 cm	15 cm	20 cm	25 cm
T1 [1]	3.42 ± 0.18 b [2]	4.42 ± 0.31 c	6.45 ± 0.25 b	7.31 ± 0.21 c
T2	3.53 ± 0.29 a	4.81 ± 0.33 a	6.54 ± 0.29 a	7.91 ± 0.21 a
T3	3.48 ± 0.22 ab	4.80 ± 0.39 a	6.52 ± 0.27 a	7.72 ± 0.24 b
T4	3.45 ± 0.29 b	4.76 ± 0.22 b	6.43 ± 0.30 b	7.65 ± 0.31 b

[1] Four cultivation-planned spinach, T1 (equal difference), T2 (rectangle), T3 (hexagon), and T4 (diamond), were investigated for leaf surface illuminance. [2] Different letters in each plant height indicate statistically significant differences according to Tukey's HSD test at $p < 0.05$.

3.2. Modeling of Leaf Surface Light Environment Under Different Spinach Cultivation Arrangements

Three optical simulation software programs, namely TracePro, LightTools, and Ansys Lumerical FDTD Solution, were used to perform ray tracing analysis on spinach models under four planting arrangements, with the results shown in Figures 6–8. Since lux is the output unit of the software, no other unit conversion was performed on the results of this experiment to ensure the effectiveness of data statistics.

For the T1 arrangement simulated by TracePro software, the maximum illuminance on the leaf surface was approximately 3362 lux, and the average value was about 1748.3 lux (Figure 6a); for the T2 arrangement, the maximum illuminance on the leaf surface was around 3339 lux, with an average of about 1729 lux (Figure 6b); for the T3 arrangement, the maximum illuminance on the leaf surface was roughly 3976 lux, and the average was approximately 1433.8 lux (Figure 6c); for the T4 arrangement, the maximum illuminance on the leaf surface was about 5909.9 lux, with an average of around 988.16 lux (Figure 6d).

For the T1 arrangement simulated by LightTools software, the maximum illuminance on the leaf surface is approximately 3142 lux, with an average of about 2078.3 lux (Figure 7a); for the T2 arrangement, the maximum illuminance on the leaf surface is around 2939 lux, and the average is about 2029 lux (Figure 7b); for the T3 arrangement, the maximum illuminance on the leaf surface is roughly 2786 lux, with an average of approximately 1593.3 lux (Figure 7c); for the T4 arrangement, the maximum illuminance on the leaf surface is about 2409.9 lux, and the average is around 1298.01 lux (Figure 7d).

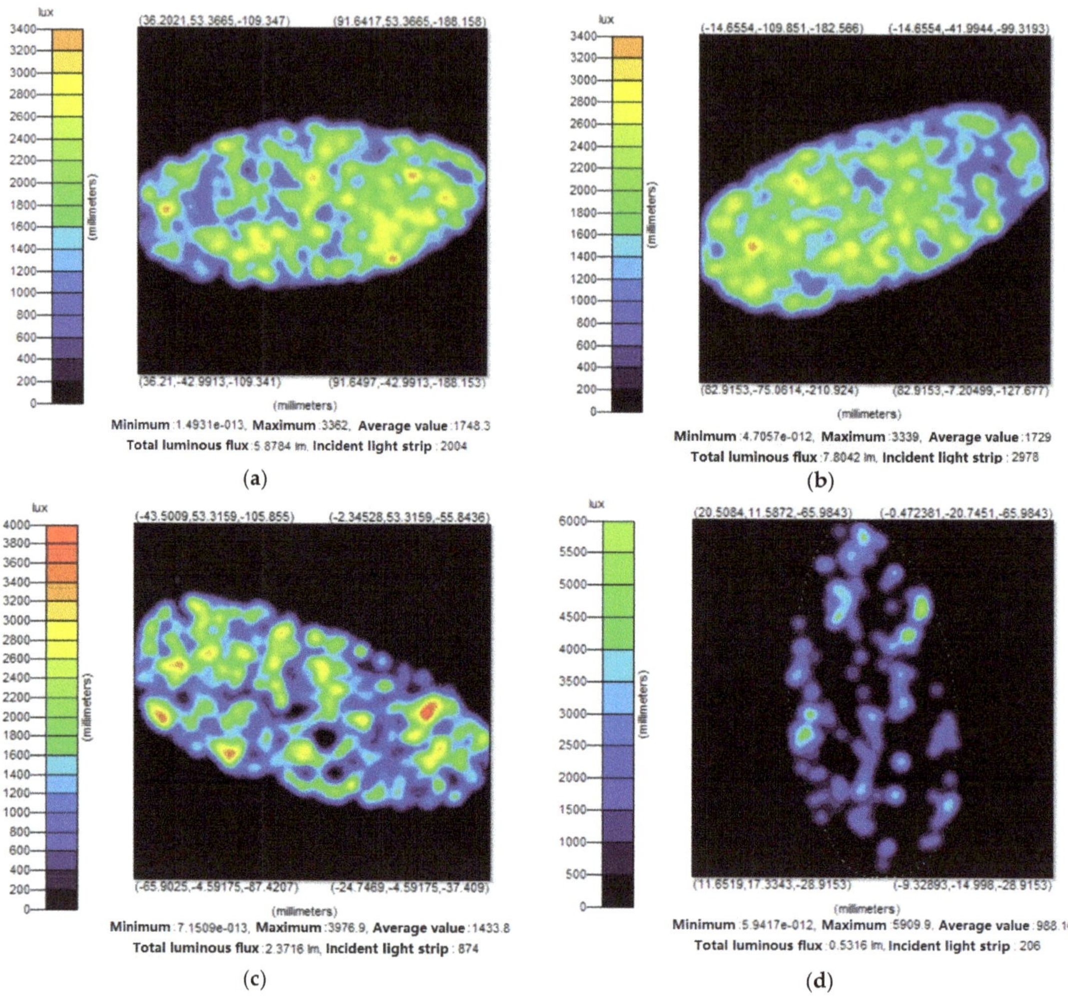

Figure 6. Illuminance on spinach leaf surfaces under different treatments simulated by TracePro: (**a**) T1 (equal difference), (**b**) T2 (rectangle), (**c**) T3 (hexagon), and (**d**) T4 (diamond).

The data heatmap output method of Ansys Lumerical FDTD Solution is slightly different from the above two software. For the T1 arrangement simulated by this software, the peak illuminance on the leaf surface is approximately 694.18 cd, and the average luminous flux is about 1758.9 lux (Figure 8a); for the T2 arrangement, the peak illuminance on the leaf surface is around 690.16 cd, with an average luminous flux of about 1731 lux (Figure 8b); for the T3 arrangement, the peak illuminance on the leaf surface is roughly 712.06 cd, and the average luminous flux is approximately 1430 lux (Figure 8c); for the T4 arrangement, the peak illuminance on the leaf surface is about 694.18 cd, with an average luminous flux of around 1030.8 lux (Figure 8d).

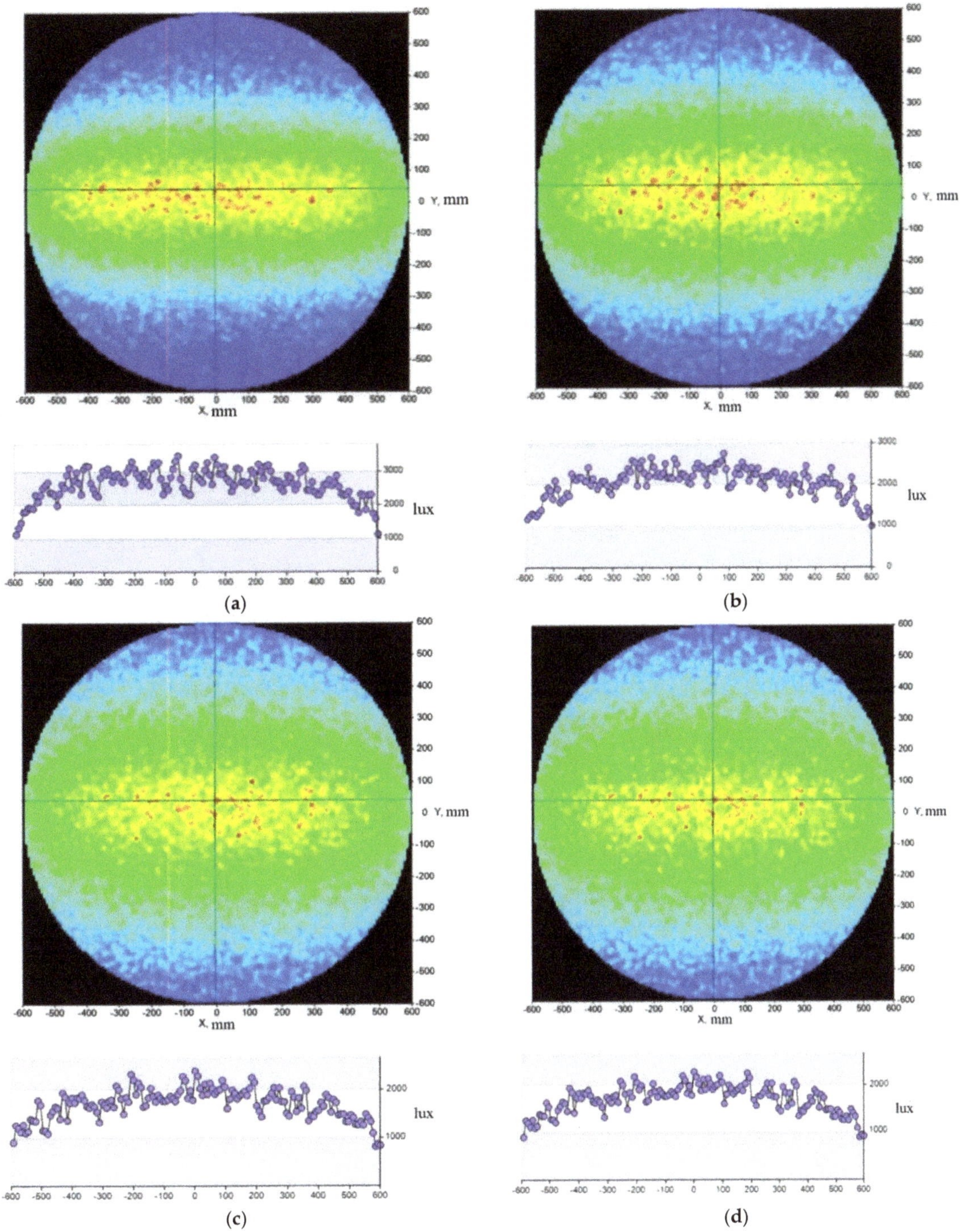

Figure 7. Illuminance on spinach leaf surfaces under different treatments simulated by Light Tools: (**a**) T1 (equal difference), (**b**) T2 (rectangle), (**c**) T3 (hexagon), and (**d**) T4 (diamond).

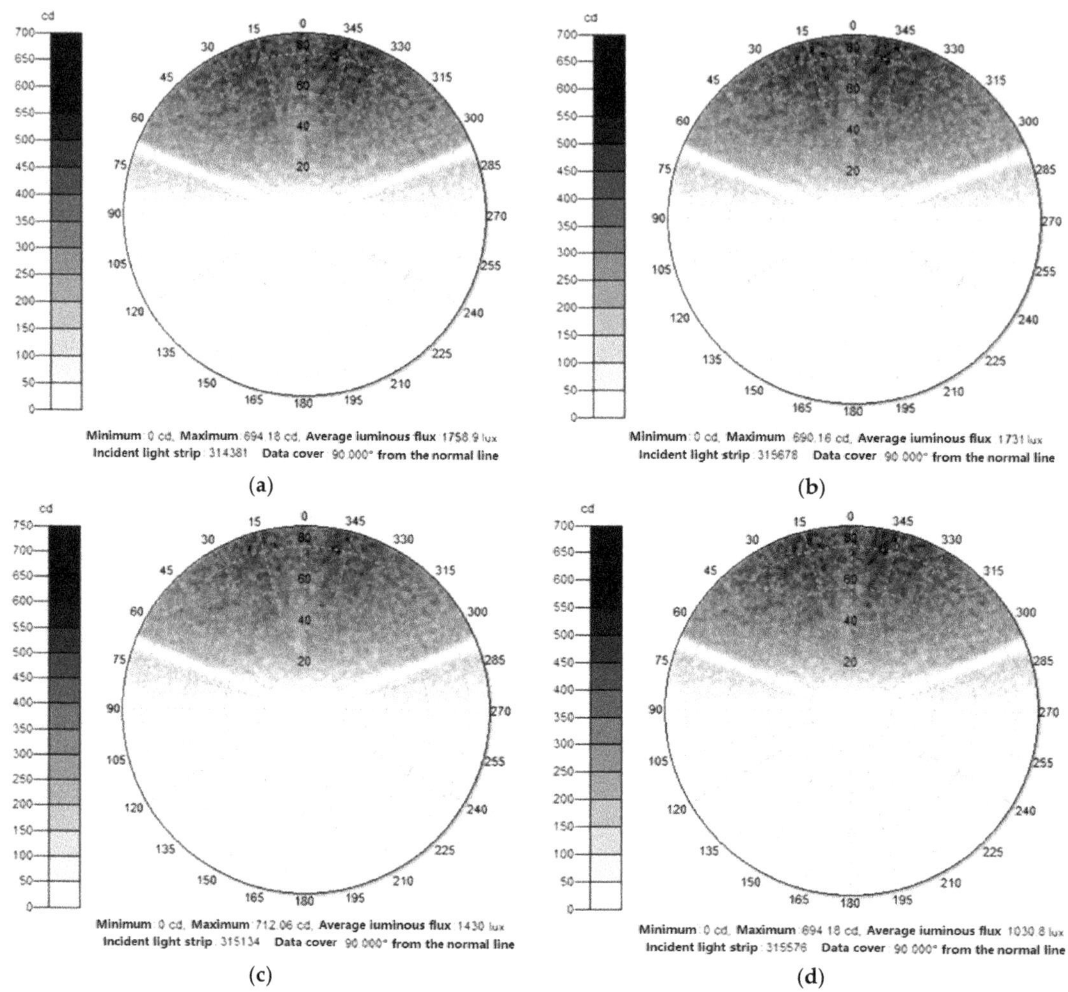

Figure 8. Illuminance on spinach leaf surfaces under different treatments simulated by Ansys Lumerical FDTD Solution: (**a**) T1 (equal difference), (**b**) T2 (rectangle), (**c**) T3 (hexagon), and (**d**) T4 (diamond).

Although the output units of illuminance among the three software may vary, it can be known through conversion that after unifying to lux, there is no cross-order-of-magnitude difference, indicating that their simulated values can be compared. In general, for the same software, the 3D heatmaps of illuminance analysis for spinach models under different planting arrangements tend to be consistent, which shows that the differences in planting arrangements have little impact on the software modeling itself. However, under different planting arrangements, there are differences in the 3D heatmaps of illuminance analysis, indicating that there are differences in the modeling results of different software.

Table 4 shows the comparison between the simulated values from the three software and the measured values of leaf surface illuminance under the four cultivation arrangements. Among all treatments, only the data from TracePro consistently showed no significant difference from the measured values. Meaning among them, the difference between the simulated values and the measured values under various cultivation arrangements obtained by TracePro simulation software is the smallest.

Table 4. Simulation and actual measurement of light intensity on spinach leaf surfaces under four cultivation arrangements (Klux).

Treatment	Measured Value	TracePro	Light Tools	Ansys Lumerical FDTD Solution
T1	1.71 ± 0.12 ab [1]	1.75 ± 0.18 a	1.65 ± 0.03 b	1.66 ± 0.41 b
T2	1.77 ± 0.23 a	1.77 ± 0.31 a	1.57 ± 0.30 c	1.74 ± 0.19 b
T3	1.41 ± 0.11 b	1.43 ± 0.05 b	1.50 ± 0.24 a	1.43 ± 0.22 b
T4	1.00 ± 0.53 bc	0.99 ± 0.06 c	1.29 ± 0.32 a	1.03 ± 0.84 b

[1] Different letters in each treatment indicate statistically significant differences according to Tukey's HSD test at $p < 0.05$.

3.3. Modeling of the Light Environment on Spinach Leaf Surfaces Under Different Light Source Distances

Based on the results from Sections 3.1 and 3.2, we investigated the illuminance on the leaf surfaces of spinach models using TracePro under a rectangular planting arrangement at distances of 10 cm, 20 cm, and 30 cm from the LED lamps (Figure 9 and Table 5). When the distance from the lamps was 10 cm, the average illuminance of the leaf model was 3068.3 lux, with a total of 3815 incident light rays on the model surface, and the measured illuminance was 3112.8 lux. At a distance of 20 cm from the lamps, the average illuminance of the leaf model was 2365.2 lux, with 2838 incident light rays on the model surface, and the measured illuminance was 2417.8 lux. When the distance was 30 cm from the lamps, the average illuminance of the leaf model was 1775.8 lux, with 2130 incident light rays on the model surface, and the measured illuminance was 1812.2 lux.

Table 5. Simulated and measured values of average illuminance on spinach leaf surfaces under different light source distances (Klux).

	10 cm	20 cm	30 cm
Measured values	3.11 ± 0.12	2.42 ± 0.11	1.81 ± 0.35
Simulated values	3.07 ± 0.23	2.37 ± 0.27	1.78 ± 0.04

A one-sample *t*-test was conducted between the measured values at the three distances and the simulated values from TracePro software, with the results shown in Table 6. It can be seen that since the *p*-value is greater than the significance level of 0.05, there is insufficient evidence to reject the null hypothesis, indicating no statistically significant difference between the measured and simulated values. Moreover, when the light source distance is 30 cm, the resulting *p*-value of 0.977 is the highest.

Table 6. One-sample *t*-test of TracePro simulated values under different light source distances.

	10 cm	20 cm	30 cm
Confidence interval	−0.225, 0.145	−0.144, 0.257	−0.342, 0.441
Significance	0.884	0.928	0.977

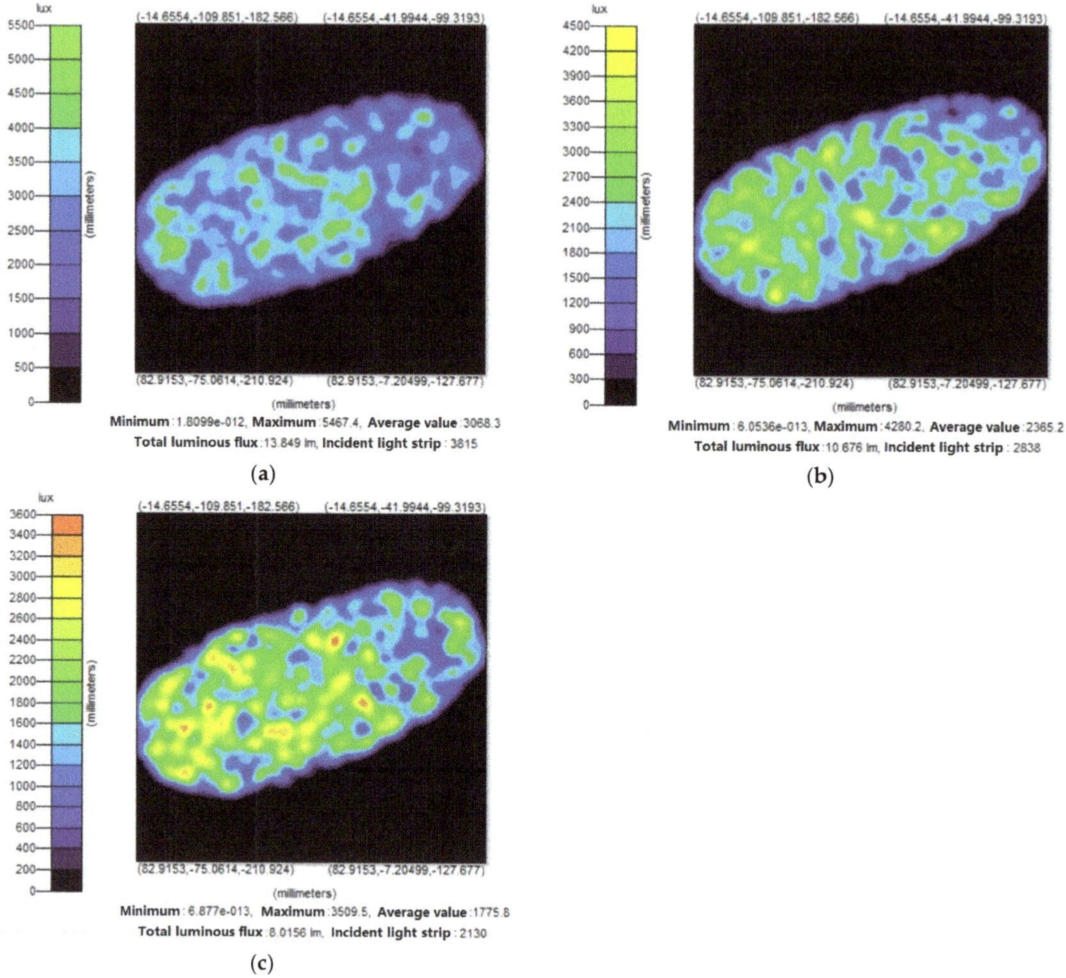

Figure 9. Analysis chart of illuminance on spinach leaf surfaces under different distances from light sources: (**a**) 10 cm, (**b**) 20 cm, (**c**) 30 cm.

4. Discussion

In this study, under the T2 rectangular cultivation arrangement, spinach leaves received more light, and the plants grew faster. This indicates that for strip-arranged LED light sources, the rectangular, i.e., evenly distributed, planting method is more conducive to the effective light energy reception by plant leaves. Under this mode, it is more beneficial for leaf photosynthesis, thereby achieving more material accumulation. This is consistent with previous research results [6].

When examining different plant planting arrangements, it was found that they had little impact on the simulation results of the same software. However, under the same planting arrangement, there were significant differences in the simulation results among the three software, among which the difference between the simulated values of TracePro and the measured values was the smallest. This should be caused by the differences in the degree of fuzzy calculation and robustness of different software for the light-receiving surface of plants. Previous studies have reported that when using different lighting analysis software and various algorithms to optimize the design of such items as LED lamp beads and array light sources, it was found that the square array arrangement under the particle swarm optimization algorithm is more suitable for plant lighting compared with circular and triangular arrays [31,32]. However, the results are only applicable to surface light sources. For mixed-color light sources, low illumination uniformity can lead to light spots on the illuminated surface at close range [32], thereby resulting in inconsistent growth light environments of the same batch of plants calculated by different software that mainly rely on direct light intensity calculation.

Further tracking and simulation of the leaf surface illuminance of spinach under the rectangular planting arrangement at different distances from the lamps using TracePro showed high accuracy, indicating that TracePro simulation software is most suitable for monitoring the changes in the light environment during spinach growth in plant factories. In our study, although the confidence intervals suggest that the difference may be close to zero (indicating no significant difference), a paired t-test was conducted to verify statistical significance, and the resulting p-value of 0.977 (when the light source distance is 30 cm) indicates no significant difference between the measured and simulated values. This demonstrates the strong application potential of this software. Based on a previous report, TracePro adopts the Monte Carlo ray tracing method, which can ensure that light rays on the surface of objects can be correctly analyzed [28]. In addition, he also pointed out that TracePro can not only edit multiple models in one program but also import model data from CAD software, providing conditions for the establishment of complex models [29]. Combined with its relatively simple operation interface design, the optical analysis function of TracePro is more suitable for application scenarios requiring intuitive simulation and rapid monitoring and should have good adaptability to the intelligent monitoring of various factors in the production environment of plant factories. In recent years, with the rapid development of smart agricultural production, high-fidelity imaging of the production environment has become an indispensable part of the monitoring process in intelligent production. When optical software is combined with new high-fidelity imaging transmission composite materials, it can maintain high imaging quality while transmitting a large amount of image data [33–36], which meets the requirements for imaging detail accuracy and data volume in various aspects such as crop phenotype monitoring and environmental data visualization in smart horticultural crop production.

In recent years, the digital plant revolution, represented by breakthroughs in high-definition imaging technology, has quietly arrived [37–47]. Various research achievements not only involve real-time monitoring and early warning of many aspects such as the growth [38], pollination [39], and diseases [40] of field crops like corn [41], rice [42,43], and wheat [44], but also their applications in the field of horticultural plants have gradually become popular [45–47]. In practical production applications, to find the most suitable illuminance for spinach growth, corresponding plant models and lamp models can be created using software, and a corresponding model library can be established for easy call-up at any time. Ray tracing analysis can be carried out with models created on demand, and the illuminance of spinach can be observed and analyzed through intuitive heatmaps. This study shows that this digital plant analysis method is fast and accurate, can intuitively present the illuminance on spinach leaf surfaces, and can provide rapid environmental monitoring data for spinach production in plant factories. Meanwhile, with the rapid development of smart agricultural production, high-fidelity imaging of the production environment has become an indispensable part of the monitoring process in intelligent production. When optical software is combined with new high-fidelity imaging transmission composite materials, it can maintain high imaging quality while transmitting a large amount of image data, which meets the requirements for imaging detail accuracy and data volume in various aspects such as crop phenotype monitoring and environmental data visualization in smart horticultural crop production. In this study, we investigated the surface light irradiation with software simulation. Though the software has considered the possible shade effect, most light rays were defined as direct rays, which might decrease the actual irradiation amount with the missing of scattered radiation. This means we need extra consideration of the light state in different canopies with digital optical technology developed. Meanwhile, the concept of light ray visualization attempts also can be applied to other plant production in plant factories, greenhouses, plastic tunnels, and even field production scenes. For example, when a greenhouse-produced high-quality cherry tomato turns to the third or fourth truss, irradiation exposure on both function leaves (mainly contributing to fruit enlargement) and fruit (quality formation) decreases significantly. Growers can adjust artificial light sources more intelligently (including supplemental lighting position, cycle, quantity, spectrum, etc.), thereby ensuring the high-quality production of products and high economic returns. The application of this ray tracing technology and optical simulation software in vegetable production will reduce labor input and time costs in production and promote the intelligentization of horticultural production. At the same time, during the modeling process of this study, it was found that none of the three software has a plant database, and the simulation process often takes a lot of time. This suggests that future research can focus on the establishment of general databases for various common horticultural crops so as to provide references for the standardization of digital technologies in smart horticultural plants.

5. Conclusions

In this paper, LightTools, TracePro, and Ansys Lumerical FDTD Solution are selected as the research objects to investigate their performance in simulating the light environment of a spinach leaf surface under different planting arrangements and different lamp source distances. Under the rectangular planting arrangement, the absolute value of illuminance was highest, indicating the leaves received more light, and the plants grew faster. Different plant planting arrangements had different impacts on the simulation results of the same software; there were significant differences between simulated value and measured value under rectangular planting arrangements and hexagonal planting arrangements with Light Tools. In addition, there were significant differences in the simulation results among the three software under the same planting arrangement, among which the difference between the simulated values of TracePro and the measured values was the smallest. When using TracePro, there was no significant difference between the simulated values of the software and the measured values, and the simulation accuracy was the highest when the distance from the light source was 30 cm.

Author Contributions: Conceptualization, C.J. and Y.S.; methodology, M.L. and T.P.; software, K.Z. and Y.M.; validation, K.Z. and Y.M.; formal analysis, M.L.; investigation, K.Z.; resources, Y.Z.; data curation, K.Z.; writing—original draft preparation, C.J. and Y.Z.; writing—review and editing, K.Z., Y.M. and T.P.; project administration, C.J. and W.L.; funding acquisition, Y.S. and W.L. All authors have read and agreed to the published version of the manuscript.

Funding: This research was funded by the Major Science and Technology Project of Xinjiang Uygur Autonomous Region (2022A02005-2), the National Science Foundation of China (32202581), and Independent Cultivation Project of Research Institute of Crop Germplasm Resources, Xinjiang Academy of Agricultural Sciences: PZ202302.

Institutional Review Board Statement: Not applicable.

Data Availability Statement: The original contributions presented in this study are included in the article. Further inquiries can be directed to the corresponding authors.

Conflicts of Interest: The authors declare no conflicts of interest.

References

1. Zou, T.; Huang, C.; Wu, P.; Ge, L.; Xu, Y. Optimization of artificial light for spinach growth in plant factory based on orthogonal test. *Plants* **2020**, *9*, 490. [CrossRef] [PubMed]
2. Cai, X.; Ge, C.; Wang, X.; Xu, C.; Wang, Q. Advances and perspectives in research of spinach breeding technology in China. *Jiangsu J. Agric. Sci.* **2019**, *35*, 996–1005. (In Chinese)
3. Akbari, F.; Mollaei, M.; Argani, P.; Daneshfard, B.; Derakhshan, A.R. *Spinacia Oleracea*: Exploring the therapeutic potential in persian medicine and modern pharmacology. *Curr. Drug Discov. Technol.* **2024**, *21*, e150224227025. [CrossRef] [PubMed]
4. Pan, Y.; Xu, X.; Lang, Q.; Liao, S.; Li, Y. Three different fertilizers enhance spinach growth and reduce spinach Cd concentration in Cd contaminated Alkaline Soil. *Horticulturae* **2023**, *9*, 445. [CrossRef]
5. Vatistas, C.; Avgoustaki, D.D.; Monedas, G.; Bartzanas, T. The effect of different light wavelengths on the germination of lettuce, cabbage, spinach and arugula seeds in a controlled environment chamber. *Sci. Hortic.* **2024**, *331*, 113118. [CrossRef]
6. Semenova, N.A.; Proshkin, Y.A.; Smirnov, A.A.; Dorokhov, A.S.; Ivanitskikh, A.S.; Burynin, D.A.; Dorokhov, A.A.; Uyutova, N.I.; Chilingaryan, N.O. The influence of the spectral composition and light intensity on the morphological and biochemical parameters of spinach (*Spinacia oleracea* L.) in vertical farming. *Horticulturae* **2023**, *9*, 1130. [CrossRef]
7. Ming, Z.; Lyu, Q.; Ming, Y.; Zeng, W.; Lyu, H.; Zhang, J. Optimization of LED light source array based on improved particle swarm algorithm. *J. Appl. Opt.* **2022**, *43*, 524. [CrossRef]
8. Fang, Y.C.; Tzeng, Y.F.; Li, S.X. Multi-objective design and extended optimization for developing a miniature light emitting diode pocket-sized projection display. *Opt. Rev.* **2008**, *15*, 241–250. [CrossRef]
9. Kozai, T. Plant factory in Japan-current situation and perspectives. *Chron. Horticult.* **2013**, *53*, 8–11.
10. Bourget, C.M. An introduction to light-emitting diodes. *Hortscience* **2008**, *3*, 1944–1946. [CrossRef]
11. Morrow, R.C. LED lighting in horticulture. *Hortscience* **2008**, *43*, 1947–1950. [CrossRef]
12. Kim, K.C.; Lee, J.; Kim, H.J.; Koo, D.H. Multiobjective optimal design for interior permanent magnet synchronous motor. *IEEE Trans. Magn.* **2009**, *45*, 1780–1783. [CrossRef]
13. Hwang, C.C.; Lyu, Y.; Liu, C.T.; Li, P.L. Optimal design of a spm motor using genetic algorithms and Taguchi method. *IEEE Trans. Magn.* **2008**, *44*, 4325–4328. [CrossRef]
14. Silva, M.M.A.; Oliveira, A.L.B.D.; Oliveira-Filho, R.A.; Gouveia-Neto, A.S.; Camara, T.J.R.; Willadino, L.G. Effect of blue/red LED light combination on growth and morphogenesis of saccharum officinarum plantlets in vitro. *SPIE* **2014**, *8947*, 342–348.
15. Zhang, S.; Ma, J.; Zou, H.; Zhang, L.; Li, S.; Wang, Y. The combination of blue and red LED light improves growth and phenolic acid contents in *Salvia miltiorrhiza Bunge*. *Ind. Crops Prod.* **2020**, *158*, 112959. [CrossRef]
16. Mitsanis, C.; Hurst, W.; Tekinerdogan, B. A 3D functional plant modelling framework for agricultural digital twins. *Comput. Electron. Agric.* **2024**, *218*, 108733. [CrossRef]
17. Onwude, D.; North, J.; Cronje, P.; Schouten, R.; Defraeye, T. Digital twins to quantify the impact of growing variability on the harvest quality of orange. *Sci. Hortic.* **2024**, *331*, 113129. [CrossRef]
18. Ganapathysubramanian, B.; Sarkar, S.; Singh, A.; Singh, A.K. Digital twins for the plant sciences. *Trend Plant Sci.* **2025**, *30*, 576–577. [CrossRef]
19. Teek, S.L.; Chen, S.X.; Gao, X.K. Robusttorque optimization for BLDC spindle motors. *IEEE Trans. Ind. Electron.* **2001**, *48*, 656–663.
20. Moreno, I. Illumination uniformity assessment based on human vision. *Opt. Lett.* **2010**, *35*, 4030–4032. [CrossRef]
21. Sun, C.C.; Moreno, I.; Lo, Y.C.; Chiu, B.C.; Chien, W.T. Collimating lamp with well color mixing of red/green/blue LEDs. *Opt. Express* **2012**, *20*, 75–84. [CrossRef]
22. Liu, P.; Wang, H.; Wu, R.; Yang, Y.; Zhang, Y.; Zheng, Z.; Li, H.; Liu, X. Uniform illumination design by configuration of LEDs and optimization of LED lens for large-scale color-mixing applications. *Appl. Opt.* **2013**, *52*, 3998–4005. [CrossRef]
23. Wang, H.C.; Chiang, Y.T.; Lin, C.Y.; Lu, M.Y.; Lee, M.K.; Feng, S.W.; Kuo, C.T. All-reflective RGB LED flashlight design for effective color mixing. *Opt. Express* **2016**, *24*, 266–271. [CrossRef] [PubMed]
24. Zheng, Z.; Ma, Z.; Li, R.; Wei, H.; Chu, X.; Liu, H. Design method for improving illumination distribution uniformity of LED plant growth light source. In Proceedings of the 2021 18th China International Forum on Solid State Lighting & 2021 7th International Forum on Wide Bandgap Semiconductors (SSLChina: IFWS), Shenzhen, China, 6–8 December 2021; pp. 202–210.
25. Wang, D.; Zou, J.; Luo, S.; Liu, W.; Li, Y.; Shi, M.; Li, Y.; Wang, Z.; Wang, Y.; Chen, Q.; et al. Study on uniformity and thermal stability of strip LED lamps in pea growing applications. *Measurement* **2025**, *245*, 116604. [CrossRef]
26. Zhang, S.; Wen, S.; Ma, B.; Pang, P.; Chen, H.; Cai, M.; Fu, M.; Hou, Y.; Zou, X. High uniformity LED panel-light for plant lighting. *Chin. J. Lumin.* **2018**, *39*, 403–413. [CrossRef]
27. Wen, Z.; Luo, S.; Zhu, L.; Wei, H.; Cai, S.; Li, M.; Ma, Z. Optimization and verification of the arrangement of lamp beads for LED light source for plant growth based on TracePro. *J. Zhongkai Univ. Agric. Eng.* **2020**, *33*, 44–49, (In Chinese with English Abstract).
28. Pan, G.; Wu, K.; Zhou, J. Application study of TracePro in LED lighting design. *J. Chizhou Univ.* **2023**, *37*, 16–18. (In Chinese)

29. Zhang, H.; Jiang, M.; Jia, D.; Shi, Y.; Zhang, L. Simulation design for teaching multimode fiber imaging features based on TracePro modeling. *Intel. Comp. Appl.* **2025**, *7*, 1–11. (In Chinese with English Abstract)
30. Zhang, Y.; Zhu, C.; Zhang, R. Simulation experiment of electromagnetic properties in metallic rectangle waveguide based on FDTD Solutions. *Exp. Tech. Manag.* **2020**, *37*, 161–165.
31. Deng, Y.; Xu, X.; Su, Y.; Song, T.; Wang, L.; Fang, Z. Study on design and uniformity of LED plant tissue culture array. *Jiangsu Agri. Sci.* **2017**, *45*, 225–228. (In Chinese)
32. Tang, H.; Wen, S.; Fu, M.; He, G.; Zhang, H.; Liao, S.; Kang, L. Design of LED plant lighting source based on particle swarm optimization algorithm under photons system. *Chin. J. Lumi.* **2019**, *40*, 340–348. (In Chinese)
33. Psaltis, D.; Moser, C. Imaging with multimode fibers. *Opt. Photonics News* **2016**, *27*, 24–31. [CrossRef]
34. Rehan, M.; Chowdhury, R.; Biswas, P.; Kang, M.S.; Varshney, S.K. Low-threshold cascaded Raman scattering and intermodal four-wave mixing in cascaded multimode fiber system. *J. Light. Technol.* **2024**, *42*, 4636–4642. [CrossRef]
35. Rahmani, B.; Oguz, I.; Tegin, U.; Hsieh, J.; Psaltis, D.; Mose, C. Learning to image and compute with multimode optical fibers. *Nanophotonics* **2022**, *11*, 1071–1082. [CrossRef] [PubMed]
36. Liu, Z.; Wang, L.; Meng, Y.; He, T.; He, S.; Yang, Y.; Wang, L.; Tian, J.; Li, D.; Yan, P.; et al. All-fiber high-speed image detection enabled by deep learning. *Nat. Commun.* **2022**, *13*, 1433. [CrossRef] [PubMed]
37. Ren, M.N.; Yu, X.J.; Mujumdar, A.S.; Yagoub, A.A.; Chen, L.; Zhou, C.S. Visualizing the knowledge domain of pulsed light technology in the food field: A scientometrics review. *Innov. Food Sci. Emerg. Technol.* **2021**, *74*, 102823. [CrossRef]
38. Tang, L.D.; Syed, A.U.; Otho, A.R.; Junejo, A.R.; Tunio, M.H.; Hao, L.; Ali, M.N.H.A.; Brohi, S.A.; Otho, S.A.; Channa, J.A. Intelligent papid asexual propagation technology-a novel aeroponics propagation approach. *Agronomy* **2024**, *14*, 2289. [CrossRef]
39. Wu, S.; Liu, J.Z.; Lei, X.J.; Zhao, S.Y.; Lu, J.J.; Jiang, Y.X.; Xie, B.B.; Wang, M. Research progress on efficient pollination technology of crops. *Agronomy* **2022**, *12*, 2872. [CrossRef]
40. Wan, L.; Li, H.; Li, C.S.; Wang, A.C.; Yang, Y.H.; Wang, P. Hyperspectral sensing of plant diseases: Principle and methods. *Agronomy* **2022**, *12*, 1451. [CrossRef]
41. Niu, Y.X.; Han, W.T.; Zhang, H.H.; Zhang, L.Y.; Chen, H.P. Estimating maize plant height using a crop surface model constructed from UAV RGB images. *Biosyst. Eng.* **2024**, *241*, 56–67. [CrossRef]
42. Sun, J.; Lu, X.Z.; Mao, H.P.; Wu, X.H.; Gao, H.Y. Quantitative determination of rice moisture based on hyperspectral imaging technology and BCC-LS-SVR algorithm. *J. Food Process Eng.* **2017**, *40*, e12446. [CrossRef]
43. Lu, X.Z.; Sun, J.; Mao, H.P.; Wu, X.H.; Gao, H.Y. Quantitative determination of rice starch based on hyperspectral imaging technology. *Int. J. Food Prop.* **2017**, *20*, S1037–S1044. [CrossRef]
44. Chen, X.X.; Tang, Y.L.; Duan, Q.F.; Hu, J.P. Phenotypic quantification of root spatial distribution along circumferential direction for field paddy-wheat. *PLoS ONE* **2023**, *18*, e0279353. [CrossRef] [PubMed]
45. Li, M.Q.; Li, J.Y.; Mao, H.P.; Wu, Y.Y. Diagnosis and detection of phosphorus nutrition level for *Solanum lycopersicum* based on electrical impedance spectroscopy. *Biosyst. Eng.* **2016**, *143*, 108–118. [CrossRef]
46. Yan, T.; Zhu, H.; Sun, L.; Wang, X.; Ling, P. Investigation of an experimental laser sensor-guided spray control system for greenhouse variable-rate applications. *Trans. ASABE* **2019**, *62*, 899–911. [CrossRef]
47. Wang, J.Z.; Zhang, Y.; Gu, R.R. Research status and prospects on plant canopy structure measurement using visual sensors based on three-dimensional reconstruction. *Agriculture* **2020**, *10*, 462. [CrossRef]

agriculture

MDPI

Article

Effects of Seed Priming with *Talaromyces ruber* Extracts on Tomato (*Solanum lycopersicum*) Growth

Álvaro Iglesias-Ganado [1,2], Jorge Poveda [1,2], Oscar Santamaría [1,2], Sara Rodrigo [3], María I. Pozo [1,4] and Jorge Martín-García [1,2,*]

1 Recognized Research Group Agrobiotech, UIC-370 (JCyL), Higher Technical School of Agricultural Engineering of Palencia, University of Valladolid, 34004 Palencia, Spain; alvaro.iglesias.ganado@uva.es (Á.I.-G.); jorge.poveda@uva.es (J.P.); oscar.santamaria@uva.es (O.S.); mariaisabel.pozo@uva.es (M.I.P.)
2 Department of Plant Production and Forest Resources, University Institute for Research in Sustainable Forest Management (iuFOR), University of Valladolid, Avda. Madrid 57, 34004 Palencia, Spain
3 Dehesa Research Institute (INDEHESA), University of Extremadura, Badajoz, Avda. Elvas s/n, 06006 Badajoz, Spain; saramoro@unex.es
4 Department of Agroforestal Sciences, University of Valladolid, Avda. Madrid 57, 34004 Palencia, Spain
* Correspondence: jorge.marting@uva.es

Abstract: Modern agriculture requires effective and sustainable tools to enhance crop performance while minimizing the environmental impact. In this context, the application of fungal-derived bioactive compounds directly onto seeds represents a promising alternative. In this study, tomato seeds (*Solanum lycopersicum*) were subjected to mycopriming treatment using two fungal extracts obtained from the mycelium and culture filtrate of *Talaromyces ruber*. Two independent greenhouse trials were conducted to assess germination dynamics, morphometric traits, and physiological parameters (chlorophyll content, flavonol index, and anthocyanin index). Although germination rates were not significantly affected, root development was consistently enhanced by the treatments compared with the control group in both experiments. In contrast, no clear improvement was observed in shoot growth or leaf physiological parameters. Overall, the application of *T. ruber* extracts via seed priming proved to be a feasible strategy to stimulate early-stage root development in tomatoes, potentially contributing to improved seedling vigor and agronomic performance. These findings support the potential use of fungal extracts as practical tools for improving seedling quality in commercial nursery production.

Keywords: plant growth promotion; fungal secondary metabolites; mycopriming; sustainable agriculture; nursery production

1. Introduction

Modern agriculture faces the complex challenge of, on the one hand, maintaining the levels of quality required by the market yet responding to the rapidly growing global food demand [1]. On the other hand, the environment poses additional pressures [2] and threats that are magnified by ongoing climatic change [3]. Given these challenges, the scientific community and the agri-food sector are seeking innovative strategies that can maintain or improve crop productivity without compromising environmental sustainability [4].

One of the most valuable crops affected by these constraints is the tomato, the world's most widely cultivated vegetable, with 5.4 million hectares of cultivated area and a global production of 192 million tons [5]. This crop faces global agricultural challenges, along with specific threats such as the proliferation of diseases caused by fungi, bacteria, nematodes, and viruses [6], possibly driven by the intensification of agricultural systems [7]. These biotic stresses are further aggravated by abiotic factors, including extreme summer temperatures that negatively affect tomato growth and fruit quality [8]. Moreover, soil salinity, caused by irrigation with saline water, reduces the plant's ability to absorb water and nutrients, ultimately leading to reduced plant growth and yield [9]. Altogether, these conditions may compromise the sustainability of tomato cultivation in traditional production areas, potentially leading to significant economic losses [10].

In this worrisome scenario, improving plant fitness during the early stages of development, particularly while seedlings are still in the horticultural nursery, might relieve most of the current production threats listed above. The rationale behind this improvement is that more vigorous plants may exhibit greater tolerance to the previously described biotic and abiotic stresses once they leave the protection of the nursery to be transplanted in the field [11]. There are, however, several biotechnological strategies that might be applied to plant reproductive material to improve plant development in the early stages. Among these strategies, seed priming stands out in terms of novelty. It consists of a pre-sowing treatment in which seeds are partially hydrated to activate metabolic processes prior to germination, with the primary goal of enhancing germination performance and seedling establishment, ultimately contributing to improved crop productivity [12].

Seed priming can be amended by using plant growth-promoting microorganisms (PGPMs) [13]. When fungi are used, the technique is referred to as mycopriming [14]. However, the application of PGPMs in agriculture remains challenging under realistic field conditions. Inconclusive field trials occur sometimes due to the difficulty to control external variables that affect the establishment of microorganisms in the soil or in the plant, and, therefore, the expected benefits are not consistently observed [15]. Another issue associated with the direct inoculation of living microorganisms is their interaction with the pre-existing microbiome in the environment. In certain cases, introduced microorganisms alter the structure and composition of soil or rhizosphere microbial communities, displacing beneficial species or disrupting key ecological functions [16,17]. Additionally, some endophytic fungi may express pathogenicity in crops [18], and their capacity to cause infections in humans has also been documented [19]. All these issues have led regulators, such as those in the European Union, to impose strict controls on the use of microorganisms in agriculture [20–22].

It has been demonstrated that the beneficial effects of PGPMs are often attributed to the production of a series of secondary metabolites [23]. Secondary metabolites (SMs) are bioactive organic compounds not directly involved in the microorganism's primary metabolism but that play crucial roles in the interaction of the plant subject with its environment, e.g., by affecting plant–microbe interactions, enhancing stress tolerance, or providing protection against pathogens [24]. These compounds can be produced under laboratory conditions by extracting them from a fungal culture grown in a nutrient medium. Consequently, an alternative approach to mycopriming could involve the in vitro production of these bioactive compounds and their direct application to seeds [25].

The application of fungal extracts and, in particular, SMs, from endophytic fungi has been mainly explored via foliar spraying [26], but the main practical and economical limitation of this method is the large quantity of extract that is required [25]. In contrast, their application as a seed treatment has not been extensively investigated, but it represents a promising approach. This is particularly relevant in crops like the tomato, where the small size of the seeds allows a large number of them to be treated with a minimal amount of bioactive compounds, making this technique both economically and technically feasible [27].

The mycopriming technique using SMs extracted from endophytic fungi has been tested successfully as a plant growth promoter. This technique involves soaking seeds in a solution containing SMs produced by fungi. Through this process, the seeds absorb these compounds, which can produce beneficial effects on germination and early development [28]. Extracts from endophytic fungi have been shown to improve germination rates in various species [29,30] and have also been successful in enhancing early growth in several crops [31,32]. There is, however, wide variation in the production of bioactive compounds among fungal taxa. The genus *Talaromyces* stands out as a producer of bioactive compounds with potential interest in agriculture, including indole-3-acetic acid (IAA), phosphate-solubilizing metabolites, siderophores, and lytic enzymes, such as chitinases and β-glucanases, which contribute to plant growth promotion and biocontrol activity [33–35]. Within the genus, *Talaromyces ruber*—a species previously classified as *Penicillium rubrum* [36]—is a proven source of agriculturally relevant bioactive metabolites such as *Rubratoxin B* [37].

The aim of this study was to evaluate the PGP effect of seed treatments using extracts obtained from the endophytic fungus *T. ruber* on tomato seedlings during their early developmental stages. The hypothesis is that enhancing seedling fitness during these stages may increase their tolerance to various biotic and abiotic stresses encountered throughout the crop cycle which should be confirmed in future trials. This approach could provide a sustainable, fully organic solution for seed producers and nurseries to increase the value of their products, aligning with current agricultural policies that promote the use of environmentally friendly inputs.

2. Materials and Methods

2.1. Fungal and Plant Material

The endophyte *T. ruber* strain EN3 belongs to the microbial collection of the GIR AGROBIOTECH (Palencia, Spain). The taxonomic identity of the fungal isolate *T. ruber* EN3 was identified by sequencing the EF1α and ITS regions. The isolate was routinely cultured onto potato dextrose agar plates (PDA, Scharlab, Spain) and incubated in dark conditions at 25 °C for 7 days. For long-term preservation, cultures were stored at 4 °C and periodically subcultured to maintain viability.

Commercial seeds of tomatoes (*S. lycopersicum*), cultivar Optima F1, were supplied by Seminis (Oxnard, CA, USA). The seeds were kept in heat-sealed multilayer aluminum foil pouches designed to safeguard the viability of the seeds during storage, protecting them against moisture, light, and oxygen. Seeds were stored at 4 °C until use. All experimental procedures were performed in 2025.

2.2. Obtaining Fungal Extracts from T. ruber

Four 5 mm mycelial disks were excised from an actively growing colony of *T. ruber* on potato dextrose agar (PDA, Scharlab) and used to inoculate 400 mL of potato dextrose broth (PDB, Scharlab) in 500 mL Erlenmeyer flasks. The flasks were kept at room temperature (20–25 °C), in darkness, and shaken at 140 rpm using an orbital shaker (Edmund Bühler KS-15, Bodelshausen, Germany). During incubation, glucose concentration in each flask was measured daily using semi-quantitative rapid test strips with a detection range of 50–2000 mg·L^{-1} (Quantofix®, Macherey-Nagel, Düren, Germany). When glucose was nearly depleted (<100 mg·L^{-1}), the extraction of SMs was carried out.

Two different sources were used for the extraction: the liquid filtrate and the mycelium, in order to recover a wider range of metabolites. For the extraction, the method described by García-LaTorre [28] was followed. The fungal culture was gravity-filtered using a qualitative filter, pore size 10–12 μm (Prat Dumas), to separate the mycelium from the liquid filtrate. The filtered solution was mixed with an equal volume of ethyl acetate 99.8% GLR (ETAC-POP-5K0, Labkem, Barcelona, Spain) in a separation funnel and shaken for 2 min. Once phase separation was complete, the aqueous phase was discarded. To remove residual water, sodium sulfate anhydrous EPR Reag. (SOSU-00T-1K0, Labkem, Barcelona, Spain) was added to the ethyl acetate phase, and the same filters used previously were employed to remove sodium sulfate residues. The organic solvent phase was then evaporated using a rotary evaporator (BUCHI R-80, Flawil, Switzerland). The resulting crude extract was resuspended in methanol 99.8% GLR (MTOL-POP-1K0, Labkem, Barcelona, Spain) and concentrated using a vial evaporator (DNA 120 Speed Vac, Eppendorf, Hamburg, Germany). Finally, to quantify the amount of fungal extract obtained, the empty vials were first weighed and subsequently reweighed after the evaporation process in the SpeedVac using a four-decimal precision analytical balance. The final volume was then adjusted with sterile water to obtain a concentration of 1 mg/L.

The mycelial part was mixed with acetone 99.6% GLR (ACET-POP-1K0, Labkem, Barcelona, Spain) and submerged in an ultrasonic bath at 40 °C and 35 kHz for 10 min to separate the metabolites from the mycelium. The content was subsequently filtered again using qualitative filters with a pore size of 10–12 μm, and the acetone was evaporated with a rotary evaporator. The resulting solid was resuspended in 50 mL of distilled water and mixed with an equal volume of ethyl acetate. From this point, the same extraction steps described above for the filtrate were followed, including the aqueous phase removal, drying with sodium sulfate, filtration, ethyl acetate evaporation, and methanol resuspension (Figure 1).

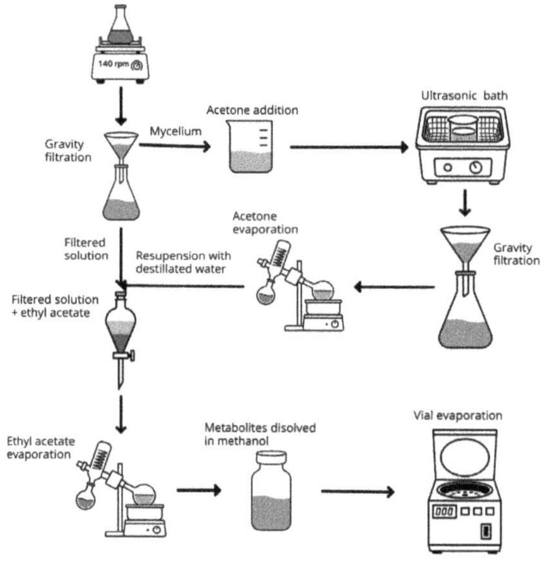

Figure 1. Schematic workflow of the extraction of SMs from culture filtrate and mycelial biomass of *T. ruber*.

2.3. Seed Treatment

Seeds were surface sterilized by immersion in 70% ethanol for 1 min, followed by immersion in 2% sodium hypochlorite for 1 min. They were then rinsed three times with sterile distilled water.

SMs were resuspended in a 2.5% (*v*/*v*) dimethyl sulfoxide (DMSO; Thermo Fisher Scientific, Waltham, MA, USA) solution prepared with sterile distilled water, due to the high compatibility of DMSO with plant tissues [38]. The fungal extract suspensions were adjusted to a final concentration of 1 mg mL^{-1}. The same protocol was applied to both the culture filtrate and the mycelium extracts.

A 2.5% DMSO solution (without SMs) was used as the control to account for any effects of the solvent itself. Tomato seeds were soaked in the treatment solutions for 12 h at room temperature under continuous shaking using an orbital shaker to ensure the uniform uptake of metabolites by the seed tissues. After soaking, the seeds were dried on sterile filter paper for 24 h in a laminar flow hood to maintain sterile conditions (Figure 2).

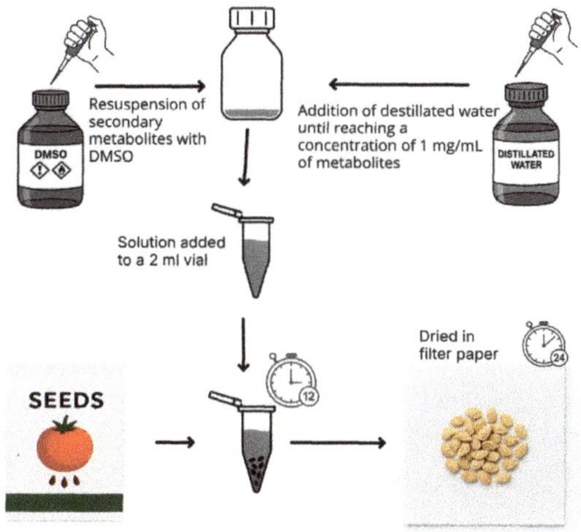

Figure 2. Schematic representation of the seed treatment process (mycopriming) using SMs extracts obtained from *T. ruber*.

2.4. In Plant Experiments

The effect of the *T. ruber* extracts obtained from both the filtrate and the mycelium was evaluated in two independent greenhouse trials conducted twice under the same methodological conditions (greenhouse temperatures ranged between 15 °C and 25 °C). Tomato seeds primed with extracts were sown in pots (44 × 54 × 49 mm) in seed trays and placed in a greenhouse. Twenty-four biological replicates for each treatment were used with a total of three treatments evaluated. The substrate used was Terrahum (Klassman-Deilmann, Geeste, Germany) with an organic matter = 90%, electrical conductivity (EC) = 0.4 dS m^{-1}, and pH = 5.5–6.5. Irrigation was applied weekly to maintain the substrate at field capacity, and no nutrient solution was added during the experiment. The greenhouse was equipped with an automatic temperature control system that maintained conditions with a minimum of 15 °C and a maximum of 25 °C. Additionally, artificial lighting (estimated photosynthetic photon flux density, PPFD, of 200–300 $\mu mol \cdot m^{-2} \cdot s^{-1}$) was used and regulated to provide a 16 h photoperiod. The climatic data of humidity, temperature, and radiation recorded in the greenhouse are shown in Table S1.

2.5. Morphometric and Physiological Characteristics Analyzed

Seedling emergence was monitored daily to determine the germination rate and to evaluate whether the treatments had any effect on the timing of emergence. Shoot length was measured at 14, 21, 28, 35, and 42 Days After Sowing (DAS), corresponding to BBCH stages 09, 10, 11, 12, and 13 in the first trial and 09, 10, 11, 13, and 14 in the second trial, according to the BBCH scale for tomatoes [39]. Shoot length was recorded from the substrate surface to the insertion point of the last visible leaf. At 42 DAS, the experiment was concluded, and root length, shoot dry weight, and root dry weight were measured.

Additionally, at 42 DAS, chlorophyll content, flavonol index, and anthocyanin index were measured using a portable multiparameter optical sensor Dualex-4 (Dualex Scientific+™ Polyphenol & Chlorophyll Meter, FORCE-A, Orsay, France). This device is equipped with a leaf clip that enables non-destructive measurements to be taken directly from the plant. To standardize the results, measurements were conducted on the last fully expanded leaf. All measurements were carried out in blocks to minimize potential errors, as the sensor is highly sensitive to environmental conditions, such as solar radiation and humidity, which may vary over time.

2.6. Statistical Analysis

A Kaplan–Meier survival analysis was applied to assess differences in germination dynamics among treatments. Survival curves were generated in R software (version 4.4.0, R Core Team, Vienna, Austria) using the Survfit function from the "Survival" package [40]. Differences among treatments were evaluated using the log-rank test applied to Kaplan–Meier germination curves. A one-way analysis of variance (ANOVA) was performed to evaluate the effects of treatments on plant height, biomass, and physiological parameters. When significant differences were detected, Tukey's test (HSD) was used for multiple comparisons at a 95% confidence level ($p < 0.05$). The assumptions of normality and homogeneity of variances were verified using the Shapiro–Wilk and Bartlett tests, respectively. All statistical analyses were performed using R software environment (R Foundation for Statistical Computing, Vienna, Austria).

3. Results

3.1. Effect of the Seed Treatments on Germination Rates

In all treatments, seedling emergence was mostly concentrated between days 5 and 9. The survival curves did not reveal significant differences among treatments neither in the first trial ($\chi^2 = 2.784$; $p = 0.25$) nor in the second ($\chi^2 = 4.753$; $p = 0.093$).

3.2. Effect of the Seed Treatments on the Morphometric and Physiological Characteristics Analyzed

To gain deeper insight into the PGP potential of the treatments, the analysis was extended to shoot and root development parameters during the seedling stage. Weekly measurements of the shoot length were conducted. During the first three assessments, no statistically significant differences were observed among treatments in either trial (Table 1). However, in the first trial, the seeds primed with mycelium extracts showed a significant increase in shoot length during weeks 5 and 6 compared with the control, an effect that could not be replicated in the second trial (Figure 3). Significant differences in the shoot dry weight were also detected among treatments (Table 1), and the use of filtrate extract led to a significant improvement relative to the control (Figure 4). Regarding root traits, both the length and dry weight showed significant variation among treatments in both trials (Table 1). Seedlings derived from seeds treated with both types of extract (the first derived from the filtrate and the second from the mycelium extract) developed longer roots than the control treatment. In terms of root biomass, both treatments produced significant increases in the first trial (Figure 5), while in the second trial, only the mycelium extract remained effective (Figure 6). No significant differences were observed in the chlorophyll content, flavonol index, or anthocyanin index among treatments (Table 1). This indicates that the application of *T. ruber* extracts did not affect these physiological parameters in the aerial part of the seedlings, at least in the variables measured and within the duration of the experiment (Table S2).

Table 1. Statistical information from ANOVA tests corresponds to the influence of seed treatments on shoot length (14, 21,28, 35, and 42 DAS), and the root length, shoot dry weight, root dry weight, chlorophyll content, flavonol index, and anthocyanin index of tomato seedlings at 42 days. DAS = Days After Sowing.

Analyzed Parameters	Degrees of Freedom	First Trial F-Value	First Trial *p*-Value	Second Trial F-Value	Second Trial *p*-Value
Shoot length 14 DAS	2	0.78	0.466	5.03	0.099
Shoot length 21 DAS	2	0.67	0.515	3.08	0.224
Shoot length 28 DAS	2	1.16	0.323	2.38	0.21
Shoot length 35 DAS	2	3.79	0.03	3.34	0.165
Shoot length 42 DAS	2	3.64	0.034	1.23	0.753
Shoot dry weight	2	3.50	0.04	1.05	0.355
Root length	2	12.73	<0.001	11.22	<0.001
Root dry weight	2	34.78	<0.001	3.48	0.037
Chlorophyll content ($\mu g/cm^2$)	2	0.02	0.979	0.33	0.722
Flavonol Index	2	0.76	0.472	0.97	0.386
Anthocyanins Index	2	0.58	0.564	0.08	0.923

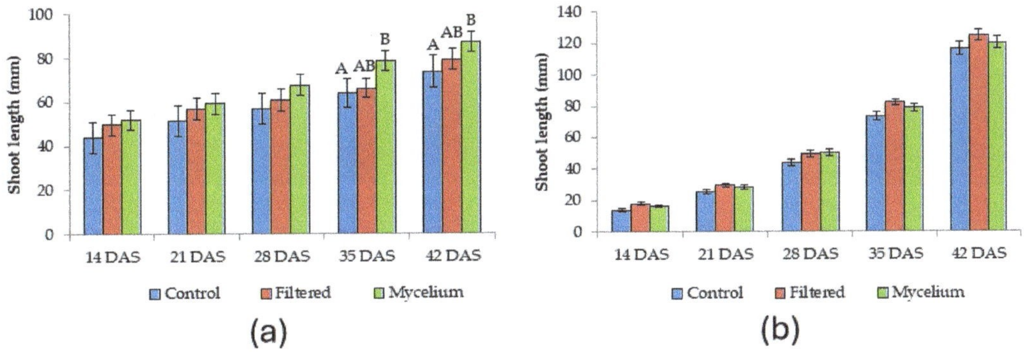

Figure 3. Effect of seed treatments on the development of shoot length of tomato seedlings. Bars depict mean ± standard error. Different letters indicate significant differences according to Tukey's HSD test ($p < 0.05$). (**a**) First trial; (**b**) second trial. DAS = Days After Sowing.

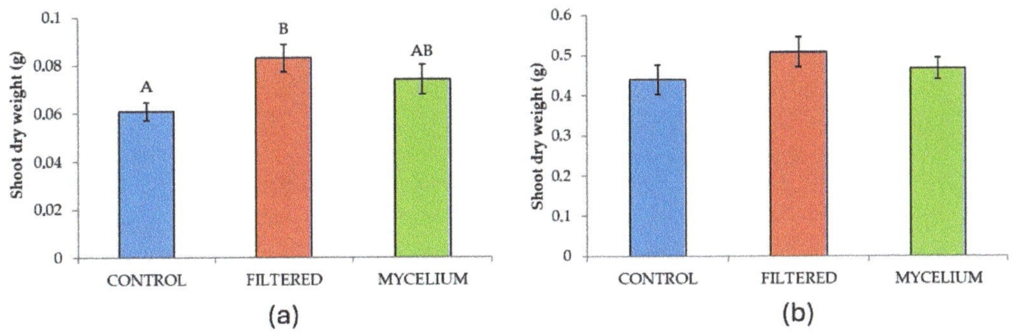

Figure 4. Effect of seed treatments on the shoot dry weight at 42 DAS. Values are means ± standard error. Different letters indicate significant differences according to Tukey's HSD test ($p < 0.05$). (**a**) First trial; (**b**) second trial.

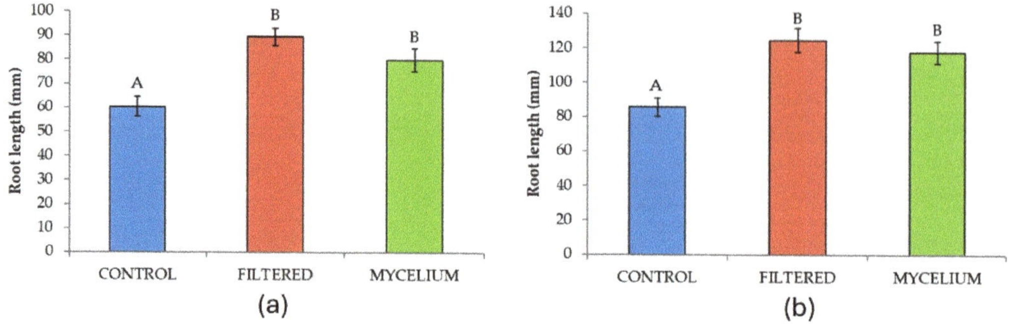

Figure 5. Effect of seed treatments on the root length at 42 DAS. Values are means ± standard error. Different letters indicate significant differences according to Tukey's HSD test ($p < 0.05$). (**a**) First trial; (**b**) second trial. DAS = Days After Sowing.

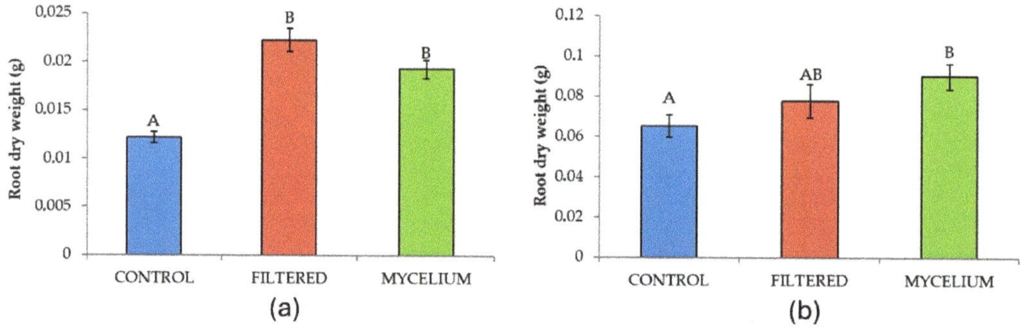

Figure 6. Effect of seed treatments on the root dry weight at 42 DAS. Values are means ± standard error. Different letters indicate significant differences according to Tukey's HSD test ($p < 0.05$). (**a**) First trial; (**b**) second trial. DAS = Days After Sowing.

4. Discussion

We evaluated the potential plant growth-promoting effects of extracts derived from the mycelium and liquid filtrate of *T. ruber*, obtained through organic solvent extraction and rotary evaporation. The extracts were applied to tomato seeds via mycopriming, and the effects were subsequently assessed in seedlings grown under controlled greenhouse conditions.

The Kaplan–Meier survival curves showed no statistically significant differences in germination rates. The use of high-quality certified seeds with a naturally high germination rate may have masked the potential beneficial effects of the treatments. Therefore, we could not confirm the positive influence reported in previous studies for seed priming with fungal extracts [29,30,41]. It is worth noting, however, that no signs of germination inhibition were observed in our trial, in contrast to previous findings that reported negative effects on the germination percentage [42,43].

We were interested in exploring treatment effects that lay beyond germination. Therefore, we analyzed the morphological traits of seedlings. In the first trial, the shoot length at 35 and 42 DAS was significantly increased by the mycelium treatment. This effect was not observed in the second trial, suggesting that the response to this treatment may vary under different conditions. The final measurements revealed the most pronounced outcomes. Focusing on the shoot, the increase in biomass observed in both treatments remained a trend that did not reach statistical significance. The consistency of the numerical increase across replicates may suggest that we should not rule out the possibility of a Type II statistical error. The greater shoot elongation observed in this second trial may have led to biomass dilution, potentially attenuating differences in the dry weight. It should be noted that improvements in root development do not necessarily lead to increases in shoot length or shoot dry weight, as previously reported in tomatoes [44,45]. This lack of correlation arises from the fact that they are generally governed by different physiological mechanisms. While shoot elongation is mainly driven by cell expansion, which is controlled by phytohormones such as auxins and gibberellins [46,47], shoot dry weight reflects the net accumulation of biomass, resulting from the balance between photosynthetic activity and assimilate allocation [48,49]. Consequently, these two parameters may respond differently to treatments. This distinction aligns with the results observed in the present study, where different treatments had varying effects on elongation and biomass accumulation. The results of this experiment are consistent with previous studies reporting improvements in the shoot length and biomass using similar techniques to those employed in this study, such as the application of *Aspergillus fumigatus* extracts in *Glycine max* [30], *Epicoccum sorghinum* extracts in tomatoes [50], or *Aspergillus niger* extracts in various *Vigna* species [51].

The most promising results of this trial were observed in the root system, as both the mycelium and filtrate treatments significantly improved the root length and root dry weight compared to the control. These findings are consistent with previous studies that also reported enhanced root development following fungal-based seed treatments, with comparable results in tomatoes using extracts of *Alternaria leptinellae* [28] or *Glutamicibacter halophytocota* [28,52], as well as in other crops such as the *Vigna* species with *Aspergillus niger* [51] or *Pennisetum glaucum* [53] with various species of *Trichoderma*. While the use of SMs extracted from the genus *Talaromyces* for seed treatment had not been studied so far, the direct inoculation of seeds with *Talaromyces* spp. has previously been tested, producing positive effects such as increased root length [54], which is consistent with the findings of the present work. It appears that the SMs derived from *T. ruber* may induce a shift in resource allocation [55], promoting root development without altering shoot phenology, which is regulated by thermal time accumulation [56]. This physiological adjustment may enable the production of seedlings at an optimal phenological stage for transplanting, while also enhancing root system traits that may benefit subsequent crop performance. This improvement could positively influence several key agronomic parameters in tomato cultivation, such as strengthening the physical anchorage of the plant to the soil, thereby reducing the risk of lodging or stem mechanical stress and improving overall plant stability [57]. Moreover, a larger root system increases the efficiency of water and nutrient uptake, potentially contributing to an improved tolerance to drought stress [58].

Regarding the physiological parameters evaluated, no significant differences were detected in chlorophyll, flavonol, or anthocyanin content. Similar results were also obtained by García-La Torre [32], who observed no significant changes in chlorophyll content despite an increase in root length following seed treatment with SMs extracted from *Fusarium avenaceum*, *Sarocladium terricola*, and *Xylariaceae* sp. A comparable pattern was observed in *Phaseolus vulgaris* [59] by foliar application of the SMs of *Alternaria sorghi*.

Our results, taken together, suggest that mycopriming with *T. ruber* extracts can positively influence vegetative development in tomato seedlings during the early stages prior to transplanting, particularly by enhancing root growth, which is associated with several agronomic advantages previously described. The reproducibility of this response was supported by the consistent results obtained in two independent trials. The hypothesis that seedlings with improved early vigor may exhibit a greater tolerance to subsequent biotic and abiotic stresses will be addressed in future research. Among the two formulations tested, the filtered extract appears to be the most viable option for practical implementation, as it involves a simpler extraction process and requires fewer inputs, thereby facilitating scalability and integration into industrial applications. Although it is highly probable that the growth-promoting effects of fungal extracts are due to the presence of secondary metabolites, further studies are needed to identify which specific metabolites are actually responsible for this biostimulant effect. Overall, this approach represents a sustainable strategy, both environmentally and economically, with a strong potential for integration into commercial nursery practices. Its simplicity of application and cost-effectiveness make it a promising tool for seedling producers, contributing to the production of more vigorous transplants aligned with the principles of sustainable agriculture.

5. Conclusions

Our study revealed the potential of the mycopriming technique using fungal extracts from endophytic fungi, specifically from the mycelium and filtrate of *T. ruber*, as a plant growth-promoting strategy in tomato cultivation. Although no significant improvements in the germination rate were observed, the treatments consistently showed a trend toward greater root system development in both independent trials. The reproducibility of this effect reinforces the robustness of the observed response. In contrast, no clear significant improvements were detected in shoot-related traits, including shoot length, aerial biomass, and leaf physiological parameters such as chlorophyll, flavonol, or anthocyanin content. This suggests the targeted stimulation of root development without substantial effects on the shoot during the early stages. These findings appear to support the hypothesis that seed treatments based on fungal extracts can effectively enhance early plant fitness. The application of these compounds directly to seeds may represent a viable alternative to the use of direct live microorganisms, which often face challenges related to field establishment, environmental variability, and regulatory constraints. This approach could offer a simple, scalable, and economically viable tool for commercial nurseries aiming to produce more vigorous tomato seedlings. Future studies should explore whether the improvements observed at the seedling stage translate into a better crop performance, greater resilience to biotic and abiotic stresses, and yield benefits after transplanting.

Supplementary Materials: The following supporting information can be downloaded at: https://www.mdpi.com/article/10.3390/agriculture15171868/s1, Table S1. Effect of seed treatments on the chlorophyll content, flavonol index, and anthocyanins index of tomato seedlings at 42 DAS. Values are means ± standard error. DAS = Days After Sowing. Table S2. Daily mean temperature, humidity, and radiation inside the greenhouse during the experiments. DAS = Days After Sowing.

Author Contributions: Conceptualization, Á.I.-G. and J.M.-G.; methodology, Á.I.-G., S.R. and J.M.-G.; formal analysis, Á.I.-G. and J.M.-G.; investigation, Á.I.-G. and J.M.-G.; writing—original draft preparation, Á.I.-G., J.P. and J.M.-G.; writing—review and editing, Á.I.-G., J.P., O.S., S.R., M.I.P. and J.M.-G.; funding acquisition, J.M.-G. and M.I.P. All authors have read and agreed to the published version of the manuscript.

Funding: This publication is part of the R&D project "PID2022-142403OA-I00 (BIOCROPPING)", funded by MCIN/AEI/10.13039/501100011033/FEDER, UE and the Proof of Concept "TCUE10 067/2300003 Technical validation for the use of mycopriming as a sustainable strategy for the crop production and protection improvement" funded by the U. of Valladolid Foundation.

Data Availability Statement: The raw data supporting the conclusions of this article will be made available by the authors on request.

Acknowledgments: Thanks to María Felicidad López-Sainz and Evelio Alonso for all their help in the development of the work.

Conflicts of Interest: The authors declare that they have no known competing financial interests or personal relationships that could have appeared to influence the work reported in this paper.

References

1. van Dijk, M.; Morley, T.; Rau, M.L.; Saghai, Y. A meta-analysis of projected global food demand and population at risk of hunger for the period 2010–2050. *Nat. Food* **2021**, *2*, 494–501. [CrossRef]
2. European Commission. *Farm to Fork Strategy*; European Commission: Luxembourg, 2019.
3. Smirnov, O.; Lahav, G.; Orbell, J.; Zhang, M.; Xiao, T. Climate Change, Drought, and Potential Environmental Migration Flows Under Different Policy Scenarios. *Int. Migr. Rev.* **2023**, *57*, 36–67. [CrossRef]
4. Boros, A.; Szólik, E.; Desalegn, G.; Tőzsér, D. A Systematic Review of Opportunities and Limitations of Innovative Practices in Sustainable Agriculture. *Agronomy* **2025**, *15*, 76. [CrossRef]
5. Food and Agriculture Organization of the United Nations. FAOSTAT Database. Available online: https://www.fao.org/statistics/en/ (accessed on 8 June 2025).
6. Karačić, V.; Miljaković, D.; Marinković, J.; Ignjatov, M.; Milošević, D.; Tamindžić, G.; Ivanović, M. Bacillus Species: Excellent Biocontrol Agents against Tomato Diseases. *Microorganisms* **2024**, *12*, 457. [CrossRef]
7. Janssen, D.G.C.; Ruiz, L.; de Cara-García, M.; Simón, A.; Martínez, A. Disease resistance in tomato crops produced in Spain. In Proceedings of the V International Symposium on Tomato Diseases: Perspectives and Future Directions in Tomato Protection, Malaga, Spain, 13–16 June 2016; pp. 63–68.
8. Litskas, V.; Migeon, A.; Navajas, M.; Marie-Stephane, T.; Stavrinides, M. Impacts of climate change on tomato, a notorious pest and its natural enemy: Small scale agriculture at higher risk. *Environ. Res. Lett.* **2019**, *14*, 084041. [CrossRef]
9. El-Banna, M.F.; Mosa, A. Exogenous application of proline mitigates deteriorative effects of salinity stress in NFT closed-loop system: An ultrastructural and physio-biochemical investigation on hydroponically grown tomato (*Solanum lycopersicum* L.). *Sci. Hortic.* **2024**, *330*, 113061. [CrossRef]
10. Ullah, A.; Bano, A.; Khan, N. Climate Change and Salinity Effects on Crops and Chemical Communication Between Plants and Plant Growth-Promoting Microorganisms Under Stress. *Front. Sustain. Food Syst.* **2021**, *5*, 618092. [CrossRef]
11. Cañizares, E.; Giovannini, L.; Gumus, B.O.; Fotopoulos, V.; Balestrini, R.; González-Guzmán, M.; Arbona, V. Seeds of Change: Exploring the transformative effects of seed priming in sustainable agriculture. *Physiol. Plant.* **2025**, *177*, e70226. [CrossRef] [PubMed]
12. Sher, A.; Sarwar, T.; Nawaz, A.; Ijaz, M.; Sattar, A.; Ahmad, S. Methods of Seed Priming. In *Priming and Pretreatment of Seeds and Seedlings: Implication in Plant Stress Tolerance and Enhancing Productivity in Crop Plants*; Hasanuzzaman, M., Fotopoulos, V., Eds.; Springer Singapore: Singapore, 2019; pp. 1–10.
13. Singh, P.; Vaishnav, A.; Liu, H.; Xiong, C.; Singh, H.B.; Singh, B.K. Seed biopriming for sustainable agriculture and ecosystem restoration. *Microb. Biotechnol.* **2023**, *16*, 2212–2222. [CrossRef]
14. García-Latorre, C.; Rodrigo, S.; Santamaría, O. Evaluation of the Extract of Pseudopithomyces chartarum to be used as Biocontrol Agent Against Phytophthora cinnamomi in *Lupinus luteus*. *J. Soil Sci. Plant Nutr.* **2024**, *24*, 6325–6337. [CrossRef]
15. Dwisandi, R.F.; Miranti, M.; Widiastuti, A.; Prismantoro, D.; Awal, M.A.; Mispan, M.S.; Joshi, R.C.; Doni, F. Microbial secondary metabolites for modulating plant biotic stress resistance: Bridging the lab-field gap. *Plant Stress* **2025**, *15*, 100720. [CrossRef]
16. Jangir, M.; Sharma, S.; Sharma, S. *Non-target Effects of Trichoderma on Plants and Soil Microbial Communities*; Springer: Cham, Switzerland, 2019; pp. 239–251.
17. Köhl, J. Use of beneficial microorganisms in crop production: Do current regulatory frameworks in the EU fit for purpose? *BioControl* **2025**, *70*, 433–450. [CrossRef]
18. Pozo, M.I.; Herrero, B.; Martín-García, J.; Santamaría, Ó.; Poveda, J. Evaluating potential side effects of Trichoderma as biocontrol agent: A two-edges sword? *Curr. Opin. Environ. Sci. Health* **2024**, *41*, 100566. [CrossRef]
19. Naeimi, S.; Hatvani, L.; Marik, T.; Balázs, D.; Dóczi, I.; Cai, F.; Vágvölgyi, C.; Druzhinina, I.S.; Kredics, L. Trichodermosis: Human Infections Caused by Trichoderma Species. In *Advances in Trichoderma Biology for Agricultural Applications*; Amaresan, N., Sankaranarayanan, A., Dwivedi, M.K., Druzhinina, I.S., Eds.; Springer International Publishing: Cham, Switzerland, 2022; pp. 607–634.
20. European Commission. *Commission Delegated Regulation (EU) 2022/1439 of 31 August 2022 Amending Regulation (EU) No 283/2013 as Regards the Information to Be Submitted for Active Substances and the Specific Data Requirements for Microorganism*; European Commission: Luxembourg, 2022; pp. 8–37.
21. European Parliament and of the Council. Regulation (EC) No 1107/2009 of 21 October 2009 concerning the placing of plant protection products on the market and repealing Council Directives 79/117/EEC and 91/414/EEC. *Off. J. Eur. Union* **2009**, *L309*, 1–50.

22. European Parliament and of the Council. Regulation (EU) 2019/1009 of 5 June 2019 laying down rules on the making available on the market of EU fertilising products and amending Regulations (EC) No 1069/2009 and (EC) No 1107/2009 and repealing Regulation (EC) No 2003/2003. *Off. J. Eur. Union* **2019**, *L170*, 1–114.
23. James, D.; Gleena Mary, C.F.K.; Mathew, S. Evaluation of Culture Filtrates of Endophytic Microorganisms from Tomato against *Ralstonia solanacearum*. *Int. J. Environ. Clim. Change* **2022**, *12*, 1476–1481. [CrossRef]
24. Sánchez-Gómez, T.; Santamaría, O.; Martín-García, J.; Poveda, J. Fungal Metabolites as Plant Growth Promoters in Crops. In *Fungal Metabolites for Agricultural Applications*; Poveda, J., Santamaría, Ó., Martín-García, J., Eds.; Springer Nature Switzerland AG: Cham, Switzerland, 2025; pp. 59–84.
25. Rodrigo, S.; García-Latorre, C.; Santamaria, O. Metabolites Produced by Fungi against Fungal Phytopathogens: Review, Implementation and Perspectives. *Plants* **2022**, *11*, 81. [CrossRef]
26. da Silva, F.M.R.; Paggi, G.M.; Brust, F.R.; Macedo, A.J.; Silva, D.B. Metabolomic Strategies to Improve Chemical Information from OSMAC Studies of Endophytic Fungi. *Metabolites* **2023**, *13*, 236. [CrossRef] [PubMed]
27. El-Nagar, D.; Salem, S.H.; El-Zamik, F.I.; El-Basit, H.M.I.A.; Galal, Y.G.M.; Soliman, S.M.; Aziz, H.A.A.; Rizk, M.A.; El-Sayed, E.S.R. Bioprospecting endophytic fungi for bioactive metabolites with seed germination promoting potentials. *BMC Microbiol.* **2024**, *24*, 200. [CrossRef]
28. García-Latorre, C.; Rodrigo, S.; Santamaria, O. Biological Control of *Pseudomonas syringae* in Tomato Using Filtrates and Extracts Produced by *Alternaria leptinellae*. *Horticulturae* **2024**, *10*, 334. [CrossRef]
29. Mathur, P.; Chaturvedi, P.; Sharma, C.; Bhatnagar, P. Improved seed germination and plant growth mediated by compounds synthesized by endophytic *Aspergillus niger* (isolate 29) isolated from *Albizia lebbeck* (L.) Benth. *3 Biotech* **2022**, *12*, 271. [CrossRef] [PubMed]
30. Waqas, M.; Khan, A.; Hamayun, M.; Kang, S.-M.; Kim, Y.-H.; Lee, I.-J. Assessment of endophytic fungi cultural filtrate on soybean seed germination. *Afr. J. Biotechnol.* **2012**, *11*, 15135–15143. [CrossRef]
31. Pablo, C.H. Plant growth-promoting characteristics of root fungal endophytes isolated from a traditional Cordillera rice landrace. *Stud. Fungi* **2020**, *5*, 536–549. [CrossRef]
32. García-Latorre, C.; Rodrigo, S.; Marin-Felix, Y.; Stadler, M.; Santamaria, O. Plant-growth promoting activity of three fungal endophytes isolated from plants living in dehesas and their effect on *Lolium multiflorum*. *Sci. Rep.* **2023**, *13*, 7354. [CrossRef]
33. Hashem, A.H.; Attia, M.S.; Kandil, E.K.; Fawzi, M.M.; Abdelrahman, A.S.; Khader, M.S.; Khodaira, M.A.; Emam, A.E.; Goma, M.A.; Abdelaziz, A.M. Bioactive compounds and biomedical applications of endophytic fungi: A recent review. *Microb. Cell Factories* **2023**, *22*, 107. [CrossRef]
34. Zhai, M.-M.; Li, J.; Jiang, C.-X.; Shi, Y.-P.; Di, D.-L.; Crews, P.; Wu, Q.-X. The Bioactive Secondary Metabolites from Talaromyces species. *Nat. Prod. Bioprospecting* **2016**, *6*, 1–24. [CrossRef]
35. Lei, L.-R.; Gong, L.-Q.; Jin, M.-Y.; Wang, R.; Liu, R.; Gao, J.; Liu, M.-D.; Huang, L.; Wang, G.-Z.; Wang, D.; et al. Research advances in the structures and biological activities of secondary metabolites from Talaromyces. *Front. Microbiol.* **2022**, *13*, 984801. [CrossRef]
36. Houbraken, J.; Visagie, C.M.; Meijer, M.; Frisvad, J.C.; Busby, P.E.; Pitt, J.I.; Seifert, K.A.; Louis-Seize, G.; Demirel, R.; Yilmaz, N.; et al. A taxonomic and phylogenetic revision of *Penicillium* section Aspergilloides. *Stud. Mycol.* **2014**, *78*, 373–451. [CrossRef]
37. Gupta, V.; Sharma, A.; Jamwal, G.; Gupta, S.; Razdan, V. Penicillium Genus as a Source of Metabolites for Agricultural Applications. In *Fungal Metabolites for Agricultural Applications*; Poveda, J., Santamaría, O., Martín-García, J., Eds.; Springer Nature: Cham, Switzerland, 2025; pp. 181–198.
38. Hao, L.; Zheng, X.; Wang, Y.; Li, S.; Shang, C.; Xu, Y. Inhibition of Tomato Early Blight Disease by Culture Extracts of a Streptomyces puniceus Isolate from Mangrove Soil. *Phytopathology* **2019**, *109*, 1149–1156. [CrossRef]
39. Feller, C.; Bleiholder, H.; Buhr, L.; Hack, H.; Heß, M.; Klose, R.; Meier, U.; Stauß, R.; Boom, T.v.d.; Weber, E. Phänologische Entwicklungsstadien von GemüsepflanzenII. Fruchtgemüse und Hülsenfrüchte: Codierung und Beschreibung nach der erweiterten BBCH-Skala—Mit Abbildungen. *Heft 9* **1995**, *47*, 217–232.
40. Therneau, T. A Package for Survival Analysis in R. Available online: https://cran.r-project.org/package=survival (accessed on 14 April 2025).
41. Kagithoju, S.; Sk, A.H.; Godishala, V.; Nanna, R.S. Role of Fungi and Fungal Extract on *Strychnos potatorum* Seed Germination. *Res. J. Biotechnol.* **2013**, *8*, 32–36.
42. Parveen, S.; Wani, A.H.; Yaqub, M. Effect of culture filtrates of pathogenic and antagonistic fungi on seed germination of some economically important vegetables. *Braz. J. Bot.* **2019**, *6*, 133–139. [CrossRef]
43. Ogórek, R.; Przywara, K.; Piecuch, A.; Cal, M.; Lejman, A.; Matkowski, K. Plant–Fungal Interactions: A Case Study of *Epicoccoum nigrum* Link. *Plants* **2020**, *9*, 1691. [CrossRef]
44. Zheng, Y.; Zou, J.; Lin, S.; Jin, C.; Shi, M.; Yang, B.; Yang, Y.; Jin, D.; Li, R.; Li, Y.; et al. Effects of different light intensity on the growth of tomato seedlings in a plant factory. *PLoS ONE* **2023**, *18*, e0294876. [CrossRef]

45. Abou Chehade, L.; Al Chami, Z.; De Pascali, S.A.; Cavoski, I.; Fanizzi, F.P. Biostimulants from food processing by-products: Agronomic, quality and metabolic impacts on organic tomato (*Solanum lycopersicum* L.). *J. Sci. Food Agric.* **2018**, *98*, 1426–1436. [CrossRef]
46. Miceli, A.; Vetrano, F.; Moncada, A. Effects of Foliar Application of Gibberellic Acid on the Salt Tolerance of Tomato and Sweet Pepper Transplants. *Horticulturae* **2020**, *6*, 93. [CrossRef]
47. Katsumi, M.; Ishida, K. The Gibberellin Control of Cell Elongation. In *Gibberellins*; Springer: New York, NY, USA, 1991; pp. 211–219.
48. Ericsson, T. Growth and shoot: Root ratio of seedlings in relation to nutrient availability. *Plant Soil* **1995**, *168*, 205–214. [CrossRef]
49. Wang, L.; Chen, X.; Yan, X.; Wang, C.; Guan, P.; Tang, Z. A response of biomass and nutrient allocation to the combined effects of soil nutrient, arbuscular mycorrhizal, and root-knot nematode in cherry tomato. *Front. Ecol. Evol.* **2023**, *11*, 1106122. [CrossRef]
50. Balasubramaniam, S.; Sakthivel, A.; Ramesh, K.; Manisseeri, C.; Ganeshan, S.; Subramani, M.; Gnanajothi, K. Bioprospecting of exopolysaccharides from the endophytic fungi *Epicoccum sorghinum* AMFS4, for its structure, composition, bioactivities and application in seed priming. *Nat. Prod. Res.* **2024**, 1–11. [CrossRef] [PubMed]
51. Mathur, P.; Agrawal, P.; Ghosh, U.; Sharma, C.; Bhatnagar, P.; Chaturvedi, P. Growth promotion ability of endophytic *Aspergillus niger* on different species of *Vigna*. *Vegetos* **2024**, *37*, 192–201. [CrossRef]
52. Chen, S.-M.; Zhang, C.-M.; Peng, H.; Qin, Y.-Y.; Li, L.; Li, C.-G.; Xing, K.; Liu, L.-L.; Qin, S. Exopolysaccharides from endophytic *Glutamicibacter halophytocota* KLBMP 5180 functions as bio-stimulants to improve tomato plants growth and salt stress tolerance. *Int. J. Biol. Macromol.* **2023**, *253*, 126717. [CrossRef] [PubMed]
53. Nandini, B.; Hariprasad, P.; Shankara, H.N.; Prakash, H.S.; Geetha, N. Total crude protein extract of Trichoderma spp. induces systemic resistance in pearl millet against the downy mildew pathogen. *3 Biotech* **2017**, *7*, 183. [CrossRef] [PubMed]
54. Patel, D.; Patel, A.; Patel, M.; Goswami, D. *Talaromyces pinophilus* strain M13: A portrayal of novel groundbreaking fungal strain for phytointensification. *Environ. Sci. Pollut. Res.* **2021**, *28*, 8758–8769. [CrossRef] [PubMed]
55. Pusztahelyi, T.; Holb, I.J.; Pócsi, I. Secondary metabolites in fungus-plant interactions. *Front. Plant Sci.* **2015**, *6*, 573. [CrossRef]
56. Flores-Velázquez, J.; Rojano, F.; Aguilar-Rodríguez, C.E.; Villagran, E.; Villarreal-Guerrero, F. Greenhouse Thermal Effectiveness to Produce Tomatoes Assessed by a Temperature-Based Index. *Agronomy* **2022**, *12*, 1158. [CrossRef]
57. Stubbs, C.; Cook, D.; Niklas, K. A general review of the biomechanics of root anchorage. *J. Exp. Bot.* **2019**, *70*, 3439–3451. [CrossRef]
58. Campobenedetto, C.; Mannino, G.; Beekwilder, J.; Contartese, V.; Karlova, R.; Bertea, C.M. The application of a biostimulant based on tannins affects root architecture and improves tolerance to salinity in tomato plants. *Sci. Rep.* **2021**, *11*, 354. [CrossRef]
59. Ismail, M.A.; Amin, M.A.; Eid, A.M.; Hassan, S.E.; Mahgoub, H.A.M.; Lashin, I.; Abdelwahab, A.T.; Azab, E.; Gobouri, A.A.; Elkelish, A.; et al. Comparative Study between Exogenously Applied Plant Growth Hormones versus Metabolites of Microbial Endophytes as Plant Growth-Promoting for *Phaseolus vulgaris* L. *Cells* **2021**, *10*, 1059. [CrossRef]

agriculture

MDPI

Article

Assessing Nutrient Losses and Recycling in Sweet Cherry Orchards: A Yield-Based Approach

Ilias Karampatzakis [1,2], Fotis Bilias [1,2,*], Chrysanthi Polychroniadou [1], Georgia Tanou [1], Panagiotis Kekelis [1], Aphrodite Theofilidou [1], Georgios Giannopoulos [1], Athina Pavlatou-Ve [2] and Vassilis Aschonitis [1,*]

1 Soil and Water Resources Institute, Hellenic Agricultural Organization DIMITRA, 57001 Thessaloniki, Greece; ikarampat@agro.auth.gr (I.K.); polychry@agro.auth.gr (C.P.); gtanou@elgo.gr (G.T.); p.kekelis@elgo.gr (P.K.); a.theofilidou@elgo.gr (A.T.); g.giannopoulos@elgo.gr (G.G.)

2 Soil Science Laboratory, School of Agriculture, Aristotle University of Thessaloniki, 54124 Thessaloniki, Greece; ave@agro.auth.gr

* Correspondence: fbilias@agro.auth.gr (F.B.); v.aschonitis@elgo.gr (V.A.)

Abstract: Mineral nutrition management in sweet cherry orchards remains a critical challenge due to the lack of site-specific fertilization guidelines, particularly in Greece, a significant cherry-producing country. This study aimed to develop a predictive framework for total nutrient losses in sweet cherry orchards by proposing simplified estimations using fresh fruit yield as the sole input variable. Field experiments were conducted in two orchards with distinct rootstocks (MxM 14 and CAB-6P), analyzing soil properties, leaf nutrient status, and uptake patterns on different plant components. Results indicated that despite differences in soil texture and pH, nutrient availability was generally sufficient, with only Fe and Zn marginally below optimal levels in leaf tissue. Principal Component Analysis (PCA) revealed distinct nutrient distribution patterns, with N evenly distributed across fruits, peduncles, and prunings, while K was concentrated in fruits and peduncles, and Ca and Mg predominantly in fallen leaves. Notably, K was redistributed from leaves to fruits under high yields, evidenced by negative correlations between leaf biomass and K uptake. Strong relationships ($r^2 > 0.8$) were found between fresh fruit yield and uptake of N, P, K, Mg, B, and Cu, enabling reliable predictions of total nutrient losses. Estimated annual nutrient removals were 85.6 kg ha^{-1} N, 8.94 kg ha^{-1} P, 42.7 kg ha^{-1} K, and 12.0 kg ha^{-1} Mg, with significant fractions retained in prunings and fallen leaves (e.g., 51.8 kg ha^{-1} N, 6.2 kg ha^{-1} P). The developed yield-based models provide a practical tool for optimizing fertilization strategies, while our findings highlight the potential for nutrient recycling through sustainable residue management.

Keywords: sweet cherry; mineral nutrition; nutrient loss prediction; nutrient recycling; fertilization management

1. Introduction

Mineral nutrition of sweet cherry trees is an aspect of major concern among farmers and stakeholders involved in the chain of cherry production and distribution, whereas for Greece, with an annual production of 45,000–113,000 tons (7–8% of EU production), ranked at No. 12 globally in cherry production [1], the above acquires a particular economic importance. However, until now, few published studies have focused on adjusting fertilization practices for cherry orchards, both to accurately define the mineral elements that are taken up by trees each year [2] as well as to account for the various soil parameters that regulate the buffering capacities of these elements in terms of their availability rates within the soil–soil solution–rhizosphere system [3,4].

In the absence of such information, and for Greece particularly, fertilization recommendations for cherry orchards are mainly based on empirical estimations either extrapolated from data specific to different deciduous stone-fruit tree species or data derived from studies on cherry trees grown under various environmental and soil conditions in other countries [5]. Similar gaps and limitations in implementing adjusted fertilization recommendation strategies have also been documented by other studies on cherry orchards grown across the broader Mediterranean region [4].

Current fertilization recommendations for deciduous fruit trees in general, and particularly for cherry orchards, are largely based on the combination of two distinct approaches: (i) leaf tissue analysis, which provides a snapshot of nutrient status in the tree at a specific point in time; and (ii) soil nutrient availability assessments, which estimate the nutrient supply capacity of the soil [6]. However, both approaches have inherent limitations if not sufficiently calibrated [7]. Leaf tissue analysis captures plant nutrient status only at the time of sampling and needs to be addressed in comparison with valid critical sufficiency thresholds of each nutrient that correspond to the specific tree species and variety [8]. Accordingly, soil nutrient analysis measures the availability of macro- and micronutrients in the soil but does not directly account for plant uptake dynamics if the respective soil tests that are to be used for the assessment are not adequately calibrated to reflect direct relationships between the available nutrient forms in the soil and their respective uptake by trees. Moreover, in fruit trees, soil analysis alone is not a satisfactory guide for fertilizer recommendations, mainly because of the difficulty of accurately determining the root zones from which most of the nutrients are taken up [7].

In view of the above, a preliminary necessary step in bypassing these limitations is to develop a nutrient budgeting framework to elucidate specific patterns of relations between yield-based data and the corresponding nutrient uptake potential. In this approach and for cherry orchards particularly, the quantification of nutrient removal through harvested fruits, leaf litter, and prunings, if linked and correlated with fresh biomass yield, could help reliably assess nutrient losses per year and, thus, establish an assessment framework for nutrient uptake [2].

Among different theoretical models that are used to describe specific relationships between yield parameters and nutrient uptake and are cited in the literature, Mitscherlich's law of diminishing returns, and Liebscher's theory of variable endurance, which is formalized by the Michaelis–Menten model, are mostly discussed, whereas their hyperbolic nature is based on the assumption that as nutrient uptake increases, yield also increases but at a decreasing rate [9].

However, the above relationships are the yield response curves, which connect yield response with a specific nutrient, a lack of which is regarded as a constraint factor for plant growth, and the supply of which forms the specific form of the exponential curve [10].

We hypothesize that if total nutrient losses can be estimated based on easily obtained data of harvested fresh fruit parameters, and respective relationships could be established, these estimates could serve as a preliminary tool for improving fertilization recommendations. Although previous studies, including those from the broader Mediterranean region, have explored nutrient budgets in cherry orchards [2], the main novelty of our work lies in predicting total annual nutrient losses directly from the fresh fruit yield. By establishing empirical relationships between fruit yield and the nutrients removed in harvested and residual plant parts (leaves and prunings), we offer a yield-based modeling approach that can be readily used by growers to adjust fertilization practices without relying exclusively on periodic soil or tissue testing. This approach is particularly valuable for Greek cherry production, where such calibrated models are currently lacking. In addition, the assessment of nutrients accumulated in plant residues that remain on the soil surface and can potentially be reused (such as fallen leaves and prunings) could provide further insights into nutrient cycling and efficiency in cherry orchards.

Based on that, the objectives of this study are to develop a predictive framework for nutrient removal and uptake in sweet cherry orchards by quantifying the relationships between fresh biomass yield and nutrient concentrations in different plant compartments, including fruits, peduncles, prunings, and fallen leaves. Specifically, this study aims to (i) assess the total annual nutrient losses from cherry orchards through harvested fruits and removed biomass, (ii) evaluate the contribution of plant residues, such as fallen leaves and prunings, to nutrient recycling within the orchard system, and (iii) to establish prediction models that can assess total nutrient losses based on the harvested fresh fruits biomass. By establishing yield-based nutrient removal estimates, this study seeks to provide a preliminary but reliable framework for refining fertilization recommendations, improving nutrient use efficiency, and promoting sustainable soil fertility management in cherry orchards.

2. Materials and Methods

2.1. Description of the Experimental Site

The experiment was conducted in 2021 on two cherry orchards located at the Achladochori farm, Giannitsa, Pella region, Greece. Both orchards featured intercropped cherry varieties, but they differed in location, tree age, and rootstocks used for grafting.

The first orchard is a commercial 0.4-hectare cherry orchard in the "Anadasmos" area of the Achladochori farm. It was established in January 2012 with the varieties Early Bigi, Bigarreau Burlat, and Lapins, all grafted onto the MxM 14 rootstock. The planting design was linear, with 5 m spacing between rows and 4 m spacing between trees within rows. The orchard comprised 8 rows, each containing 26 trees per row, pruned into a vase-like shape with approximately 4 central branches per tree.

In the odd-numbered rows (1, 3, 5, 7), the Early Bigi variety was planted, while in the even-numbered rows (2, 4, 6, 8), the Bigarreau Burlat and Lapins varieties were alternated in a 2-2 pattern. Early Bigi was the primary cultivar of the experimental farm, yielding the bulk of the orchard's production, while the other two served as pollinators. Despite its excellent flavor, Bigarreau Burlat has diminished commercial value, whereas Lapins remains attractive to growers due to its marketability and desirable traits. Therefore, the study of this specific variety is of greater interest.

The second orchard was a commercial 0.2-hectare cherry orchard, half the size of the first, established in December 2013. Although the same varieties were used as in the first orchard, this time, they were grafted onto the CAB-6P rootstock. Planting distances were 5 m between rows and 5 m within rows, resulting in 4 rows of 19 trees each. Trees were pruned into a vase-like shape with approximately 4 central branches per tree. In the odd-numbered rows (1, 3), the Early Bigi variety was planted, while in the even-numbered rows (2, 4), the Bigarreau Burlat and Lapins varieties alternated in a 3-3 pattern.

In each orchard, 5 trees of the Lapins variety were selected. The selected trees were representative of typical management and yield conditions in each orchard, while sampling was stratified by tree vigor and canopy uniformity within each orchard to minimize variability due to tree condition. Soil sampling and analysis were conducted once per tree, while leaf sampling and analysis were performed twice—once in July and once during leaf fall. Additional measurements included production volume, leaf litter weight (collected in autumn), and pruning weight (collected during winter pruning). Finally, chemical analyses were performed on fruits, peduncles, leaves, and prunings to quantify macro-

and micronutrient removal and evaluate the feasibility of nutrient recycling to minimize resource waste.

2.2. Fruit Harvest and Production Measurement

One advantage of the Lapins variety is its ability to ripen in one or two stages, depending on climatic conditions, compared to the three-stage ripening of the other two varieties. In 2021, all cherries ripened almost simultaneously, allowing for harvest completion in two days—one day per orchard. The harvest took place on 13 and 14 June 2021, in the Achladochori farm area.

Harvesting was performed manually using 0.45 kg tare-weight containers. Weighing was performed with a small professional scale (max weight capacity: 150 kg, precision: 2 decimal places in multiples of 5). This allowed for calculating the total production per tree and the average yield per Lapins tree in the two orchards.

2.3. Collection and Weighing of Leaves and Prunings

Leaf sampling and analysis were conducted twice, once in July 2021 and once during leaf fall. Leaf collection involved spreading nets under the trees. Net placement varied between orchards due to tree size differences. The orchard with MxM 14 rootstock had larger trees, necessitating a setup that would not obstruct routine orchard management activities, such as soil mechanical weed control or post-harvest spraying. In contrast, the CAB-6P orchard was regularly visited by wild boars, which could damage equipment. Consequently, nets were adjusted to prevent damage from wildlife. Leaf collection occurred monthly during the summer when natural leaf drop was observed. Leaves were collected into bags and weighed using the same professional scale used for fruit production.

2.4. Soil and Plant Tissue Analyses

Soil sampling was conducted in July 2021. Surface soil samples (0–30 cm) were collected around the tree trunk within the projection of the canopy's perimeter. After transport to the laboratory, samples were air-dried, crushed, and sieved through a 2 mm mesh. The following analyses were performed on this material:

The soil texture was determined using the hydrometer method [11]. Soil pH and electrical conductivity (EC) were measured in a saturated paste extract [12]. For samples with pH $\geq$ 6.5, total calcium carbonate ($CaCO_3$) was quantified using a digital calcimeter (FOG L, bd INVENTIONS, Thessaloniki, Greece). Soil organic matter (SOM) was determined via the wet oxidation method [13].

Soil total N was measured using the Kjeldahl method [14], while available NO_3-N was determined via UV–visible spectrophotometry after extraction with 2 M KCl [15]. Available P (Olsen-P) was extracted using 0.5 M $NaHCO_3$ at pH 8.5 and analyzed using the ascorbic acid method [16]. Exchangeable cations (K, Ca, Mg) were extracted with 1 M NH_4OAc at pH 7 and analyzed via ICP-OES [17]. Micronutrients (Fe, Mn, Cu, Zn) were determined after extraction with DTPA [18] and analyzed using ICP-OES, whereas available B was measured spectrophotometrically using the azomethine-H method after hot water extraction [19].

The harvested aboveground biomass was subdivided into different plant parts (fruits, peduncles, prunings, and fallen leaves), dried at 75 °C until a constant weight was reached, and then weighed to determine the respective dry biomass yield. The dried samples were subsequently ground for chemical analysis. Total N was determined using the Kjeldahl method [14], while the concentrations of other nutrients (P, K, Ca, Mg, Fe, Mn, Cu, Zn, and B) were analyzed following dry combustion at 500 °C for at least four hours. The resulting ash was dissolved in 6 M HCl, and nutrient determination in plant tissues was performed using the same methods as those applied for soil analysis.

2.5. Statistical Analyses

Linear or linearized regression models were applied between yield parameters and nutrient uptake among different plant parts to investigate relationships that could describe and predict nutrient outputs during the growing season. The empirical criterion of standardized skewness and kurtosis below 2 was applied to test the normality of the data. Transformations of variables were performed in the context of linearized modeling. The best models were selected based on the coefficient of determination criterion r^2, between observed and simulated data. Differences among mean values were assessed using ANOVA—One Way Analysis of Variance, single factor, at a probability level of $p < 0.05$. Significant differences between treatments were determined using Tukey's HSD test with a significance level of $p \leq 0.05$, whereas principal component analysis (PCA) was also performed on centered and standardized data. All analyses, along with corresponding visualization of graphs, were conducted using R software (V.4.2.2), installing "ggplot2", "datasets", "multcompView", "dplyr", and "tidyr" packages.

3. Results

3.1. Physicochemical Properties of Studied Soils

The physicochemical properties of the two soils used in this experiment, along with their differences from the ANOVA analysis, are summarized in Figure 1. The soils were classified as Loam (L), where the MxM14 rootstock was established, and Sandy Loam (SL), in the case of CAB-6P rootstock, respectively. They exhibited differences in clay, silt, and sand content, pH, and bulk density. In contrast, no significant differences were observed in their $CaCO_3$ content, electrical conductivity (EC), or organic matter (OM) content.

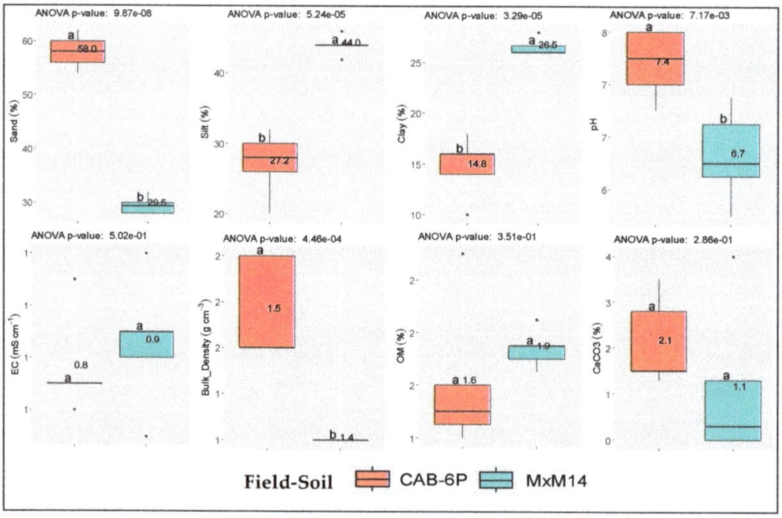

Figure 1. Mean values ($n = 5$ per field) and respective differences in the studied physicochemical properties between the two selected soils. Dots (·) indicate outliers; EC: electrical conductivity; OM: soil organic matter. Soil coding (MxM14 and CAB-6P) refers to the respective rootstock established at each field.

3.2. Available Forms of Studied Macro- and Micronutrients in Soil and Respective Concentrations from Leaf Analysis Between Two Fields

The mean values of the available forms of each studied macro- and micronutrient from soil samples taken from the two fields, along with their differences, are presented in Figure 2. Results indicated that among the three major macronutrients, only soil NO_3-N levels were marginally below sufficiency for the CAB-6P rootstock and marginally sufficient for the MxM14 rootstock, based on reference sufficiency levels reported in the literature [20]. In contrast, P-Olsen and available K values (K-ex) were within the respective sufficiency ranges and did not differ significantly between the two soils [21,22]. Regarding micronutrients, significant differences were observed between the two soils for available Fe, Mn, and Cu, whereas Zn and B showed no significant variation. Notably, the soil supporting the CAB-6P rootstock exhibited micronutrient levels of Fe, Mn, and Cu below the reported sufficiency thresholds [20] (Supplementary Table S1).

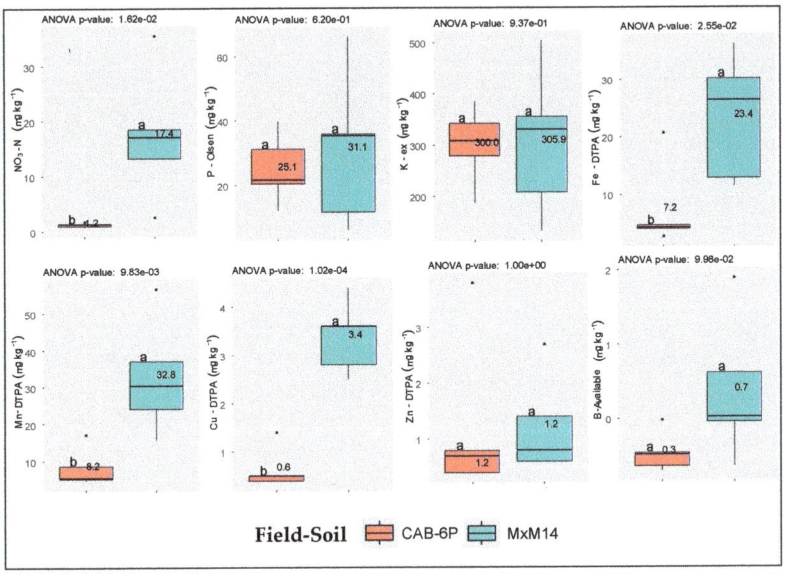

Figure 2. Mean values (n = 5 per field) and respective differences in the available forms of macro- and micronutrients between the two selected soils. Dots (·) indicate outliers; Soil coding (MxM14 and CAB-6P) refers to the respective rootstock established at each field.

The mean concentrations of macro- and micronutrients derived from the regular leaf tissue analysis of samples collected from the two fields are presented in Figure 3. The results indicate that the concentrations of the major macronutrients (N, P, and K) fall within the sufficiency ranges reported in the literature for cherry orchards [23]. Similarly, among the micronutrients, Mn, Cu, and B were also within adequate levels. However, Fe concentrations were below the sufficiency range of 100–250 mg kg^{-1}, while Zn levels were slightly below the corresponding sufficiency threshold of 20–50 mg kg^{-1} (Supplementary Table S1).

3.3. Distribution Patterns of Nutrient Concentration and Their Total Uptake Among Different Plant Parts

Principal Component Analysis (PCA) was employed to assess the overall variability patterns of macro- and micronutrient concentration among different plant parts. Results showed that a significant proportion of the total variance (93.4%) was clustered within four components, with two of them accounting for 68.4% of the variance (Figure 4).

Accordingly, as indicated by the PCA biplot, higher concentrations of N, Cu, and Zn are closely associated with prunings, while K concentration shows a strong alignment with peduncles. On the other hand, Mg and Mn concentrations are more closely aligned with fallen leaves, whereas Ca and B appear to be equally balanced between prunings and fallen leaves.

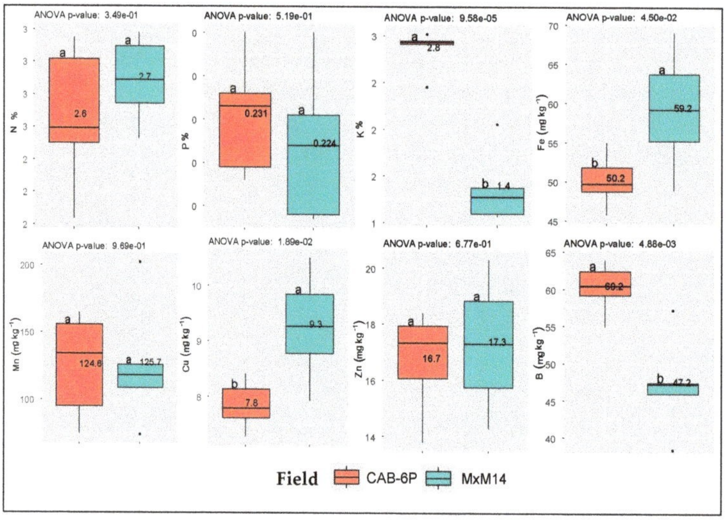

Figure 3. Mean values (n = 5 per field) and respective differences in the leaf tissue concentrations of macro- and micronutrients between the two selected fields. Dots (·) indicate outliers; Coding MxM14 and CAB-6P refer to the respective rootstock established at each field.

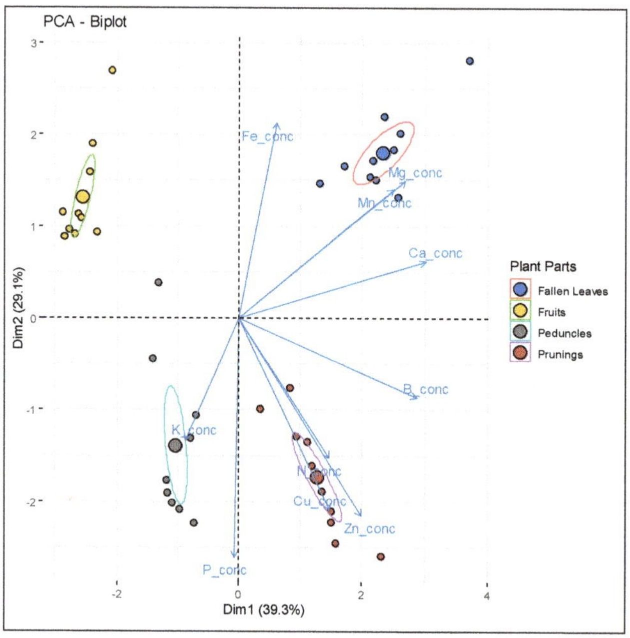

Figure 4. PCA biplots (95% confidence ellipses) of plant tissue concentrations of macro- and micronutrients scattered in different plant parts (n = 10, representing 40 total plant part measurements).

As expected, the total uptake percentage profile among the plant components further corroborates this distribution, as indicated in Figure 5. Notably, the majority of N uptake is evenly distributed between fruits and peduncles and prunings, while P uptake is primarily associated with prunings. On the other hand, K uptake is predominantly concentrated in fruits and peduncles, and Ca uptake is mostly attributed to fallen leaves. Similarly, Mg and Mn show the highest uptake in fallen leaves, at 64.8% and 72.2%, respectively, whereas Fe uptake is relatively balanced, with substantial portions found in fruits and peduncles (44.6%) and fallen leaves (40.6%).

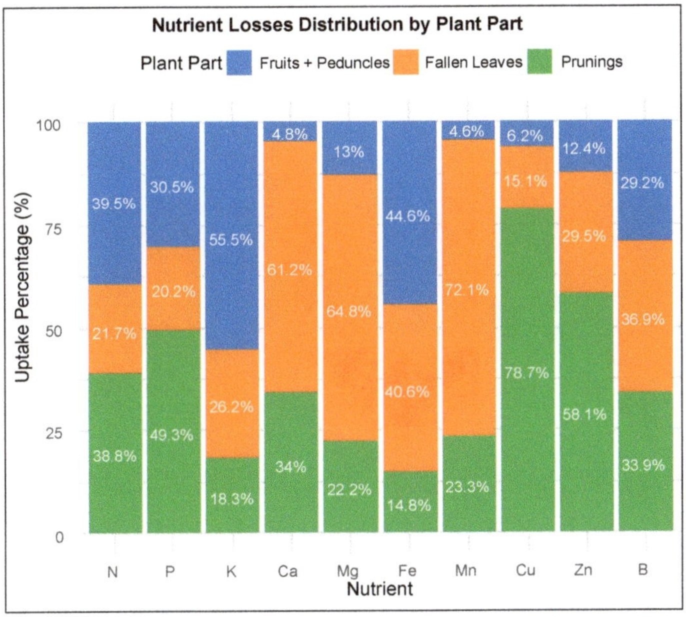

Figure 5. Stacked bar charts showing the percentage contribution of each plant part to the total losses of accumulated macro- and micronutrients.

Accordingly, a strong association with prunings was found for Cu and Zn, accounting for 78.7% and 58% of their total uptake, whereas B uptake appeared more evenly distributed, with significant contributions from all three components.

3.4. Relations Between Fresh Yield Parameters of Different Plant Parts and Respective Nutrient Uptake Magnitudes

The datasets of fresh weight and uptake magnitudes were normalized by dividing each value by the maximum within its respective plant part group, thereby generating scaled values ranging from 0 to 1 for easier comparison. The results are presented in Figure 6. Linear regressions were then applied, and the intercept and slope variables, along with the coefficients of determination (r^2) and *p*-values, are shown in Supplementary Table S2.

The acquired linear models show that for nutrients such as N, Ca, Mg, B, and Cu, there are strong relationships (high $r^2 > 0.8$ and significant p-values < 0.05) between fresh weight yield and nutrient uptake across most plant parts, whereas for nutrients such as P, K, Mn, Zn, and Fe, the relationships are moderate to weak ($r^2 < 0.7$ in some cases, e.g., P in peduncles, K in peduncles, Zn in fruits). Regarding plant-specific trends, fruits generally show the strongest relationships for most nutrients (e.g., N, P, K, Mg, B), whereas, in contrast, fallen leaves exhibit weaker relationships for some other nutrients (e.g., P, K, Mn, Zn). Notably, K uptake in fallen leaves and Fe uptake in fruits showed negative slopes and nonlinear patterns, indicating unique uptake dynamics.

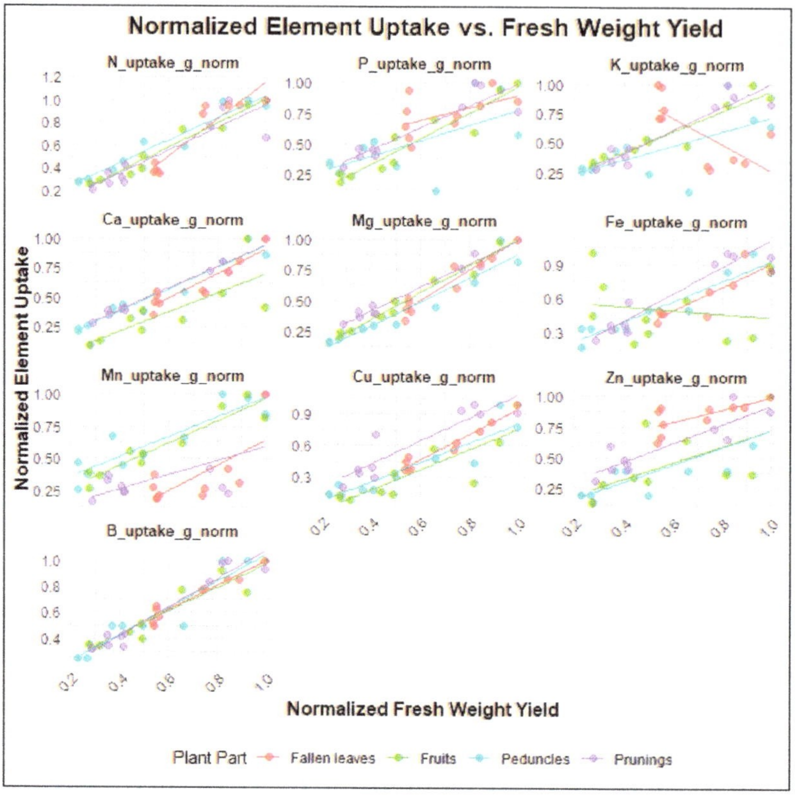

Figure 6. Relationships between uptake of each studied macro- or micronutrient and fresh weight yield of respective plant part that was removed during the year ($n = 10$ per plant part). Fresh weight and uptake values were normalized by dividing each value by the maximum within its respective plant part group (scale 0–1).

3.5. Relations Between Harvested Fresh Weight of Fruits and Peduncles and Total Nutrient Losses

Linear or linearized regression models were used to analyze the relationship between the fresh weight yield parameters of harvested fruits and peduncles (x) and the total nutrient losses, which include the amounts of nutrients taken up by prunings, fallen leaves, fruits, and peduncles (y). The best-fitting models are presented in Table 1 and Figure 7. Additionally, simplified linear models, excluding the intercept, represented by the equation y = ax, were applied to assess their prediction accuracy (Table 2 and Figure 8). This approach was adopted to enhance practical field applications, as the simplified models better support grower decision-making through their straightforward calculations.

Table 1. Linearized equation models between the fresh weight yield parameters of harvested fruits and peduncles (x) and the total nutrient losses, which include the amounts of nutrients taken up by prunings, fallen leaves, fruits, and peduncles (y). The approximate nutrient loss per hectare was assessed by multiplying the fresh weight threshold, set at 50 kg per tree, by 400, assuming 400 trees ha^{-1}.

Nutrient	Equation	r^2 (%)	*p*	Nutrient Loss ha^{-1} (kg)
N	$y = \left(-0.779532 + 2.17971 * \sqrt{x}\right)^2$	90.71	0.000	85.6
P	$y = \sqrt{42.8664 + 0.182855 * x^2}$	92.19	0.000	8.94

Table 1. *Cont.*

Nutrient	Equation	r^2 (%)	*p*	Nutrient Loss ha^{-1} (kg)
K	$y = \sqrt{4190.43 + 2.881 * x^2}$	82.33	0.001	42.7
Ca	$y = \sqrt{10002.2 + 8.7779 * x^2}$	50.99	0.006	-
Mg	$y = \sqrt{180.638 + 0.285702 * x^2}$	78.63	0.000	12.0
B	$y = \sqrt{0.0366927 + 0.0000591254 * x^2}$	86.97	0.000	0.172
Mn	$y = \sqrt{0.350896 + 0.000260695 * x^2}$	9.63	0.291	-
Zn	$y = \sqrt{0.0218186 + 0.0000153912 * x^2}$	75.84	0.017	0.100
Fe	$y = \sqrt{0.540711 + 0.000116915 * x^2}$	30.80	0.159	-
Cu	$y = 0.00561566 + 0.00499962 * x$	78.76	0.001	0.104

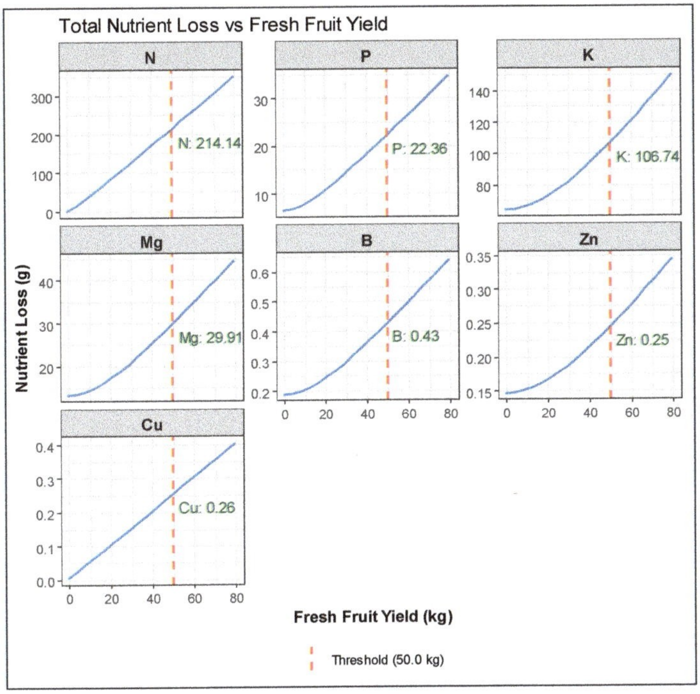

Figure 7. Models of total nutrient loss prediction for each studied macro- and micronutrient (n = 10), as derived from equations of Table 1. Fresh weight threshold was set at 50 kg per tree.

Table 2. Simplified linear equation models, excluding the intercept between the fresh weight yield parameters of harvested fruits and peduncles (x) and the total nutrient losses, which include the amounts of nutrients taken up by prunings, fallen leaves, fruits, and peduncles (y). Nutrient loss refers to a fresh weight threshold at 20,000 kg ha^{-1}.

Nutrient	Equation	rs [1]	*p*	Nutrient Loss ha^{-1} (kg)
N	$y = 4.31441 * x$	0.976	0.003	86.3
P	$y = 0.449012 * x$	0.927	0.005	9.0
K	$y = 2.17082 * x$	0.685	0.039	43.4
Ca	$y = 3.59362 * x$	0.830	0.012	-
Mg	$y = 0.605055 * x$	0.915	0.006	12.1
B	$y = 0.00870624 * x$	0.964	0.004	0.100
Mn	$y = 0.0177537 * x$	0.273	0.413	-

Table 2. *Cont.*

Nutrient	Equation	rs [1]	*p*	Nutrient Loss ha^{-1} (kg)
Zn	$y = 0.00500335 * x$	0.758	0.023	0.100
Fe	$y = 0.0183554 * x$	0.333	0.317	-
Cu	$y = 0.00511409 * x$	0.794	0.017	0.102

[1] rs: Spearman's rank correlation coefficient.

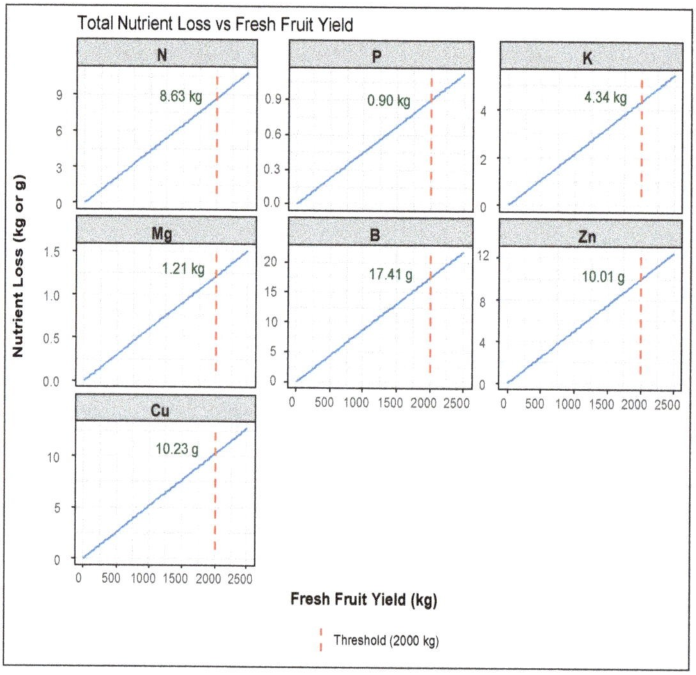

Figure 8. Models of total nutrient loss prediction for each studied macro- and micronutrient ($n = 10$), as derived from the simplified 1:1 equations of Table 2. Fresh weight threshold was set at 2000 kg of fresh-weight fruit production.

In the first group of equations, a fresh weight threshold was set at 50 kg per tree, and then, the approximate nutrient loss per hectare was assessed by multiplying by 400, assuming 400 trees ha^{-1}. As shown in Table 1, nutrients with an acceptable fit ($r^2 > 0.75$) included N, P, K, Mg, B, Zn, and Cu, with N, P, K, and B demonstrating the strongest relationships. In contrast, Fe and Mn uptake did not fit any of the tested models, preventing the prediction of total loss. In the second group of equations (Table 2 and Figure 8), the calculated nutrient loss refers to a fresh weight threshold at 20,000 kg ha^{-1}. For primary nutrients, including N and P, the simplified $y = ax$ models maintained high predictive accuracy while yielding nutrient removal estimates that closely matched those from the complex equations.

4. Discussion

4.1. Soil and Regular Plant Analysis Interpretation

While soil NO_3-N, Cu, and Fe levels were below sufficiency thresholds in the MxM14 rootstock field, leaf analysis showed that most nutrients were within sufficiency ranges, except for Fe and Zn. Nevertheless, micronutrient deficiencies in the leaves of deciduous trees may not always be associated with chlorosis symptoms or have a negative impact on yield and fruit quality, as demonstrated by Milošević et al. (2014) in the case of Fe deficiency [24]. Given the absence of visible chlorotic symptoms, our findings suggest that in both fields, cherry trees grew under soil conditions that provided sufficient nutrient availability to support adequate uptake. Therefore, no major nutrient limitation affecting cherry tree nutrition could be assumed.

Parameters that may influence the vegetative growth and yield of fruit trees are either controlled by orchard management or are genotype-dependent [25]. Among them, factors such as climate, soil, cultivar, rootstock, training, pruning, plant protection, weed control, irrigation, and fertilization are considered the most crucial [6]. Given that both orchards followed the same management practices and were located in close proximity, minimizing potential climatic influences, the variation in yield parameters can be primarily attributed to differences in cherry tree age. In addition, the distinct rootstocks onto which the studied cherry variety was grafted might also have a potential effect on yield variance [26].

4.2. Nutrient Distribution and Uptake Patterns

Results derived from the PCA biplots and Figure 5 revealed distinct nutrient distribution patterns among different plant parts, as well as respective allocation strategies. In particular, the balanced distribution of N uptake between fruits, peduncles, and prunings indicates the crucial role that N plays in both reproductive and vegetative growth [27], whereas the predominant association of P with prunings and of K with fruits and peduncles could be ascribed to the limited accumulation of P in the cherry fruits [28] or the well-known function of K in cherry fruit quality [29], respectively.

In addition, Mg, despite its phloem mobility, was predominantly partitioned to leaves, which is consistent with its role in chlorophyll and photosynthetic enzymes, while Cu, in contrast, accumulated strongly in woody tissues (prunings), likely due to its essential role in lignification processes [7]. These distribution patterns, along with the trends observed for the other macro- and micronutrients studied, are generally consistent with findings reported in previous studies on cherry trees and other deciduous species [2,30].

The equations derived from this study (Tables 1 and 2) provide a predictive framework for assessing total nutrient loss based on the fresh biomass yield of the harvested fruits and peduncles.

In contrast to yield response models such as the Michaelis–Menten or Mitscherlich's equation, which relate yield response to a specific nutrient whose lack is regarded as a constraint factor for plant growth, thus shaping the characteristic exponential curve, our approach differs. Here, the dependent variable is total nutrient uptake, representing the overall nutrient loss due to the removal or harvest of plant parts. The independent variable, on the other hand, is the harvested fresh yield weight, chosen as a readily accessible and easily measurable predictor for estimating nutrient loss.

From this point of view, the shape of the curves obtained for assessing the total nutrient loss for each element (Figure 7) is influenced by the factors governing the individual relationships that define the partial uptake patterns from the corresponding plant part yield parameters used to estimate total nutrient loss (Figure 6 and Supplementary Table S2). Consequently, the linearity or non-linearity of these individual relationships, variations in slope, and their predictive power or statistical significance play a key role in determining the overall pattern of total nutrient uptake removed, as plotted against the fresh weight yield of the harvested fruits and peduncles. The above could be attributed to specific uptake patterns caused by the specific effects of soil properties [7], translocation and redistribution patterns from different plant organs [30,31], or dilution effects [32].

The case of K is an indicative example in this context and a particularly important finding of this study. The observed negative and nonlinear relationship between fallen leaf yield and its corresponding K uptake (Figure 6), along with the relationship between fresh fruit weight, peduncle yield, and the K uptake of fallen leaves, strongly suggests a redistribution of K between leaves and fruits. This pattern is particularly evident under high fruit yield conditions, where nutrient demand is elevated. Supporting our findings, Roeva et al. (2022) also reported that in years with higher fruit loads on sour cherry trees, potassium concentrations in leaves were significantly reduced [29]. Similarly, comparable trends have been observed in other deciduous trees, such as pear and apple, where foliar potassium levels were negatively correlated with increasing yield [33,34].

Conversely, the limited predictive power of models using fresh fruit and peduncle yield to estimate total Fe and Mn loss stems from the inability to establish well-fitted individual 1:1 relationships in respective plant parts. Specifically, this limitation arises from weak correlations between fresh fruit and peduncle yield with Fe loss and between fallen leaves and prunings with Mn loss.

Fe, for instance, is predominantly found in soils as poorly soluble oxides and hydroxides, particularly under neutral to alkaline pH conditions, which is the case of the soils used in this study [35]. This limited solubility restricts Fe availability for plant uptake, often resulting in highly variable Fe uptake that is influenced by soil conditions and root physiology. Consequently, Fe homeostasis in plant cells must be tightly regulated to prevent deficiency [36]. This complex regulation, combined with the inconsistent availability of Fe in soil, may explain the weak predictive power of models attempting to estimate total Fe loss based solely on fresh fruit and peduncle yield. A similar rationale may also account for the respective incapability to predict Mn loss in soil alkaline conditions.

On the other hand, macronutrients such as Ca and Mg followed a distinct pattern with a more pronounced reduction in total uptake at higher fresh fruit yields (a trend that was also well described by a third-order polynomial curve). In the case of Ca, the above could be attributed to its limited mobility in the phloem, which prevents efficient translocation [31], a fact which is also evidenced in our study from respective low levels of accumulation in fruits and peduncles shown in Figure 5. At elevated yields, the plant prioritizes phloem-fed fruit expansion, but since Ca remains locked in vegetative tissues (61.2% in fallen leaves), total uptake trends may show a sharper decline than for other more mobile nutrients. In a similar manner, although Mg is more mobile in the phloem than Ca, its accumulation in fruits may also be limited at high yields, leading to patterns similar to those observed for Ca.

4.3. Integrating Fresh Fruit Yield Data into Nutrient Uptake Predictions

Our final predictions of total nutrient loss are generally consistent with findings from previous studies. Roversi and Monteforte (2006) reported total P losses ranging from 6.20 to 10.77 kg ha^{-1} for different sweet cherry varieties, while K losses ranged from 16.47 to 47.09 kg ha^{-1} [2]. These values align with our results, which estimate P and K losses at 8.94 kg ha^{-1} and 42.7 kg ha^{-1}, respectively. However, our predicted total N (85.6 kg ha^{-1}) and Mg losses (12.0 kg ha^{-1}) exceeded the ranges reported by the same authors (38.56–64.60 kg ha^{-1} for N and 5.08–8.79 kg ha^{-1} for Mg). In contrast, Baghadi and Sadowski (1998), who assessed the nutrient requirements of sour cherry trees, reported a total N loss of 95.0 kg ha^{-1}, a value comparable to our findings [37].

The same authors analyzed the annual nutrient uptake of sour cherry trees by distinguishing between nutrients that are permanently removed through harvested fruits, those retained within the trees, and those present in plant residues that remain on the soil surface and can potentially be reused (such as fallen leaves and prunings). Their findings indicate that 47% of total nitrogen, 55% of total potassium, 38% of total phosphorus, and 83% of total magnesium are retained in these residues and may contribute to nutrient recycling. These percentages correspond to estimated recoverable amounts of 55.2 kg ha^{-1} N, 2.2 kg ha^{-1} P, 34.38 kg ha^{-1} K, and 12.1 kg ha^{-1} Mg, respectively.

By comparing these values with the total nutrient losses predicted by our models (Table 1) and the nutrient distribution among plant parts (Figure 5), we found similar trends. Specifically, in our study, the nutrients potentially available for reuse—assuming plant residues remain on the soil surface and are not removed—amount to 51.8 kg ha^{-1} N, 6.2 kg ha^{-1} P, 19.0 kg ha^{-1} K, and 10.4 kg ha^{-1} Mg.

In contrast to prior nutrient-budget studies for cherry trees [2,37], which required laborious organ-specific sampling, our models simplify the estimation of total nutrient losses by using fresh fruit yield as the sole input variable—a practical metric readily available to growers. This approach eliminates the need to quantify partial uptake from individual plant parts (e.g., leaves, prunings) while maintaining predictive accuracy (Figures 7 and 8).

Moreover, our findings provide valuable insight into the fraction of nutrients that can be recycled, which is essential for evaluating soil fertility and tree nutrition within a comprehensive fertilization strategy [30]. Incorporating these estimates into fertilization planning can help minimize external chemical inputs and promote sustainable nutrient management under an integrated orchard management approach, refining existing fertilization guidelines for cherries in Greece [5]. In addition, the predictive models developed in this study could serve as a basis for practical applications in nutrient management planning, such as mobile decision-support tools, regional nutrient management guidelines, or farmer advisory platforms. The use of two different rootstocks in the orchards included in our study allowed the prediction models to capture variability potentially associated with rootstock type. This suggests that future recalibration of the models could also accommodate additional sources of variation, such as cultivar differences.

5. Conclusions

Our study provides a predictive framework for nutrient uptake in cherry orchards, offering valuable insights into nutrient distribution, uptake dynamics, and total nutrient losses associated with fruit production. Despite variations in soil properties, nutrient availability appeared sufficient to support cherry tree growth, while tree age was the main factor that affected the variability in yield between the two orchards. The relationships between fresh fruit yield and total nutrient uptake were found to be nonlinear in the studied yield range in most cases due to the complexity of soil properties, redistribution, and retention in different plant parts, as well as dilution effects in the models. Our prediction models estimate total nutrient losses of 85.6 kg ha^{-1} N, 8.94 kg ha^{-1} P, 42.7 kg ha^{-1} K, and 12.0 kg ha^{-1} Mg, with a significant portion retained in plant residues. Specifically, 51.8 kg ha^{-1} N, 6.2 kg ha^{-1} P, 19.0 kg ha^{-1} K, and 10.4 kg ha^{-1} Mg remain in fallen leaves and prunings, potentially contributing to nutrient recycling. Respective results were also obtained for micronutrients such as B, Zn, and Cu. These findings underscore the need for integrated fertilization strategies that account for both nutrient removal and reuse, supporting sustainable orchard management and reducing dependence on external inputs. While our models offer a preliminary framework for site-specific nutrient recommendations based on fresh yield parameters, further validation across a broader set of orchards and seasons will be necessary to confirm the robustness and generalizability of the proposed approach.

Supplementary Materials: The following supporting information can be downloaded at https://www.mdpi.com/article/10.3390/agriculture15121312/s1, Table S1: Sufficiency thresholds for interpreting soil and plant tissue nutrient concentrations; Table S2: Linear regression equation components between uptake of each studied macro- or micronutrient and fresh weight yield of respective plant part that was removed during the year. Fresh weight and uptake values were normalized by dividing each value by the maximum within its respective plant part group (scale 0–1).

Author Contributions: Conceptualization, V.A. and I.K.; methodology, F.B., I.K. and V.A.; software, F.B. and I.K.; validation, I.K., F.B. and V.A.; formal analysis, F.B., V.A. and I.K.; investigation, I.K., C.P., A.T., P.K. and G.G.; resources, V.A.; data curation, I.K., C.P., A.T., P.K. and G.G.; writing—original draft preparation, F.B.; writing—review and editing, F.B., V.A., G.T., I.K. and A.P.-V.; visualization, F.B.; supervision, V.A. and A.P.-V.; project administration, V.A. and F.B.; funding acquisition, V.A. All authors have read and agreed to the published version of the manuscript.

Funding: This research received no external funding.

Institutional Review Board Statement: Not applicable.

Data Availability Statement: All data supporting the results of this study are included in the manuscript, and datasets are available upon request.

Conflicts of Interest: The authors declare no conflicts of interest.

References

1. Papapetros, S.; Louppis, A.; Kosma, I.; Kontakos, S.; Badeka, A.; Kontominas, M.G. Characterization and Differentiation of Botanical and Geographical Origin of Selected Popular Sweet Cherry Cultivars Grown in Greece. *J. Food Compos. Anal.* **2018**, *72*, 48–56. [CrossRef]
2. Roversi, A.; Monteforte, A. Preliminary Results on the Mineral Uptake of Six Sweet Cherry Varieties. *Acta Hortic.* **2006**, *721*, 123–128. [CrossRef]
3. Ughini, V.; Roversi, A. Application of the Szűcs Algorithm as an Aid for Orchard Mineral Fertilization. *Acta Hortic.* **2006**, *721*, 299–306. [CrossRef]
4. Ughini, V.; Roversi, A. Estimation of Sweet Cherry Fertilizer Requirements by the Szűcs' Method Varies by Cultivar. *Acta Hortic.* **2008**, *795*, 733–738. [CrossRef]
5. Stylianidis, D.C.; Simonis, A.D.; Syrgiannidis, G.D. *Nutrition, Fertilization of Deciduous Fruit Trees: Deficiencies, Toxicities, Physiological Disorders of Fruits*; Stamoulis Publications: Athens, Greece, 2002. (In Greek)
6. Milošević, T.; Milošević, N. Soil fertility: Plant nutrition vis-à-vis fruit yield and quality of stone fruits. In *Fruit Crops*; Elsevier: Amsterdam, The Netherlands, 2020; pp. 583–606.
7. Marschner, H. *Marschner's Mineral Nutrition of Higher Plants*; Academic Press: London, UK, 2012; Volume 89, p. 651.
8. Smith, F.W.; Loneragan, J.F. Interpretation of Plant Analysis: Concepts and Priciples. In *Plant Analysis: An Interpretation Manual*; Reuter, D.J., Robinson, J.B., Eds.; CSIRO Publishing: Collingwood, Australia, 1997; pp. 3–33.
9. Nijland, G.O.; Schouls, J. *The Relation Between Crop Yield, Nutrient Uptake, Nutrient Surplus and Nutrient Application*; Wageningen Agricultural University: Wageningen, The Netherlands, 1997.
10. Dhanoa, M.S.; Sanderson, R.; Cardenas, L.M.; Shepherd, A.; Chadwick, D.R.; Powell, C.D.; Ellis, J.L.; López, S.; France, J. Overview and Application of the Mitscherlich Equation and Its Extensions to Estimate the Soil Nitrogen Pool Fraction Associated with Crop Yield and Nitrous Oxide Emission. In *Advances in Agronomy*; Elsevier: Amsterdam, The Netherlands, 2022; Volume 174, pp. 269–295, ISBN 978-0-323-98957-2.
11. Bouyoucos, G.J. Hydrometer Method Improved for Making Particle Size Analyses of Soils. *Agron. J.* **1962**, *54*, 464–465. [CrossRef]
12. Rhoades, J.D. Salinity: Electrical Conductivity and Total Dissolved Solids. In *Methods of Soil Analysis: Part 3 Chemical Methods*; Sparks, D.L., Page, A.L., Helmke, P.A., Loeppert, R.H., Soltanpour, P.N., Tabatabai, M.A., Johnston, C.T., Sumner, M.E., Eds.; SSSA Book Series; Soil Science Society of America, American Society of Agronomy: Madison, WI, USA, 1996; pp. 417–435.
13. Walkley, A.J.; Black, I.A. Estimation of Soil Organic Carbon by the Chromic Acid Titration Method. *Soil Sci.* **1934**, *37*, 29–38. [CrossRef]
14. Bremner, J.M. Nitrogen-Total. In *Methods of Soil Analysis: Part 3 Chemical Methods*; Sparks, D.L., Page, A.L., Helmke, P.A., Loeppert, R.H., Soltanpour, P.N., Tabatabai, M.A., Johnston, C.T., Sumner, M.E., Eds.; SSSA Book Series; Soil Science Society of America, American Society of Agronomy: Madison, WI, USA, 1996; pp. 1085–1121.
15. Mulvaney, R.L. Nitrogen-Inorganic Forms. In *Methods of Soil Analysis: Part 3 Chemical Methods*; Sparks, D.L., Page, A.L., Helmke, P.A., Loeppert, R.H., Soltanpour, P.N., Tabatabai, M.A., Johnston, C.T., Sumner, M.E., Eds.; SSSA Book Series; Soil Science Society of America, American Society of Agronomy: Madison, WI, USA, 1996; pp. 1123–1184.
16. Kuo, S. Phosphorus. In *Methods of Soil Analysis: Part 3 Chemical Methods*; Sparks, D.L., Page, A.L., Helmke, P.A., Loeppert, R.H., Soltanpour, P.N., Tabatabai, M.A., Johnston, C.T., Sumner, M.E., Eds.; SSSA Book Series; Soil Science Society of America, American Society of Agronomy: Madison, WI, USA, 1996; pp. 869–919.

17. Thomas, G.W. Exchangeable Cations. In *Agronomy Monographs*; Page, A.L., Ed.; American Society of Agronomy, Soil Science Society of America: Madison, WI, USA, 2015; pp. 159–165.
18. Lindsay, W.L.; Norvell, W.A. Development of a DTPA Soil Test for Zinc, Iron, Manganese, and Copper. *Soil Sci. Soc. Am. J.* **1978**, *42*, 421–428. [CrossRef]
19. Keren, R. Boron. In *Methods of Soil Analysis: Part 3 Chemical Methods*; Sparks, D.L., Page, A.L., Helmke, P.A., Loeppert, R.H., Soltanpour, P.N., Tabatabai, M.A., Johnston, C.T., Sumner, M.E., Eds.; SSSA Book Series; Soil Science Society of America, American Society of Agronomy: Madison, WI, USA, 1996; pp. 603–626.
20. Siatwiinda, S.M.; Ros, G.H.; Yerokun, O.A.; De Vries, W. Options to reduce ranges in critical soil nutrient levels used in fertilizer recommendations by accounting for site conditions and methodology: A review. *Agron. Sustain. Dev.* **2024**, *44*, 9. [CrossRef]
21. Bai, Z.; Li, H.; Yang, X.; Zhou, B.; Shi, X.; Wang, B.; Li, D.; Shen, J.; Chen, Q.; Qin, W.; et al. The critical soil P levels for crop yield, soil fertility and environmental safety in different soil types. *Plant Soil* **2013**, *372*, 27–37. [CrossRef]
22. Bilias, F.; Barbayiannis, N. Evaluation of sodium tetraphenylboron (NaBPh4) as a soil test of potassium availability. *Arch. Agron. Soil Sci.* **2017**, *63*, 468–476. [CrossRef]
23. Reuter, D.J.; Edwards, D.G.; Wilhelm, N.S. Temperate and tropical crops. In *Plant Analysis: An Interpretation Manual*; Reuter, D.J., Robinson, J.B., Eds.; CSIRO Publishing: Collingwood, Australia, 1997; pp. 83–278.
24. Milošević, T.; Milošević, N.; Milivojević, J.; Glišić, I.; Nikolić, R. Experiences with Mazzard and Colt sweet cherry rootstocks in Serbia which are used for high density planting system under heavy and acidic soil conditions. *Sci. Hortic.* **2014**, *176*, 261–272. [CrossRef]
25. Milošević, T.; Moreno, M.Á.; Milošević, N.; Milinković, M. Regulation of Yield, Fruit Size, and Leaf Mineral Nutrients of the 'Šumadinka' Sour Cherry Cultivar with Help of Rootstocks. *J. Plant Growth Regul.* **2023**, *42*, 5587–5599. [CrossRef]
26. Cantín, C.M.; Pinochet, J.; Gogorcena, Y.; Moreno, M.Á. Growth, yield and fruit quality of 'Van' and 'Stark Hardy Giant' sweet cherry cultivars as influenced by grafting on different rootstocks. *Sci. Hortic.* **2010**, *123*, 329–335. [CrossRef]
27. Rutkowski, K.; Łysiak, G.P. Effect of Nitrogen Fertilization on Tree Growth and Nutrient Content in Soil and Cherry Leaves (*Prunus cerasus* L.). *Agriculture* **2023**, *13*, 578. [CrossRef]
28. Rowley, S.D. Phosphorous and Potassium Fertility Management for Maximizing Tart Cherry Fruit Quality and Productivity on Alkaline Soils. Master's Thesis, Utah State University, Logan, UT, USA, 2013. [CrossRef]
29. Roeva, T.; Leonicheva, E.; Leonteva, L.; Stolyarov, M. Potassium dynamics in orchard soil and potassium status of sour cherry trees affected by soil nutritional conditions. *J. Cent. Eur. Agric.* **2022**, *23*, 103–113. [CrossRef]
30. Quartieri, M.; Polidori, G.; Baldi, E.; Toselli, M. Evaluation of Removed and Recycled Mineral Nutrients in Italian Commercial Persimmon Orchards. *Horticulturae* **2023**, *9*, 374. [CrossRef]
31. Kumar, A.; Singh, U.M.; Manohar, M.; Gaur, V.S. Calcium transport from source to sink: Understanding the mechanism(s) of acquisition, translocation, and accumulation for crop biofortification. *Acta Physiol. Plant* **2015**, *37*, 1722. [CrossRef]
32. Jarrell, W.M.; Beverly, R.B. The dilution effect in plant nutrition studies. In *Advances in Agronomy*; Elsevier: Amsterdam, The Netherlands, 1981; pp. 197–224. [CrossRef]
33. Brunetto, G.; Nava, G.; Ambrosini, V.G.; Comin, J.J.; Kaminski, J. The pear tree response to phosphorus and potassium fertilization. *Rev. Bras. Frutic.* **2015**, *37*, 507–516. [CrossRef]
34. Kuzin, A.I.; Kashirskaya, N.Y.; Kochkina, A.M.; Kushner, A.V. Correction of Potassium Fertigation Rate of Apple Tree (*Malus domestica* Borkh.) in Central Russia during the Growing Season. *Plants* **2020**, *9*, 1366. [CrossRef]
35. Li, M.; Watanabe, S.; Gao, F.; Dubos, C. Iron Nutrition in Plants: Towards a New Paradigm? *Plants* **2023**, *12*, 384. [CrossRef]
36. Briat, J.-F.; Dubos, C.; Gaymard, F. Iron nutrition, biomass production, and plant product quality. *Trends Plant Sci.* **2015**, *20*, 33–40. [CrossRef] [PubMed]
37. Baghdadi, M.; Sadowski, A. Estimation of nutrient requirements of sour cherry. *Acta Hortic.* **1998**, *468*, 515–522. [CrossRef]

agriculture

MDPI

Article

Phenological Performance, Thermal Demand, and Qualitative Potential of Wine Grape Cultivars Under Double Pruning

Carolina Ragoni Maniero [1,*], Marco Antonio Tecchio [1], Harleson Sidney Almeida Monteiro [1], Camilo André Pereira Contreras Sánchez [1], Giuliano Elias Pereira [2], Juliane Barreto de Oliveira [3], Sinara de Nazaré Santana Brito [1], Francisco José Domingues Neto [4], Sarita Leonel [1], Marcelo de Souza Silva [1], Ricardo Figueira [1] and Pricila Veiga dos Santos [1]

1 São Paulo State University (UNESP), School of Agricultural Sciences, Botucatu 18610-034, SP, Brazil; marco.a.tecchio@unesp.br (M.A.T.); harleson.sa.monteiro@unesp.br (H.S.A.M.); camilo.apc.sanchez@unesp.br (C.A.P.C.S.); sinara.santana@unesp.br (S.d.N.S.B.); sarita.leonel@unesp.br (S.L.); marcelo.s.silva@unesp.br (M.d.S.S.); ricardo.figueira@unesp.br (R.F.); pricila.veiga@unesp.br (P.V.d.S.)

2 Brazilian Agricultural Research Corporation (EMBRAPA Grape and Wine), Bento Gonçalves 95701-008, RS, Brazil; giuliano.pereira@embrapa.br

3 Federal Institute of Education, Science and Technology of Rio Grande do Sul (IFRS), Bento Gonçalves 95700-206, RS, Brazil; juliane.oliveira@bento.ifrs.edu.br

4 São Paulo State University (UNESP), School of Agricultural and Veterinary Sciences, Jaboticabal 14884-900, SP, Brazil; francisco.domingues@unesp.br

* Correspondence: carolina.maniero@unesp.br

Abstract: The production of winter wines in Southeastern Brazil represents a relatively recent but expanding viticultural approach, with increasing adoption across diverse wine-growing regions. This system relies on the double-pruning technique, which allows for the harvest of grapes during the dry and cooler winter season, favoring a greater accumulation of sugars, acids, and phenolic compounds. This study aimed to characterize the phenological stages, thermal requirements, yield, and fruit quality of the fine wine grape cultivars 'Sauvignon Blanc', 'Merlot', 'Tannat', 'Pinot Noir', 'Malbec', and 'Cabernet Sauvignon' under double-pruning management in a subtropical climate. The vineyard was established in 2020, and two production cycles were evaluated (2022/2023 and 2023/2024). Significant differences in the duration of phenological stages were observed among cultivars, ranging from 146 to 172 days from pruning to harvest. The accumulated thermal demand was higher in the first cycle, with a mean of 1476.9 growing degree days (GDD) across cultivars. The results demonstrate the potential of *Vitis vinifera* L. cultivars managed with double pruning for high-quality wine production under subtropical conditions, supporting the viability of expanding viticulture in the state of São Paulo. 'Cabernet Sauvignon' and 'Sauvignon Blanc' showed the highest yields, reaching 3.03 and 2.75 kg per plant, respectively, with productivity values of up to 10.8 t ha^{-1}. 'Tannat' stood out for its high sugar accumulation (23.4 °Brix), while 'Merlot' exhibited the highest phenolic (234.9 mg 100 g^{-1}) and flavonoid (15.3 mg 100 g^{-1}) contents. These results highlight the enological potential of the evaluated cultivars and confirm the efficiency of the double-pruning system in improving grape composition and wine quality in non-traditional viticultural regions.

Keywords: grape wine; phenology; subtropical viticulture; winter wine production; bioactive compounds; grape quality

1. Introduction

Grapes (*Vitis* spp.) are among the most economically important fruit crops worldwide, encompassing a wide diversity of species and cultivars used for fresh consumption (ta-

https://doi.org/10.3390/agriculture15121241

ble grapes), raisin production, juice, and winemaking [1]. Brazilian viticulture occupies approximately 75,000 hectares of vineyards [2], with the states of Rio Grande do Sul, São Paulo, Pernambuco, Santa Catarina, Paraná, Bahia, and Minas Gerais standing out as the main wine-growing areas. In terms of fine wine production, a 9.3% increase was recorded in 2021 and 2022, with a notable growth of 19.5% in red wine production, while volumes of white and rosé wines remained virtually stable [3]. According to IBGE estimates, grape production in Brazil reached 1,450,805 tons in 2022.

In southeastern Brazil, an emerging viticulture has shown significant potential for high-quality fine wine production, driven by the implementation of the double-pruning technique [4–6]. This management strategy has enabled the cultivation of *Vitis vinifera* L. cultivars in regions previously considered unsuitable for premium wine production [7]. The double-pruning technique modifies the phenological cycle of the grapevine, allowing for harvest to occur during the dry winter season—a period characterized by mild daytime temperatures, lower nighttime temperatures, and minimal rainfall. This strategy enables the harvest of grapes under optimal health and ripening conditions, resulting in berries with improved physicochemical and biochemical attributes, and consequently, wines of superior intrinsic quality [8].

For the production of fine wines in new viticultural regions, it is essential to assess the adaptability of grapevine cultivars in relation to the specific characteristics of each terroir. The double-pruning management enhances sugar accumulation and phenolic ripening, promoting superior sensory attributes and improving wine quality [9]. In addition to their commercial relevance, grapes are rich in polyphenols, resveratrol [10], vitamins (such as C and K), and dietary fiber [11], which have been associated with antioxidant, anti-inflammatory, and cardioprotective effects. These nutritional and functional properties have contributed to the growing interest in grape consumption and value-added products [10,11].

Although winter wine production in southeastern Brazil began only in 2004, in southern Minas Gerais, the system has already expanded to over 35 distinct wine-producing regions, with a growing community of winegrowers [12]. This expansion underscores the need to evaluate the agronomic and qualitative performance of grapevine cultivars under the specific conditions imposed by double pruning in subtropical climates. Within this context, double pruning has emerged as a promising strategy for advancing viticulture in the state of São Paulo [13]. Among the *Vitis vinifera* L. cultivars grown in the region, the red cultivars 'Cabernet Sauvignon' and 'Merlot' and the white cultivar 'Sauvignon Blanc' are particularly notable due to their recognized enological potential [14].

The double-pruning technique modifies the phenological cycle of the grapevine, allowing harvest to occur during the dry winter season, a period characterized by mild daytime temperatures, lower nighttime temperatures, and minimal rainfall [14]. Such conditions favor the accumulation and preservation of sugars, organic acids, and phenolic compounds, leading to musts with superior technological and sensory quality [8]. Moreover, recent studies have demonstrated that double pruning reduces disease pressure, enhances grape balance, and increases vineyard sustainability in tropical and subtropical regions.

Winter wine production has expanded considerably since its introduction in 2004 in southern Minas Gerais, with over 35 producing regions now adopting this model [13]. Notable examples include the production of high-quality 'Syrah', 'Tempranillo', and 'Chardonnay' wines with enhanced color, acidity, and aromatic profiles. The alignment of grape ripening with cooler and drier conditions also lowers chemical inputs and production costs. Among the cultivars used in subtropical regions of Brazil, 'Cabernet Sauvignon', 'Merlot', and 'Sauvignon Blanc' stand out for their high enological potential and commercial value [13,14]. These varieties are widely used both for table grape and wine markets, especially red blends and varietal whites. Their adaptation across diverse Brazilian terroirs reinforces their economic relevance for the industry.

Given the increasing interest in double-pruning management and the expansion of vineyards in non-traditional wine-growing areas, the present study aims to evaluate the phenological development, thermal requirements, ripening pattern, yield, physicochemical composition, and biochemical quality of the cultivars Sauvignon Blanc, Merlot, Tannat, Pinot Noir, Malbec, and Cabernet Sauvignon subtropical conditions. The findings are expected to guide grape growers and winemakers in cultivar selection and vineyard planning, contributing to the sustainable development of the fine wine industry in southeastern Brazil.

2. Materials and Methods

2.1. Experimental Site

The experiment was conducted from February 2023 to August 2024 in a vineyard located at Santa Lúcia do Tietê Farm, in the municipality of Mineiros do Tietê, São Paulo State, Brazil (22°32′25″ S, 48°24′13″ W; altitude: 580 m). The experimental area is situated in a transitional zone between the tropical montane Atlantic Forest and the Cerrado biome, and is characterized by a humid subtropical climate (Cfa) according to the Köppen–Geiger classification, with hot summers and evenly distributed rainfall throughout the year.

2.2. Climatic Data

According to the climatic data obtained during the two production cycles (Figure 1), the area exhibits climatic characteristics similar to those of other winter wine-producing regions in southeastern Brazil, such as in the states of Minas Gerais and São Paulo [15,16].

Climatic conditions favorable to grape cultivation for winter wine production include reduced rainfall during the harvest period (June to August), low relative humidity, high thermal amplitude (below 30 °C), and high solar radiation incidence. These factors contribute to a slower ripening process, resulting in the greater accumulation of phenolic compounds, aroma compounds, and aroma precursors. Such conditions are essential for the production of both red and white wines with high enological potential [5,17].

During the 2023 growing cycle, total annual rainfall was 2135.4 mm, with an average temperature of 23.4 °C and a relative humidity of 76%. In 2024, a reduction in these climatic indices was observed, with total precipitation of 989.0 mm, a mean annual temperature of 19.6 °C, and relative humidity of 75.5%. During the berry ripening period, meteorological data collection was intensified and conducted daily in both production cycles (Figure 1a,b).

2.3. Vineyard Establishment and Growing Conditions

The establishment of the vineyard began in 2020, with soil preparation, liming, and corrective fertilization based on agrochemical analyses. A vertical shoot positioning (VSP) trellis system with wooden posts was installed, with a spacing of 3.0 m between rows and 1.0 m between vines. Certified grapevine seedlings were planted in November 2020.

The vines were managed under a double-pruning system over two production cycles. Training pruning was performed on 17 August 2022, and 13 August 2023, while production pruning was carried out on 12 February 2023, and 17 February 2024, aiming for harvest during the dry winter period typical of southeastern Brazil. Following pruning, hydrogen cyanamide (Dormex®, BASF, Ludwigshafen am Rhein, Germany) was applied at a 5% concentration, combined with 0.1% silicon-based surfactant, to promote uniform budbreak.

Pruning was carried out following the long pruning system, characterized by the maintenance of canes with three active buds. This type of pruning aims to promote a balance between vegetative vigor and fruit production, resulting in a more homogeneous distribution of clusters throughout the canopy and allowing for improved aeration and light interception. The decision to retain three buds seeks to optimize bud break and fruit load, particularly in cultivars that exhibit good fertility in basal to median buds. Moreover, long pruning is a recommended practice for young vines still in formation or in regions where climatic conditions may impair the sprouting of distal buds, thereby ensuring greater production consistency. This strategy also contributes to vineyard sustainability by reducing the need for corrective interventions throughout the vegetative cycle [18].

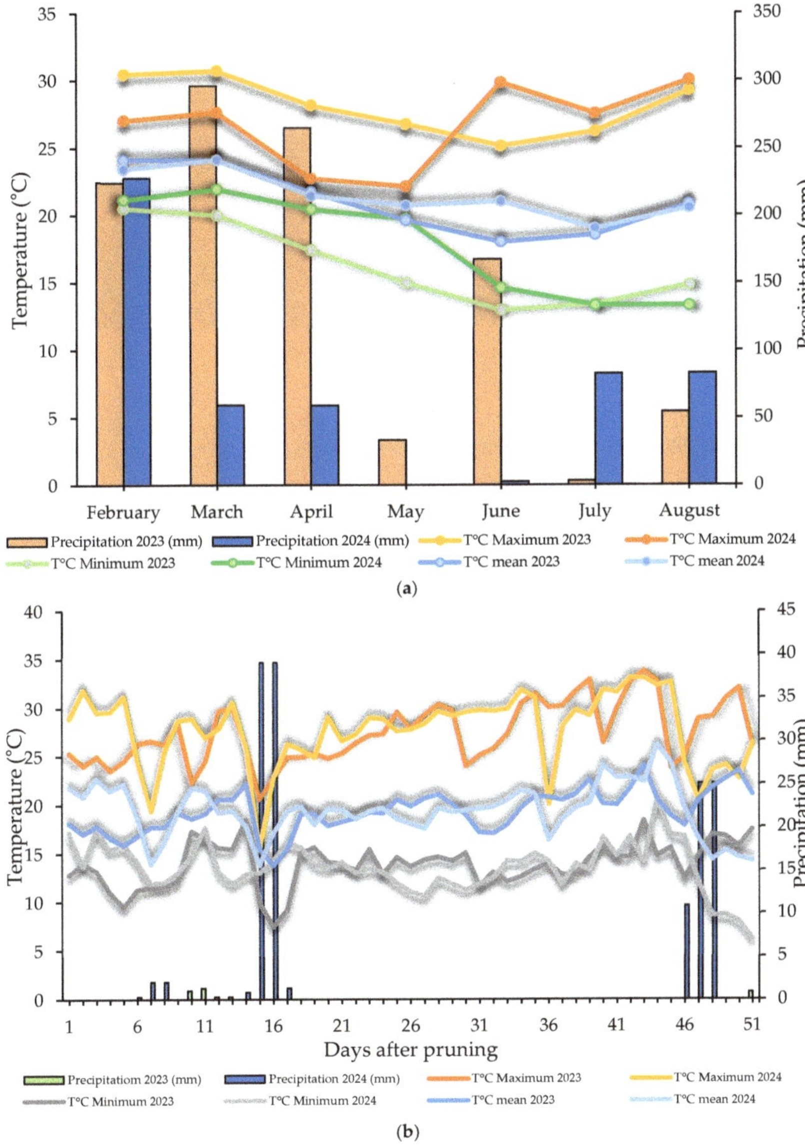

Figure 1. Annual climatic data for maximum, minimum, and mean temperatures and accumulated rainfall (**a**); and climatic data after the onset of ripening, including maximum, minimum, and mean temperatures and accumulated precipitation (**b**), during the 2023 and 2024 growing seasons in Mineiros do Tietê, São Paulo, Brazil.

Canopy management practices included the removal of leaves from the basal part of the branch, shoot thinning, tipping, tying, and berry thinning (40–70%). Top-dressing fertilizations were performed based on soil analysis and the recommendations of Bulletin 100 from the Agronomic Institute of Campinas [19]. At the onset of berry ripening (veraison), protective nets with 10% shading were installed to guard against hail, birds, and bees.

2.4. Experimental Design and Treatments

The experimental design was a randomized complete block design (RCBD), with six treatments and eight blocks, totaling 48 experimental plots. Each plot consisted of five grapevines, spaced 3.0 m between rows and 1.0 m between plants, resulting in a total area of 15 m^2 per plot. A buffer space of 1.5 m was maintained between adjacent plots to reduce edge effects and minimize interference between treatments. The treatments corresponded to six *Vitis vinifera* L. cultivars: 'Sauvignon Blanc', 'Merlot', 'Tannat', 'Pinot Noir', 'Malbec', and 'Cabernet Sauvignon'.

2.5. Evaluated Variables

2.5.1. Duration of Phenological Stages

Phenological stages were characterized according to the phenology scale by Eichhorn and Lorenz [20]. Evaluations were performed twice a week until flowering and weekly thereafter until harvest. The phenological stages observed were:

- Budburst (BB) (4): 50% of buds with visible young leaves;
- Flowering (FL) (23): 50% at full open flowering stage;
- Fruit Set (FS) (27): berries with a diameter greater than 2 mm;
- Verasion (V) (35): 50% of berries showing color change and softening;
- Harvest (H) (38): 100% of berries with appropriate color and physiological ripeness.

2.5.2. Thermal/Temperature Condition

Thermal/temperature condition was estimated using the accumulation of growing degree days (GDD), following the method proposed by Winkler [21], using the equation

$$GDD = \Sigma\,(T_m - 10\ ^\circ C) \times DAP$$

where T_m represents the daily mean temperature (°C), 10 °C is the base temperature for grapevine development, and DAP refers to days after pruning.

2.5.3. Ripening (Maturation) Dynamics

Grape ripening was monitored weekly (precisely every 7 days) from the onset of berry ripening. Soluble solids (SSs, °Brix) were determined using a digital refractometer (Reichert r2i300) (Buffalo, NY, USA); titratable acidity (TA, % tartaric acid) was measured by titrating 5 mL of must diluted in 100 mL of distilled water with 0.1 N sodium hydroxide solution, using phenolphthalein as an indicator. The results were expressed as a percentage of tartaric acid, with 0.1 N NaOH until pH 8.2–8.4; pH was assessed by direct reading using a digital pH meter (Tecnal) (Piracicaba, Brazil); and the maturity index (SS/TA) was calculated. For each sampling, 16 bunches were randomly collected per plot, and six berries were sampled from each cluster (two from each third).

2.5.4. Quantitative and Qualitative Parameters

Harvest was carried out when clusters reached physiological maturity. At this stage, all clusters were counted and weighed to determine the total number of clusters and yield per vine, and to estimate productivity in tons per hectare (t ha^{-1}).

From a sample of 15 clusters per plot, the fresh weight, length, and width of clusters, berries, and rachis were evaluated, as was the number of berries per cluster. For the physicochemical analyses of the must, 100 berries per experimental plot were used. The must was obtained by the manual pressing of the berries.

2.5.5. Bioactive Compounds and Antioxidant Activity

Biochemical Characteristics of Berries

Biochemical analyses were carried out at the Laboratory of Plant Chemistry and Biochemistry, UNESP/Botucatu. Ten berries per cluster were collected (three from the top, four from the middle, and three from the bottom). The berries were halved, immediately frozen in liquid nitrogen, and ground using mortar and pestle. The resulting samples were stored at −20 °C until analysis, which was performed in triplicate.

For supernatant extraction, 300 mg of pulp was mixed with 5 mL of acidified methanol (80%). The mixture was homogenized for 30 s in a vortex and subjected to an ultrasonic bath for 20 min. It was then centrifuged at 6000 rpm for 10 min at 5 °C. The supernatant was collected, and the extraction procedure was repeated once more. The combined extracts were stored in amber vials for further analysis.

Total Phenolic Compounds

Total phenolic content was quantified using the Folin–Ciocalteu reagent, according to Singleton and Rossi [22]. The results are expressed as mg of gallic acid equivalents per kg of fresh matter (mg GAE/kg FM). To achieve this, 0.5 mL of the extract was mixed with 0.5 mL of distilled water, 0.5 mL of diluted Folin–Ciocalteu reagent (1:4), and 2.5 mL of 4% sodium carbonate solution. The mixture was vortexed and kept in the dark for one hour before reading absorbance at 725 nm using a spectrophotometer. A gallic acid standard curve was also established.

Total Flavonoids

Total flavonoid content was determined using aluminum chloride ($AlCl_3$) and a benchtop spectrophotometer. Four milliliters of the extract were homogenized and left in the dark for 30 min, followed by centrifugation at 6000 rpm for 25 min at 5 °C. Absorbance was read at 510 nm, and the results are expressed as mg quercetin equivalent per 100 g fresh matter, according to Popova et al. [23].

Total Anthocyanin Content

Total anthocyanins (TAC) were quantified by the pH differential method proposed by Giusti and Wrolstad [24]. An aliquot of the extract was diluted in two buffer solutions—0.025 M potassium chloride (pH 1.0) and 0.4 N sodium acetate (pH 4.5). Absorbance readings were taken at 510 nm (cyanidin-3-*O*-glucoside peak) and 700 nm (turbidity correction) using a spectrophotometer. The results are expressed as mg cyanidin-3-*O*-glucoside equivalents per 100 g of whole berry grape sample.

Antioxidant Activity

Antioxidant activity was assessed using the DPPH (2,2-diphenyl-1-picrylhydrazyl) radical scavenging assay, according to Brand-Williams et al. [25]. The results were calculated using a Trolox standard curve ($y = 23.606x + 2.5805$; $R^2 = 0.9961$), and are expressed as µg Trolox equivalents per 100 g of sample.

Additionally, the FRAP (Ferric Reducing Antioxidant Power) method was applied according to Brand-Williams et al. [25], using a mixture of 300 mM acetate buffer (pH 3.6), 2.5 mL of 10 mM TPTZ (in 40 mM HCl), and 2.5 mL of $FeCl_3$. Absorbance was read at

595 nm, and the results are expressed as mmol $FeSO_4$ equivalents per 100 g of sample (y = 0.007 + 0.0044; R^2 = 0.9973).

2.6. Statistical Analysis

Data were subjected to tests for normality and homogeneity of variances. A two-factor ANOVA was performed in a split-plot design over time, with cultivars as the main plots and production cycles as the subplots. The experimental design was a randomized complete block with eight replicates and five plants per plot. Means were compared using Tukey's test at a 5% significance. Ripening curve data were analyzed by polynomial regression. All statistical analyses were performed using SISVAR 5.6 software [26].

3. Results and Discussion

3.1. Duration of Phenological Stages

Significant variations were observed in the duration of phenological stages between the 2023 and 2024 production cycles. In 2024, budburst occurred on average 13.1 days after pruning, and harvest took place at 128.0 days, while in 2023 these events occurred at 12.4 and 146.5 days, respectively. Thermal demand was also lower in 2024, with an accumulated total of 1207.6 degree days, in contrast to 1746.3 degree days recorded in the previous cycle, reflecting differences in climatic conditions and management practices between years.

The total duration from pruning to harvest varied among cultivars—Sauvignon Blanc (146 days), Merlot (159 days), Tannat (156 days), Pinot Noir (146 days), Malbec (158 days), and Cabernet Sauvignon (142 days) (Table 1). The duration of the phenological stages varied significantly among grapevine cultivars and production cycles, highlighting the influence of genetic characteristics and environmental conditions on the developmental cycle. The stages of budburst, flowering, fruit set, onset of ripening, and harvest exhibited notable differences, directly affecting vineyard management and harvest planning.

Table 1. Duration of phenological stages (days after pruning) of the grape cultivars 'Sauvignon Blanc', 'Merlot', 'Pinot Noir', 'Tannat', 'Malbec', and 'Cabernet Sauvignon' during the 2023 and 2024 production cycles under subtropical conditions in São Paulo, Brazil.

Cultivar	Production Cycle	BB	FL	FS	V	HV	GDD
Sauvignon Blanc	I	12.3 ± 0.71 Abb *	44.3 ± 2.12 Ca	47.1 ± 2.59 Ca	87.9 ± 0.64 Ba	130.9 ± 0.00 Ca	1700.3 ± 5.89 Ca
	II	13.4 ± 0.52 ABa	31.6 ± 7.44 Ab	34.8 ± 1.49 Bb	85.8 ± 4.27 Ca	121.0 ± 0.00 Cb	1122.7 ± 14.3 Db
Merlot	I	11.9 ± 0.83 Ba	48.3 ± 4.50 Ba	53.5 ± 3.25 Ba	93.8 ± 3.33 Aa	171.0 ± 0.00 Aa	1758.9 ± 1.55 Ba
	II	12.5 ± 0.76 BCa	32.1 ± 0.83 Ab	35.8 ± 2.12 ABb	90.6 ± 2.07 Bb	128.0 ± 0.00 Bb	1218.0 ± 10.1 Bb
Tannat	I	13.0 ± 0.53 Aa	53.9 ± 6.20 Aa	58.4 ± 5.53 Aa	97.0 ± 3.33 Aa	148.0 ± 0.00 Ba	1785.5 ± 2.25 Aa
	II	13.5 ± 0.53 ABa	35.0 ± 1.51 Ab	39.0 ± 0.76 Ab	95.3 ± 1.04 Aa	128.0 ± 0.00 Bb	1204.7 ± 8.45 Cb
Pinot Noir	I	12.8 ± 1.04 Aba	45.3 ± 3.45 BCa	48.1 ± 3.68 Ca	88.3 ± 0.71 Bb	130.0 ± 0.00 Da	1702.8 ± 4.39 Ca
	II	13.5 ± 1.07 ABa	32.9 ± 0.83 Ab	37.0 ± 1.69 ABb	90.8 ± 4.50 Ba	121.0 ± 0.00 Cb	1119.9 ± 18.4 Db
Malbec	I	11.8 ± 0.89 Bb	43.3 ± 1.83 Ca	45.9 ± 1.64 Ca	94.7 ± 1.39 Aa	171.0 ± 0.00 Aa	1769.8 ± 0.49 Ba
	II	13.6 ± 0.52 Aa	32.0 ± 0.76 Ab	35.6 ± 1.77 ABb	93.9 ± 2.03 Aba	128.0 ± 0.00 Bb	1202.7 ± 8.18 Cb
Cabernet Sauvignon	I	12.6 ± 1.06 Aba	43.9 ± 1.55 Ca	46.6 ± 1.77 Ca	94.0 ± 0.00 Aa	129.0 ± 0.00 Eb	1760.3 ± 0.96 Ba
	II	12.3 ± 0.46 Ca	32.0 ± 0.76 Ab	35.8 ± 1.58 ABb	92.8 ± 2.66 Aba	142.0 ± 0.00 Aa	1377.8 ± 5.46 Ab
CV(%)	I	5.95	7.06	6.17	2.54	2.93	0.56
	II	6.11	6.47	5.84	2.82	2.01	0.62
Mean		12.7	39.5	43.1	92.1	137.3	1476.90

* Means followed by different uppercase letters in columns and lowercase letters in rows differ significantly according to Tukey's test at a 5% probability level ($p > 0.05$). Values are expressed as mean ± standard deviation (n = 8). Abbreviations: BB—budburst; FL—full flowering; FS—fruit set; V—verasion; HV—harvest; GDD—growing degree days. CV: coefficient of variation.

The observed variations in the duration of phenological stages between the 2023 and 2024 production cycles under double-pruning management align with findings from recent studies conducted in Brazil. For instance, research on the 'BRS Isis' grape cultivar in São Paulo demonstrated that the duration from pruning to harvest varied significantly across different harvest seasons, ranging from 137 to 168 days, influenced by climatic conditions and rootstock combinations. The thermal demand also fluctuated, with accumulated degree days between 1651 and 1841, highlighting the impact of environmental factors on grapevine development [27]. Similarly, studies on the 'BRS Núbia' seedless grape in São Manuel, SP, reported that the total cycle duration from pruning to harvest was 155 days, with thermal requirements of 1806.4 degree days [28]. These findings underscore the importance of understanding the phenological behavior and thermal requirements of grape cultivars under double-pruning systems to optimize vineyard management and harvest planning in subtropical regions of Brazil [29,30].

The duration of phenological stages in grapevines subjected to double pruning varies considerably between production cycles and cultivars, being influenced by both genetic factors and environmental conditions. In subtropical climates, the complete cycle from pruning to harvest has been reported to range from approximately 146 to 172 days, depending on the grape cultivar and the specific climatic conditions of each growing season [31]. For instance, in 'Syrah' grapevines cultivated under the double-pruning system in Southeastern Brazil, researchers observed that the summer cycle was shorter than the winter cycle. This was primarily due to a more rapid maturation period during the summer, which led to variations in the duration of stages such as budburst, flowering, veraison, and harvest [11]. These results are consistent with observations that the thermal time requirement (expressed in growing degree days, GDD) also differs between production years, reflecting variability in climate and management practices.

Such fluctuations in the phenological timing have direct implications for vineyard management strategies and harvest scheduling. Therefore, understanding the dynamics of phenological development across cultivars and growing seasons is essential for optimizing the quality of grapes and wine produced under winter viticulture conditions in subtropical regions [16,31].

3.1.1. Budburst

Budburst showed little variation between the production cycles—an average of 12.4 days in 2023 and 13.1 days in 2024. The shortest period was recorded for 'Merlot' (11.9 ± 0.83 days) and 'Malbec' (11.8 ± 0.89 days) in the first cycle, while the longest was observed for 'Malbec' in the second cycle (13.6 ± 0.52 days). This phenological stage marks the beginning of the vegetative cycle and is crucial for the overall development rate. Early budburst may accelerate the cycle but increases the risk of damage from late frosts in regions with unstable climates [32].

3.1.2. Full Flowering

Marked differences were observed between production cycles. The period from pruning to flowering was longer in 2023 (46.5 days) compared to 2024 (32.6 days), due to higher temperatures in the latter (Figure 1), which accelerated phenological development. The cultivars 'Tannat' (53.9 ± 6.20 days) and 'Merlot' (48.3 ± 4.5 days) exhibited the longest pruning-to-flowering periods. In contrast, 'Sauvignon Blanc' had the shortest (37.9 days). Early flowering, as observed in 'Sauvignon Blanc' and 'Malbec', may be advantageous in regions requiring shorter production cycles.

This reduction is likely attributable to higher average temperatures in 2024, which accelerated phenological development. These findings are consistent with previous research conducted by Júnior and Hernandes [33], who evaluated the phenology and maturation dynamics of 'Syrah' grapes cultivated under a double pruning system during the winter season. Their study reported that the period from pruning to flowering can vary depending on climatic conditions, particularly temperature, which plays a crucial role in modulating the timing of phenological events. In our study, cultivars such as 'Tannat' and 'Merlot' showed extended durations (53.9 and 48.3 days, respectively), while 'Sauvignon Blanc' flowered earlier (37.9 days), indicating genotypic variability in developmental response. As earlier flowering may be beneficial in subtropical regions where compressed production cycles are desirable to avoid adverse weather conditions later in the season [34].

3.1.3. Fruit Set

The duration from pruning to fruit set was shorter in 2024 (36.3 days) than in 2023 (49.9 days). 'Tannat' consistently exhibited the longest cycle in both years (58.4 ± 5.53 days in 2023 and 39.0 days in 2024), while 'Sauvignon Blanc' showed the shortest durations (47.1 days in 2023 and 34.8 ± 1.49 days in 2024). This stage is highly sensitive to environmental factors such as water deficit and temperature [35], and cultivars with shorter cycles may exhibit greater resilience under climatic stress.

3.1.4. Verasion

The onset of ripening occurred slightly earlier in 2024 (91.5 days) compared to 2023 (92.1 days). 'Tannat' exhibited the latest ripening (97.0 ± 3.33 days in 2023 and 95.3 ± 1.04 days in 2024), whereas 'Sauvignon Blanc' was the earliest (87.9 ± 0.64 days in 2023 and 85.8 ± 4.27 days in 2024). This stage directly influences wine quality as it defines the balance between sugars and acids. Early ripening, as seen in 'Sauvignon Blanc', is often associated with fresher wines with higher acidity [27,36].

3.1.5. Harvest

In the 2023 production cycle, the period from pruning to harvest ranged from 121 to 185 days, with 'Pinot Noir' and 'Sauvignon Blanc' being the earliest cultivars and 'Cabernet Sauvignon' the latest. In 2024, all cultivars presented shorter cycles, with 'Sauvignon Blanc' again showing the shortest duration (121 days), while 'Cabernet Sauvignon' remained the last to be harvested (142 days).

The shortened cycle in 2024 was mainly attributed to climatic conditions between February and June, with average temperatures of 23.4 °C and 21 °C, respectively. Shorter cycles can reduce the level of biotic stress caused by pathogens, favoring early-ripening cultivars in regions with short summers [37,38]. Recent studies have highlighted that under tropical and subtropical climates, early harvests supported by short-cycle cultivars contribute significantly to reducing fungal outbreaks and ensuring technological maturity at harvest [39]. Cycle duration influences both sensory attributes and vineyard management feasibility, and is thus critical for production planning.

3.2. Thermal Demand of Grapevines

In 2023, the average thermal requirement was higher (1476.9 GDD), reflecting greater vine demand to complete the cycle. 'Tannat' showed the highest accumulation (1785.5 ± 2.25 GDD), followed by 'Malbec' (1769.8 ± 0.49) and 'Cabernet Sauvignon' (1760.3 ± 0.96). In contrast, the 2024 cycle required less thermal accumulation, with 'Cabernet Sauvignon' needing 1377.8 ± 5.46 GDD. A base temperature of 10 °C was used for GDD calculations, below which no vegetative development occurs [40]. In comparison with other regions, 'Cabernet Sauvignon' requires around 1200 GDD in Santa Catarina (Epagri, 2005) [41], and up to 1625 GDD in semiarid regions [42]. Thermal demand directly affects the pace of development, with higher temperatures shortening the intervals between phenological stages [43]. Understanding thermal requirements allows for the prediction of phenological progression and enables better harvest scheduling, optimizing both management practices and labor organization.

These results are consistent with previous research on grapevines grown under double-pruning or double-cropping systems in subtropical regions. Koyama et al. [29], for instance, reported thermal requirements of 1744.1 GDD for the summer crop and 1845.0 GDD for the off-season crop of the hybrid seedless cultivar BRS Melodia under a double-cropping system. These findings highlight the role of seasonal and environmental variations in shaping phenological dynamics and thermal accumulation patterns.

Further evidence is provided by Monteiro et al. [28], who observed that the cultivar 'BRS Núbia' required an average of 1,807.5 GDD across two production cycles, demonstrating the relative consistency in thermal requirements of some cultivars under similar environmental conditions. In addition, Sánchez et al. [27] examined the 'BRS Isis' grapevine grafted onto different rootstocks and reported thermal demands ranging from 1651 to 1841 GDD, showing how both climatic factors and rootstock genotype influence grapevine development in subtropical climates. Taken together, these studies reinforce the importance of characterizing the thermal requirements of each cultivar under specific growing conditions, particularly when innovative management systems such as double pruning are employed. Understanding these thermal patterns is critical for optimizing vine performance, harvest scheduling, and grape quality under changing climatic conditions.

3.3. Yield Components and Physical Characteristics

Significant differences were observed in the physical characteristics of clusters, berries, and rachis among cultivars (Table 2), with variations attributed to varieties and climatic conditions. 'Sauvignon Blanc' showed consistency across cycles, with cluster weights of 87.9 g (Cycle I) and 84.5 g (Cycle II), and an increase in cluster length (8.29 cm in Cycle II). 'Tannat' exhibited a marked increase in cluster weight (from 125.4 g to 143.0 g) and overall size.

In contrast, 'Pinot Noir' and 'Malbec' showed reduced cluster weight in the second cycle. Cluster width increased across all cultivars in Cycle II compared to Cycle I, while length varied, with decreases observed only in 'Sauvignon Blanc' and 'Cabernet Sauvignon'. Berry size also varied, with 'Malbec' showing notable increases in both length and width. Smaller berries (with a mass below 2.0 g) are favorable for the extraction of phenolic compounds, which are essential for red wine production [42]. The number and size of berries directly affect cluster mass [43]. Cultivars such as 'Tannat' produce robust clusters suitable for full-bodied wines, while 'Pinot Noir' tends to yield lighter, more aromatic wines.

These findings align with previous studies that have reported cultivar-specific responses to double pruning. For example, Favero et al. [17] observed that 'Syrah' grapevines under double pruning in Southeastern Brazil produced higher yields and improved fruit composition during the winter season compared to the summer, highlighting the influence of seasonal conditions on grape development. Similarly, Tecchio et al. [44] reported that 'Cabernet Franc' and 'Syrah' cultivars grafted onto 'IAC 766' rootstock exhibited superior yield performance and berry quality under subtropical conditions, emphasizing the role of rootstock-scion combinations in determining vine productivity and fruit characteristics.

Furthermore, the increase in cluster width across all cultivars in Cycle II, despite variations in cluster length, suggests an adaptive morphological response to environmental factors. Berry size also varied, with 'Malbec' showing significant increases in both length and width, which could be attributed to favorable climatic conditions or improved assimilate partitioning during the second cycle. These observations underscore the importance of considering both genetic and environmental factors in vineyard management practices.

Table 2. Physical characteristics of clusters, berries, and rachis of wine grape cultivars grown under double-pruning management in a subtropical climate during the 2023 (I) and 2024 (II) growing seasons in São Paulo, Brazil.

Cultivar	Production Cycle	FWC	CL	CW	CL/CW	FWB	BL	BW	BL/BW	NBC	FWR
Sauvignon Blanc	I	87.9 ± 16.9 bcA *	7.90 ± 1.07 Ba	4.67 ± 0.37 BCb	1.69 ± 0.16 Bca	1.41 ± 0.10 Ba	1.45 ± 0.05 ABa	1.28 ± 0.03 Ba	1.13 ± 0.02 ABa	59.3 ± 12.3 Ba	4.27 ± 0.78 Ca
	II	84.5 ± 4.05 BCa	8.29 ± 0.54 Ca	6.75 ± 0.61 BCa	1.24 ± 0.12 BCb	1.36 ± 0.10 Ca	1.42 ± 0.03 BCa	1.27 ± 0.03 Ca	1.11 ± 0.03 Aba	58.8 ± 5.61 Ba	5.00 ± 0.59 Ba
Merlot	I	56.6 ± 8.47 Da	8.25 ± 0.83 Bb	5.25 ± 0.63 BCb	1.58 ± 0.12 Ca	1.09 ± 0.04 Cb	1.14 ± 0.14 Cb	1.10 ± 0.03 Db	1.03 ± 0.14 Ba	49.4 ± 8.29 Ba	3.01 ± 0.48 Da
	II	68.0 ± 10.9 CDa	10.3 ± 0.82 Ba	7.00 ± 0.85 Ba	1.49 ± 0.08 Aba	1.27 ± 0.05 CDa	1.31 ± 0.19 Ca	1.35 ± 0.03 Ba	0.98 ± 0.15 Ba	50.8 ± 7.08 ABa	3.50 ± 0.69 Ca
Tannat	I	125.4 ± 20.4 Ab	10.3 ± 1.98 Ab	14.1 ± 1.17 Ab	0.72 ± 0.10 Da	1.41 ± 0.12 Bb	1.37 ± 0.04 Bb	1.26 ± 0.03 BCb	1.09 ± 0.02 ABa	83.2 ± 8.75 Aa	7.38 ± 0.54 Ab
	II	143.0 ± 19.8 Aa	15.3 ± 1.98 Aa	18.1 ± 1.82 Aa	0.84 ± 0.04 Da	1.64 ± 0.15 Ba	1.57± 0.06 Aba	1.39 ± 0.08 Ba	1.13 ± 0.04 Aa	83.0 ± 8.61 Aa	8.20 ± 0.36 Aa
Pinot Noir	I	66.1 ± 10.7 CDa	6.88 ± 0.74 Ba	4.22 ± 0.74 Cb	1.67 ± 0.27 BCa	1.09 ± 0.14 Ca	1.40 ± 0.20 ABa	1.19 ± 0.03 Ca	1.18 ± 0.19 Aa	58.3 ± 8.88 Ba	2.63 ± 0.80 Da
	II	46.3 ± 8.16 Db	6.18 ± 0.59 Da	5.52 ± 0.33 CDa	1.12 ± 0.11 Cb	1.21 ± 0.14 CDa	1.30 ± 0.03 Cb	1.19 ± 0.04 Da	1.09 ± 0.02 Aba	26.9 ± 7.60 Db	2.21 ± 0.60 Da
Malbec	I	137.2 ± 28.6 Aa	10.5 ± 1.21 Aa	5.53 ± 0.52 Bb	1.90 ± 0.21 Ba	2.11 ± 0.18 Aa	1.55 ± 0.06 Ab	1.45 ± 0.05 Ab	1.07 ± 0.03 ABa	62.0 ± 11.4 Ba	6.19 ± 1.54 Ba
	II	97.0 ± 22.8 Bb	11.3 ± 1.34 Ba	7.65 ± 0.83 Ba	1.48 ± 0.17 ABb	2.13 ± 0.11 Aa	1.67 ± 0.05 Aa	1.50 ± 0.03 Aa	1.12 ± 0.02 Aba	43.6 ± 8.88 CDb	3.75 ± 0.76 Cb
Cabernet Sauvi-gnon	I	99.9 ± 11.0 Ba	10.2 ± 1.10 Aa	4.14 ± 0.20 Ca	2.48 ± 0.27 Aa	1.26 ± 0.09 BCa	1.32 ± 0.02 Ba	1.22 ± 0.02 BCa	1.07 ± 0.02 ABa	75.7 ± 6.27 Aa	4.60 ± 0.49 Ca
	II	56.0 ± 11.3 Db	7.61 ± 0.88 CDb	4.70 ± 0.70 Da	1.63 ± 0.05 Ab	1.10 ± 0.03 Db	1.30 ± 0.01 Ca	1.19 ± 0.00 Da	1.09 ± 0.00 ABa	47.9 ± 8.13 BCDb	2.97 ± 0.80 CDb
CV(%)	I	21.5	17.1	14.6	13.1	9.38	8.92	3.75	2.53	16.6	16.7
	II	16	8.86	9.15	10.2	7.2	2.53	2.50	2.48	12.9	19.1
Média		88.9	9.41	7.31	1.49	1.42	1.40	1.28	1.10	50.1	4.47

* Means followed by different uppercase letters in columns and lowercase letters in rows differ significantly according to Tukey's test at a 5% probability level ($p > 0.05$). Values are expressed as mean ± standard deviation (n = 8). Abbreviations: FWC: fresh weight of clusters; CL: cluster length; CW: cluster width; CL/CW: cluster length-to-width ratio; FWB: fresh weight of berries; BL: berry length; BW: berry width; BL/BW: berry length-to-width ratio; NBC: number of berries per cluster; FWR: fresh weight of rachis. CV: coefficient of variation.

3.4. Fertility

The number of clusters per plant varied significantly among cultivars. 'Cabernet Sauvignon' recorded the highest number in the second cycle (11.8 ± 0.99 clusters plant^{-1}), while 'Pinot Noir' showed the lowest value in the first cycle (4.88 ± 1.36 clusters plant^{-1}) (Table 3). The increased yield observed in Cycle II reflects favorable environmental conditions, such as temperature and water availability, which contributed to greater biomass accumulation. This trend was consistent across cultivars, suggesting that the second cycle was more suitable for maximizing productivity.

Significant differences ($p < 0.05$) between cycles and among cultivars confirm the influence of double-pruning management under subtropical conditions. The superior performance in Cycle II may be associated with a greater number of fruitful shoots and increased bud fertility. These results indicate a strong genotype × environment interaction. 'Cabernet Sauvignon' stood out for its high yield and productivity, demonstrating good adaptability to the double-pruning system, whereas 'Malbec' and 'Pinot Noir' showed lower performance, indicating greater sensitivity to the growing conditions.

Similarly, 'Pinot Noir' has been studied under double-cropping systems—conceptually similar to double pruning—in which a second crop is forced within the same season. These studies have shown that the cultivar can respond positively, but yield components such as the number of clusters per plant and berry size may be significantly influenced by the timing of bud forcing and prevailing environmental conditions. The cultivar has been shown to exhibit a certain degree of sensitivity to modifications in phenological timing, which may impact its reproductive efficiency under non-traditional management strategies [45].

Table 3. The yield, productivity, and number of clusters per plant of wine grape cultivars grown under double-pruning management in a subtropical climate during the 2023 (I) and 2024 (II) growing seasons in Mineiros do Tietê, São Paulo, Brazil.

Cultivar	Production Cycle	Yield	Productivity	Number of Clusters per Vine
		(kg Vine^{-1})	(t ha^{-1})	(un)
Sauvignon Blanc	I	2.13 ± 0.29 Ab *	7.59 ± 1.02 Ab	9.25 ± 5.09 Ab
	II	2.75 ± 0.15 Bb	9.82 ± 0.53 Ba	11.6 ± 5.18 Aba
Tannat	I	0.88 ± 0.24 Bb	3.12 ± 0.85 Bb	5.63 ± 1.41 BCb
	II	1.09 ± 0.01 Ca	3.89 ± 0.05 Ca	7.38 ± 1.06 Ca
Pinot Noir	I	0.70 ± 0.09 BCb	2.50 ± 0.33 BCb	4.88 ± 1.36 Cb
	II	0.89 ± 0.01 CDa	3.18 ± 0.04 CDa	7.63 ± 1.19 Ca
Malbec	I	0.50 ± 0.13 Cb	1.79 ± 0.48 Cb	6.37 ± 1.41 BCb
	II	0.78 ± 0.03 Da	2.77 ± 0.11 Da	10.0 ± 1.20 Ba
Cabernet Sauvignon	I	2.13 ± 0.28 Ab	7.59 ± 0.99 Ab	7.38 ± 1.06 Bb
	II	3.03 ± 0.06 Aa	10.8 ± 0.20 Aa	11.8 ± 0.99 Aa
CV(%)	I	11.6	11.6	39.5
	II	10.6	10.7	5.2
Mean		1.49	5.3	8.2

* Means followed by different uppercase letters in columns and lowercase letters in rows differ significantly according to Tukey's test at a 5% probability level ($p > 0.05$). Values are expressed as mean ± standard deviation (n = 8). CV: coefficient of variation.

Regarding 'Malbec', late shoot pruning (LSP)—a practice related to double pruning—has been investigated as a strategy to shift phenological phases and improve grape quality. While LSP has been associated with reductions in the number of clusters and total yield, it can simultaneously enhance berry composition, particularly in terms of phenolic content. This trade-off underscores the importance of aligning management techniques with specific production goals, such as favoring quality over quantity in premium wine production [46]. These findings collectively highlight the relevance of genotype × environment interactions under double-pruning systems. They also reinforce the need for cultivar-specific management strategies to maximize both yield potential and grape quality in subtropical viticultural regions.

3.5. Yield and Productivity

Grape yield varied significantly among cultivars and between production cycles. The cultivars 'Cabernet Sauvignon' and 'Sauvignon Blanc' showed the highest yields in the second cycle (3.03 ± 0.06 kg plant^{-1} and 2.75 ± 0.15 kg plant^{-1}, respectively), whereas 'Malbec' exhibited the lowest performance (0.50 ± 0.13 kg plant^{-1} in Cycle I), likely due to poor adaptation to the local conditions. According to Steduto et al. [32], the number of clusters per plant has a direct influence on grapevine yield under this management system. All cultivars showed increased yield in Cycle II compared to Cycle I, which can be attributed to a greater formation of fruitful shoots.

These results suggest that double-pruning management under subtropical conditions benefited cultivars such as 'Sauvignon Blanc' and 'Cabernet Sauvignon', which demonstrated better adaptation and higher productivity. 'Sauvignon Blanc' has shown high productivity in different Brazilian regions, particularly when managed with the double-pruning technique for winter wine production. In contrast, cultivars such as 'Malbec' may require specific management adjustments to optimize their performance under these conditions. By specific management, we refer to tailored viticultural practices such as modifications in pruning type and timing, shoot and cluster thinning, canopy management, irrigation scheduling, and fertilization strategies, all adjusted to the physiological responses and phenological behavior of the cultivar within the regional climatic context. These interventions aim to improve fruit quality and yield stability, ensuring the cultivar reaches its full productive and enological potential.

According to Rizzon et al. [47], the factors influencing grapevine productivity can be grouped into permanent and cultural components. Permanent factors include imposed elements such as climate, soil, and biological environment, as well as selected factors such as cultivar, rootstock, planting density, and row orientation. Cultural factors encompass training systems, pruning strategies, irrigation methods, fertilization, and phytosanitary management. Additionally, Pedro junior et al. [48] emphasized that grape production is influenced by multiple factors that determine both fruit quality and productivity. The production of fine wine grapes demands extensive technical knowledge to perform the cultural practices required for high-quality grapes, as well as significant labor input for manual vineyard operations, which raises the overall production costs [49]. The cultivar 'Merlot' did not produce enough grapes to allow for the evaluation of physical characteristics under double-pruning management in subtropical conditions. This contrasts with southern Brazil, where it is considered one of the most important cultivars.

3.6. Berry Ripening

On average, berry ripening began 93 days after pruning (DAP) during the 2023 cycle, with a thermal accumulation of 1745.3 growing degree days. The cultivars 'Sauvignon Blanc', 'Merlot', 'Tannat', 'Pinot Noir', 'Malbec', and 'Cabernet Sauvignon' began ripening at 88, 94, 97, 88, 95, and 94 DAP, respectively. The ripening phenology stage showed significant variation in duration among cultivars ($p > 0.05$), although a consistent pattern of uniformity was observed within each variety. Ripening in the 2023 season began on 16 May (93 DAP), with 'Sauvignon Blanc' and 'Pinot Noir' showing early ripening (11 May 88 DAP), while 'Malbec' and 'Tannat' exhibited the latest onset at 95 and 97 DAP, respectively.

Analyses of the ripening curve revealed significant differences among cultivars for soluble solids (°Brix), pH, titratable acidity (TA), and maturity index (MI). Over the course of the production cycle, a typical ripening trend was observed—increases in SS and pH and a decrease in acidity, reflecting progressive ripening (Figures 2–4). Quadratic models provided a good fit for the variables as a function of days after the onset of ripening. In Figures 2–5, * denotes a statistically significant difference at the 5% level ($p < 0.05$), while ** indicates a highly significant difference at the 1% level ($p < 0.01$).

3.6.1. Soluble Solids Content

Soluble solids (SSs) content, expressed in °Brix, is one of the main indicators of grape maturity and grape quality, as it reflects sugar accumulation in the berries. Among the evaluated cultivars, 'Tannat' reached the highest value, with 23.4 °Brix at 128 DAP, indicating a high potential alcohol content for winemaking. This result reinforces its good adaptation to the double-pruning system in a subtropical climate, which favors sugar synthesis [50,51]. 'Pinot Noir' and 'Sauvignon Blanc' also showed consistent increases in SS content, reaching ideal levels for wines with balanced acidity and alcohol content. As noted by Falcão et al. [52], although °Brix is not a direct measurement of sugars, its increase directly reflects the physiological maturity of the grape and the quality of the must.

High SS values are directly related to the sensory profile and complexity of wines, as they influence the synthesis of phenolic compounds, aromatic precursors, and organic acids [53,54]. According to Brazilian regulations, wine grapes must reach 18–21 °Brix for optimal harvest [55]. All cultivars in this study achieved values within this range, indicating appropriate technological maturity. 'Tannat', in particular, stood out with 23.4 °Brix at

128 DAP (Figure 4), demonstrating excellent sugar accumulation and strong potential for producing structured wines.

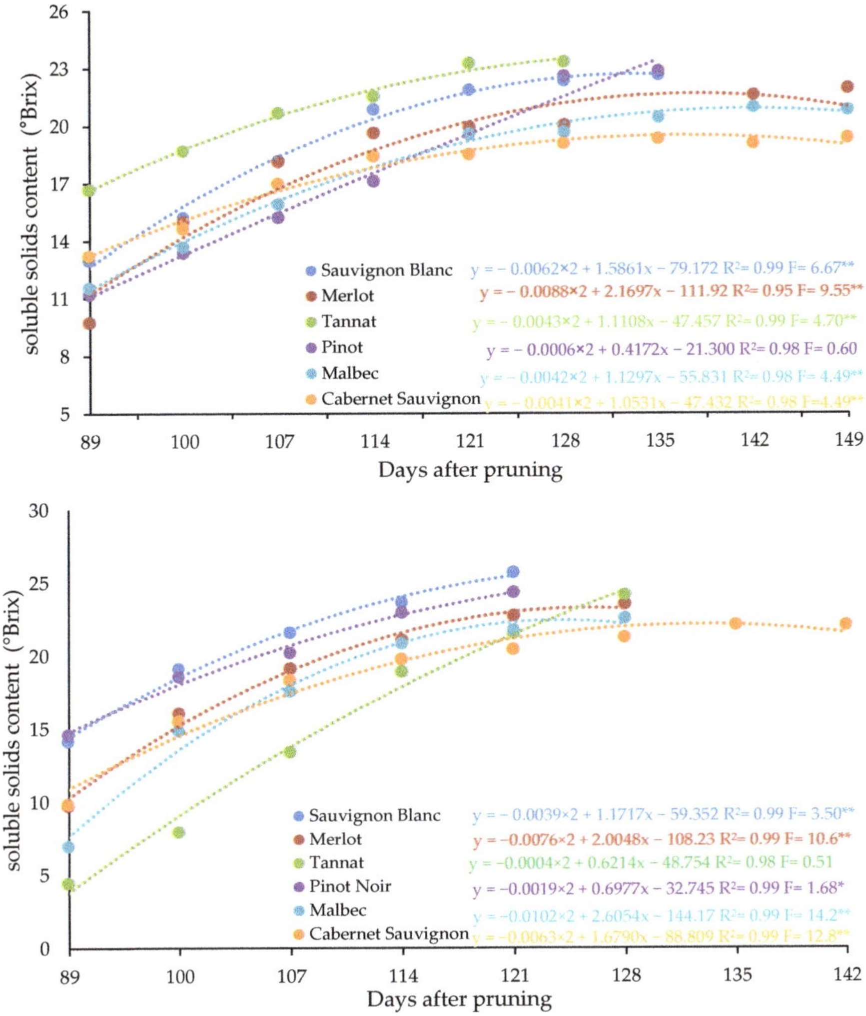

Figure 2. Evolution of soluble solids content (°Brix) in the must of wine grapes as a function of days after pruning, in grapevines managed under the double-pruning system. Mineiros do Tietê, São Paulo, Brazil.

According to Falcão et al. [52], monitoring the progression of berry ripening through variables such as soluble solids and titratable acidity is essential to characterize the chemical composition up to the point of physiological maturity. The primary criterion for determining the optimal harvest time is sugar content, measured in degrees Brix, which represents the total soluble solids—approximately 90% of which are sugars [56].

The decision on the harvest date is one of the most critical for grape growers, as it involves factors beyond physicochemical parameters, such as flavor, berry and seed integrity, sanitary status, and the intended wine style. Moreover, research has demonstrated that the timing of harvest can significantly influence the aromatic composition of wines. A study by Allamy et al. [57] highlighted that delaying harvest dates in Merlot and Cabernet-

Sauvignon grapes led to increased concentrations of compounds associated with cooked fruit aromas, thereby altering the wine's sensory profile. These findings underscore the importance of integrating sensory evaluations with traditional physicochemical analyses to make informed harvest decisions that align with the intended wine style.

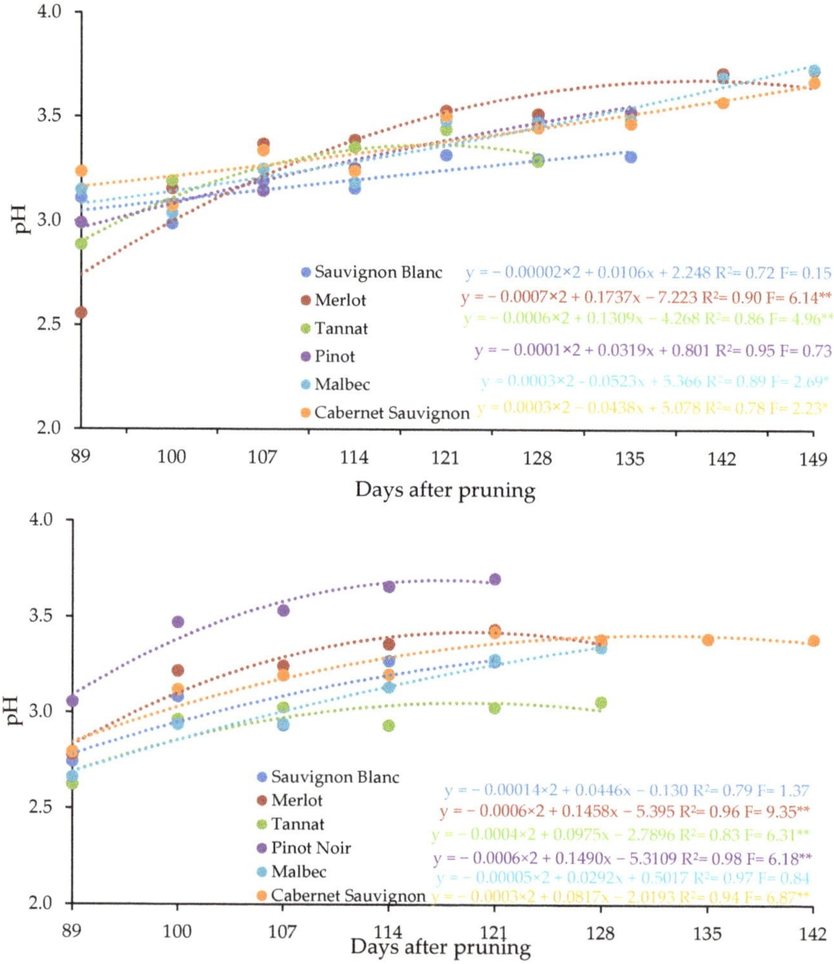

Figure 3. Evolution of must pH in wine grapes as a function of days after pruning in grapevines managed under the double-pruning system. Mineiros do Tietê, São Paulo, Brazil.

The cultivars 'Tannat', 'Pinot Noir', and 'Sauvignon Blanc' showed a quadratic increase in soluble solids over days after pruning, reflecting efficient sugar accumulation and favorable SS/TA ratios, which contribute to enological balance, similarly reported in previous studies [58,59]. For red wine production, Júnior et al. [60] recommend a minimum of 21 °Brix. Although °Brix is an indirect measure of sugars—since it also includes vitamins, phenolics, pectins, and organic acids [61], sugars still account for up to 90% of the total soluble solids.

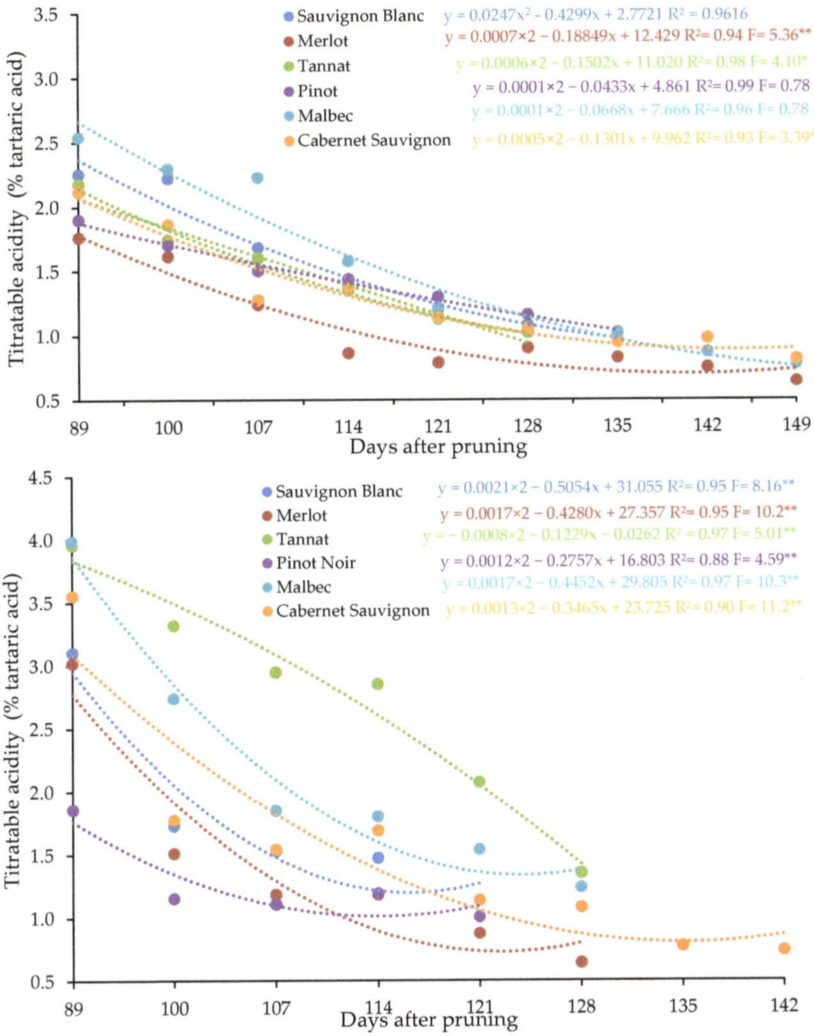

Figure 4. Evolution of titratable acidity in the must of wine grapes as a function of days after pruning in grapevines managed under the double-pruning system. Mineiros do Tietê, São Paulo, Brazil.

Harvesting for fine red wines should be performed at the peak of both technological and phenolic ripeness, with soluble solids ranging from 21 to 26 °Brix and titratable acidity between 5 and 6.5 g L^{-1} [61,62]. °Brix is also critical for estimating potential alcohol content, as higher values indicate greater fermentable sugar levels [63]. Harvest was performed when soluble solids and titratable acidity values stabilized between two consecutive samplings, as described by Falcão et al. [52]. The cultivars 'Tannat', 'Pinot Noir', and 'Sauvignon Blanc' reached satisfactory SS levels, with 'Tannat' standing out by reaching 23.4 °Brix at 128 DAP, indicating a high potential for producing wines with greater alcohol content and complexity (Figure 4).

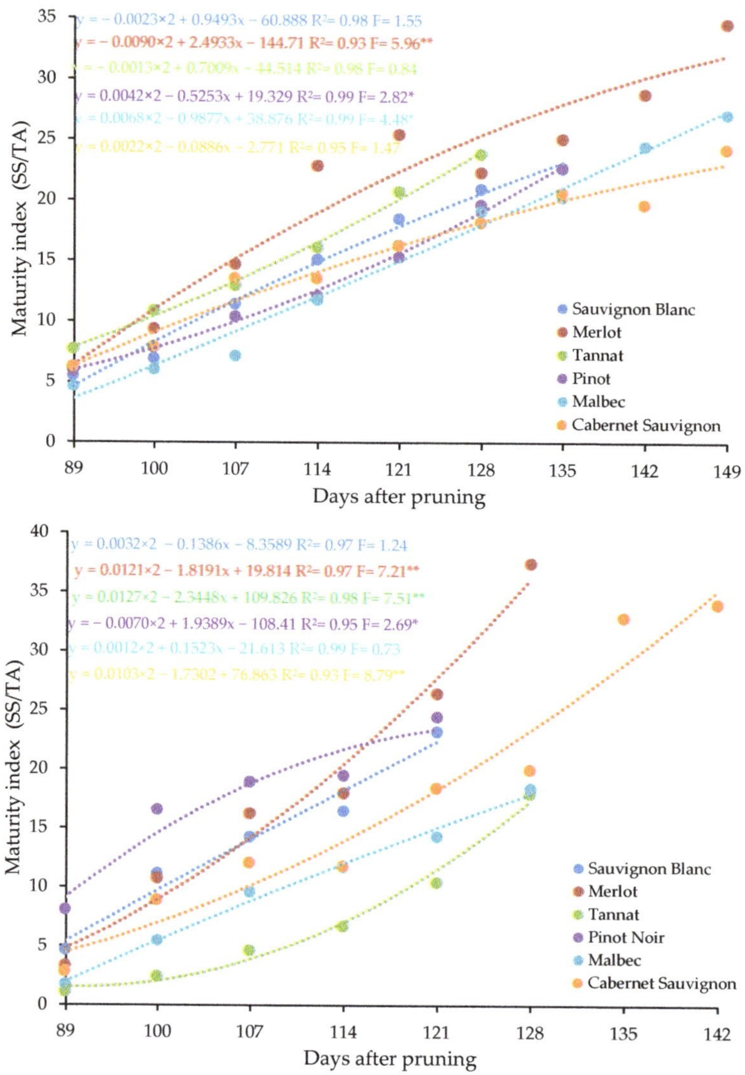

Figure 5. The evolution of the maturity index in the must of wine grapes as a function of days after pruning in grapevines managed under the double-pruning system. Mineiros do Tietê, São Paulo, Brazil.

3.6.2. pH

The analysis of pH evolution throughout the 2023 and 2024 production cycles revealed significant differences among cultivars. Statistical models fitted to the data indicated a linear trend for 'Sauvignon Blanc' and 'Pinot Noir', and a quadratic trend for 'Tannat', 'Malbec', 'Merlot', and 'Cabernet Sauvignon', reflecting distinct ripening dynamics.

In the 2023 cycle, pH values remained relatively stable, with 'Sauvignon Blanc' and 'Cabernet Sauvignon' standing out for their consistency. 'Tannat' exhibited a progressive increase in pH, indicating advanced ripening, while 'Pinot Noir' remained close to levels considered ideal for winemaking. In 2024, differences among cultivars became more pronounced. 'Cabernet Sauvignon' showed a greater increase in pH compared to the previous year, reflecting a higher accumulation of compounds associated with maturity. 'Merlot' and

'Malbec' also exhibited elevated pH values, while 'Sauvignon Blanc' maintained similar behavior to that observed in the previous cycle, demonstrating good consistency.

Quadratic pH behavior peaked at 128 days after pruning (DAP), ranging from 3.2 ('Tannat') to 3.7 ('Malbec'). 'Merlot' recorded the highest pH value at the end of ripening (3.72), indicating an advanced stage of maturity. The increase in pH, together with the reduction in titratable acidity, suggests optimal technological ripening, favoring wines with a smoother and more balanced sensory profile [54]. pH has a direct influence on wine quality, affecting microbial stability, color, and sensory characteristics Gerra and Pereira [8], reported that wines produced in southeastern Brazil from cultivars such as 'Cabernet Sauvignon', 'Malbec', and 'Merlot' usually present pH values between 3.2 and 3.5. Values between 3.2 and 3.4 are considered ideal, as they are associated with anthocyanin stability and color intensity [64].

During ripening, acidity reduction is mainly due to the degradation of malic acid via cellular respiration, which leads to an increase in pH [65,66]. Thus, double-pruning management under subtropical conditions proved effective in achieving pH levels suitable for quality wine production, ensuring a balance between freshness, acidity, and stability. The pH is an indicator directly related to berry acidity and showed significant variation among the evaluated cultivars, reaching its maximum at 128 days after pruning (DAP). Values ranged from 3.2 for 'Tannat' to 3.7 for 'Malbec' (Figure 5).

The evolution of pH throughout the ripening cycle provides relevant information regarding the degree of maturation and the acid–base characteristics of the must, being a decisive parameter for wine quality. pH affects flavor, microbial stability, and wine color, directly influencing the sensory profile [52]. Among the cultivars, 'Merlot' showed the greatest variation, reaching 3.72 at 142 DAP, indicating a sharp decrease in acidity and a more advanced stage of ripening. Elevated pH values are associated with lower perceived acidity, favoring wines with a smoother mouthfeel.

According to Guerra and Pereira [8], wines produced in southeastern Brazil from red cultivars such as 'Cabernet Sauvignon', 'Malbec', 'Merlot', 'Pinot Noir', and the white 'Sauvignon Blanc' generally present pH values ranging from 3.2 to 3.5, which supports the findings of the present study. Secondly, Rojas-Lara and Morrison [67], reported that pH tends to increase linearly during ripening, whereas total acidity decreases exponentially, mainly due to malic acid degradation. In environments with higher water availability, pH tends to be lower for a given level of acidity, highlighting the influence of environmental conditions on grape chemical composition.

3.6.3. Titratable Acidity (TA)

Titratable acidity (TA), expressed as tartaric acid equivalents, showed a quadratic decline throughout the ripening process in all cultivars. The highest acidity values were observed in 'Pinot Noir' and 'Tannat', which stood out compared to the other cultivars (Figure 4). This reduction in TA during ripening is essential for achieving a balanced wine taste. High acidity levels can result in wines with a more astringent mouthfeel, while controlled acidity contributes to freshness and sensory stability.

Cooler temperatures during the production cycle help preserve berry acidity, which in turn favors the retention of aroma compounds and enhances wine liveliness, as reported by Esteban et al. [68]. According to Brazilian regulations, acceptable acidity levels for fine wines range from 40 to 130 meq L^{-1}, with the ideal interval being between 55 and 130 meq L^{-1} [58]. The cultivars 'Pinot Noir' and 'Tannat' remained within this range, ensuring enological quality and an appropriate structure for wines with greater aging potential.

Titratable acidity (TA), expressed as tartaric acid equivalents, showed a declining trend during ripening across all evaluated cultivars, with a more pronounced reduction during the final stages of the cycle. In contrast, soluble solids content increased as ripening progressed due to sugar accumulation. This inverse relationship between SS and TA was consistent across both production cycles (Figure 4).

The cultivars 'Pinot Noir' and 'Tannat' maintained the highest acidity levels throughout the period, reflecting their genetic characteristics and greater potential for producing premium wines with aging capacity. The combined evaluation of these variables is essential for determining the optimal harvest point, as technological maturity—necessary for fine wine production—is achieved when physicochemical parameters are balanced [69]. The acidity values observed are in accordance with Brazilian regulations for fine wines, which stipulate acceptable limits between 40 and 130 meq L^{-1}, with the ideal range between 55 and 130 meq L^{-1} [58]. The high acidity recorded in some cultivars may be attributed to cooler temperatures in high-altitude regions, which slow the degradation of organic acids such as malic acid [70].

Previous studies, such as that of Brighenti et al. [71], reported similar trends for the cultivar 'Sauvignon Blanc' in high-altitude regions of Brazil, reinforcing the present findings. Several factors contribute to the reduction in TA, including physiological, genetic, and environmental variables [72–74]. The decline in acidity is mainly attributed to the respiratory degradation of malic acid, dilution resulting from berry growth, and the salt formation of organic acids [75].

3.6.4. Maturity Index (SS/TA)

The maturity index (MI), calculated as the ratio between soluble solids content and titratable acidity (SS/TA), showed significant variation among cultivars, reaching values of 20.6 for 'Cabernet Sauvignon' and 28.8 for 'Merlot' at 142 days after pruning.

This indicator is widely used to assess the balance between sugar and acidity and is fundamental in defining the wine style to be produced. According to Ribéreau-Gayon [76], determining the ideal harvest time involves integrating technological (SS/TA), aromatic, and phenolic parameters to maximize enological potential.

The highest MI values observed in 'Merlot' indicate a favorable balance between sugars and acidity, supporting the production of wines with a smoother and rounder profile. In contrast, 'Cabernet Sauvignon' presented a lower MI, a feature that may impart greater freshness and structure to the wine—desirable attributes for age-worthy red wines.

According to Caceres-Mella [77], the ratio between soluble solids and titratable acidity (SS/TA) is widely used as an indicator of grape ripeness and enological quality and potential of varieties in experimental years. However, this index should be interpreted with caution, as the increase in SS does not always occur in inverse proportion to the reduction in TA. Nonetheless, the SS/TA ratio can serve as a reliable indicator of the sugar–acid balance for a given cultivar and region, especially when compared to reference vintages with established enological benchmarks. In this study, maturity index values at 142 days after pruning ranged from 20.6 ('Cabernet Sauvignon') to 28.8 ('Merlot') (Figure 5), highlighting significant differences among cultivars regarding the ideal harvest balance point.

3.7. *Biochemical Composition and Antioxidant Potential of Grapes*

The biochemical characterization of grapes—including total phenolics, flavonoids, anthocyanins, and antioxidant activity—is essential for determining wine sensory quality, stability, and aging potential. The cultivars 'Merlot' and 'Tannat' exhibited the highest contents of total phenolics and flavonoids in grape must (Table 4), indicating a higher antioxidant potential and, consequently, greater aging capacity and complexity. Phenolic compounds such as anthocyanins and proanthocyanidins are products of the grapevine's secondary metabolism and are directly associated with the color, body, bitterness, and astringency of red wines, particularly under water deficit conditions [78].

Table 4. The bioactive compounds and antioxidant activity of wine grape cultivars grown under double-pruning management in a subtropical climate during the 2024 growing season, São Paulo, Brazil.

Cultivar	Bioactive Compounds and Antioxidant Activity				
	Phenols	Flavonoids	Anthocyanins	DPPH	FRAP
		(mg 100 g^{-1})		(μg Trolox g^{-1})	(mmol $FeSO_4$ g^{-1})
Sauvignon Blanc	101.3 c *	10.1 c	18.7 cd	12.2 a	3.49 a
Merlot	234.9 a	15.3 a	76.1 a	11.1 cd	3.34 cd
Tannat	216.6 a	14.7 a	64.6 ba	11.6 bc	3.41 bc
Pinot Noir	151.2 b	12.3 b	33.8 bc	11.7 ab	3.42 ab
Malbec	209.9 a	14.5 a	77.7 a	11.1 d	3.33 d
Cabernet Sauvignon	206.3 a	14.4 a	14.4 d	11.6 bcd	3.41 bcd
DMS	1.22	2.07	111.5	0.55	1.91
CV (%)	8.65	27.9	22.4	2.39	3.49
Mean	186.69	13.5	47.5	11.6	3.40

* Lowercase letters in rows differ significantly according to Tukey's test at a 5% probability level ($p > 0.05$). CV: coefficient of variation.

The double-pruning technique promotes ripening under ideal climatic conditions, enhancing the accumulation of polyphenols and antioxidants, which improves both the sensory and enological quality of wines [17]. This not only adds value to the final product but also enhances the health benefits associated with moderate wine consumption. The 'Merlot' cultivar, with the highest contents of phenolics and flavonoids, is highly promising for the production of age-worthy wines due to its oxidative stability and aromatic persistence. In contrast, 'Sauvignon Blanc', with lower levels of these compounds, is better suited for fresh, early-consumption wines. The high anthocyanin content observed in 'Malbec' and 'Merlot' contributes to the intense coloration of wines, a desirable trait in more robust products (Table 4).

In red wines, phenolic compound concentrations typically range from 1000 to 4000 mg L^{-1}, while in white wines they range from 200 to 300 mg L^{-1} [78]. The color of red wines is determined by multiple factors, including pre-fermentation conditions. The interaction between tannins and anthocyanins—especially malvidin-3-*O*-glucoside, the most abundant pigment in red grapes—plays a key role in defining color intensity and stability [79]. Antioxidant activity, assessed by DPPH and FRAP assays, confirmed the superior performance of 'Merlot' and 'Tannat' in terms of phenolic stability and suitability for producing complex wines. Interestingly, 'Sauvignon Blanc' and 'Pinot Noir' also showed high antioxidant activity regardless of the method used, with 'Pinot Noir' standing out in the DPPH assay. Despite having lower total phenolic content, these cultivars demonstrated excellent antioxidant performance, with similar average values.

It is worth noting that the methods employed yielded distinct sensitivities: the DPPH assay proved more responsive to the presence of antioxidant compounds than the FRAP method. Among all samples, 'Sauvignon Blanc' recorded the highest antioxidant potential, with performance comparable to red cultivars, despite its lower anthocyanin content—compounds that are exclusive to red grapes. The average antioxidant activity measured by DPPH across all cultivars was 11.6, underscoring the suitability of the evaluated grapes for high-quality wine production. The results demonstrate that, beyond their enological potential, these cultivars represent a viable and technically distinct alternative for sustainable wine production in subtropical regions.

A study conducted by Rockenbach et al. [80] evaluated the phenolic content and antioxidant activity of grape pomace from various *Vitis vinifera* L. cultivars widely produced in Brazil. The research focused on red grape varieties such as 'Cabernet Sauvignon', 'Merlot', and 'Bordeaux'. The findings revealed that 'Cabernet Sauvignon' grape pomace exhibited the highest total phenolic content, measuring 74.75 mg of gallic acid equivalents per gram. Correspondingly, this variety also demonstrated superior antioxidant activity, with DPPH and ABTS radical scavenging capacities of 505.52 and 485.42 μmol Trolox equivalents per gram, respectively. The FRAP assay further confirmed its potent antioxidant potential, recording a reducing power of 249.46 μmol Trolox equivalents per gram. These results underscore the significant role of grape variety in determining the phenolic composition and antioxidant capacity of grape by-products. The study highlights the potential of utilizing grape pomace, particularly from 'Cabernet Sauvignon', as a valuable source of natural antioxidants for food and nutraceutical applications.

4. Conclusions

The cultivars evaluated in this study exhibited distinct phenological, physicochemical, and productive performances under double pruning in a subtropical environment. Among them, 'Sauvignon Blanc', 'Merlot', 'Tannat', and 'Cabernet Sauvignon' stood out for their favorable agronomic behavior and fruit composition. The cultivation of 'Cabernet Sauvignon' and 'Sauvignon Blanc' is particularly recommended for winter wine production under double pruning, given their high yield potential and desirable technological attributes. Red cultivars, especially 'Tannat' and 'Merlot', demonstrated elevated phenolic content and antioxidant activity, reinforcing their value for premium red wine production. Double pruning proved to be an effective management strategy for enhancing grape quality in sub-tropical regions, although it requires precise technical execution throughout the production cycle. The results support informed decision-making for cultivar selection and vineyard planning, and they contribute to expanding viticulture into non-traditional areas of Brazil. Further studies are encouraged across multiple production cycles and locations to validate these findings, as well as to explore the enological potential of each cultivar and the long-term viability and economic impact of adopting the double pruning system.

This study demonstrates the relevance of double pruning as a viable strategy for optimizing grapevine performance and improving fruit quality under subtropical conditions. By shifting the production cycle to the winter season, the technique benefits from more favorable climatic conditions, contributing to the production of high-quality grapes. Given the limited research on Vitis vinifera cultivars managed under double pruning, especially in emerging terroirs such as the State of São Paulo, this work provides valuable insights into cultivar adaptability and productivity. Understanding these responses is essential for developing targeted management practices and strengthening the scientific foundation of winter viticulture. Such knowledge is crucial for advancing Brazil's position in the global wine industry and unlocking the potential of São Paulo as a competitive and distinctive wine-producing region.

Author Contributions: Conceptualization: M.A.T., C.R.M. and G.E.P.; methodology: M.A.T., C.R.M., C.A.P.C.S., J.B.d.O., R.F., M.d.S.S. and H.S.A.M.; investigation:; M.A.T., C.R.M., C.A.P.C.S., J.B.d.O., M.d.S.S., S.d.N.S.B. and F.J.D.N.; writing—original draft: M.A.T., C.R.M., F.J.D.N. and S.L.; writing—review and editing: M.A.T., C.R.M., G.E.P., J.B.d.O., F.J.D.N., S.L., H.S.A.M. and P.V.d.S.; tables and figures: M.A.T., C.R.M., M.d.S.S., H.S.A.M., F.J.D.N. and C.A.P.C.S.; supervision: M.A.T., G.E.P., S.L. and R.F.; final approval: M.A.T., C.R.M., G.E.P., S.L. and M.d.S.S. All authors have read and agreed to the published version of the manuscript.

Funding: This research was funded by the Fundação Coordenação de Aperfeiçoamento de Pessoal de Nível Superior (Coordination for the Improvement of Higher Education Personnel), CAPES process n°. 88887.805056/2023-00 through granting the scholarship for the first author. We also have the support of Conselho Nacional de Desenvolvimento Científico e Tecnológico (CNPq) for the Research Productivity Scholarship (process n°. 307377/2021-0).

Data Availability Statement: The original contributions presented in the study are included in the article material, further inquiries can be directed to the corresponding author.

Acknowledgments: To Sao Paulo State University (UNESP), School of Agriculture, especially the Graduate Program in Agronomy/Horticulture, for the opportunity to pursue the master's degree and carry out this research. To the rural property Fazenda Santa Lúcia do Tietê, for the concession of the experimental area and also to the producer, Leonardo Sgargeta Ustulin, for allowing the execution of this research.

Conflicts of Interest: Author Giuliano Elias Pereira was employed by Brazilian Agricultural Research Corporation (EMBRAPA Grape and Wine). The remaining authors declare that the research was conducted in the absence of any commercial or financial relationships that could be construed as a potential conflict of interest.

References

1. Khadatkar, A.; Sawant, C.P.; Thorat, D.; Gupta, A.; Jadhav, S.; Gawande, D.; Magar, A.P. A comprehensive review on grapes (*Vitis* spp.) cultivation and its crop management. *Discov. Agric.* **2025**, *3*, 9. [CrossRef]
2. IBGE Sistema IBGE de Recuperação Automática—SIDRA. *Produção Agrícola Municipal: Área Destinada à Colheita, Área Colhida, Quantidade Produzida, Rendimento Médio e Valor da Produção das Lavouras Permanentes*; IBGE: Rio de Janeiro, Brazil, 2022.
3. UVIBRA União Brasileira de Vitivinicultura. *Estatísticas do Setor Vitivinícola do Estado do Rio Grande do Sul*; UVIBRA União Brasileira de Vitivinicultura: Bento Gonçalves, Brazil, 2023.
4. Regina, M.A.; Mota, R.V.; Souza, C.R.; Favero, A.C. Viticulture for fine wines in Brazilian Southeast. *Acta Hortic.* **2011**, *910*, 113–120. [CrossRef]
5. Dias, F.A.N.; Mota, R.V.; Souza, C.R.; Pimentel, R.M.; Souza, L.C.; Souza, A.L.; Regina, M.D. Rootstock on vine performance and wine quality of 'Syrah' un- der double pruning management. *Sci. Agric.* **2017**, *74*, 134–141. [CrossRef]
6. Brant, L.A.C.; Figueredo, G.M.; Mota, R.V. Vinhos de Inverno do Sudeste Brasileiro. *Territ. Vin* **2018**, *9*, 1–12. [CrossRef]
7. Pereira, G.E.; Peccin, E.G.; Riste, U.S.; Deconti, G.S.; Oliveira, J.B.; Santos, F.C.; Santana, D.P.; Albuquerque Filho, M.R.; Figueiredo, A.B.A. Agronomical characterization of grapevines and enological of grapes and winter wines in Brazilian Cerrado. *Bol. Pesqui. Desenvolv.—Embrapa Uva Vinho* **2023**, *36*, 20240001781.
8. Guerra, C.C.; Pereira, G.E. A qualidade e a tipicidade dos vinhos finos tranquilos e espumantes brasileiros. *Territ. Vin* **2018**, *9*, 1–32. [CrossRef]
9. Da Mota, R.V.; de Amorim, D.A.; Favero, A.C.; Purgatto, E.; Regina, M.d.A. Effect of trellising system on grape and wine composition of Syrah vines grown in the cerrado region of Minas Gerais. *Food Sci. Technol.* **2011**, *31*, 967–972. [CrossRef]
10. Balanov, P.E.; Smotraeva, I.V.; Abdullaeva, M.S.; Volkova, D.A.; Ivanchenko, O.B. Study on Resveratrol Content in Grapes and Wine Products. In Proceedings of the International Conference on Efficient Production and Processing (ICEPP-2021), Kazan, Russia, 25–26 February 2021; Volume 247.
11. Zhou, D.-D.; Li, J.; Xiong, R.-G.; Saimaiti, A.; Huang, S.-Y.; Wu, S.-X.; Yang, Z.-J.; Shang, A.; Zhao, C.-N.; Gan, R.-Y.; et al. Bioactive Compounds, Health Benefits and Food Applications of Grape. *Foods* **2022**, *11*, 2755. [CrossRef]
12. Tonietto, J.; Pereira, G.E.; Peregrino, I.; Regina, M.D.A. Potencial Para Construção de Indicações Geográficas de Vinhos de Inverno do Sudeste brasileiro. *Inf. Agropecu.* **2020**, *41*, 91–98.
13. Costa, R.S. Zonas de Vigor Vegetativo Para Colheita Seletiva em Viticultura Irrigada com Base em Sensoriamento Proximal. Ph.D. Thesis, Universidade Estadual Paulista (Unesp), São Paulo, Brazil, 2021.
14. Tecchio, M.A.; Hernandes, J.L.; Pires, E.J.; Terra, M.M.; Moura, M.F. Cultivo da Videira Para Mesa, Vinho e Suco. In *Cultivo de Fruteiras Clima Temperado em Regiões Subtropicais e Tropicais*, 2nd ed.; Pio, R., Ed.; Lavras: Editora UFLA: Lavras, Brazil, 2018; pp. 512–584.
15. Mota, R.V.; Silva, C.P.; Favero, A.C.; Purgatto, E.; Shiga, T.M.; Regina, M.D. Composição físico-química de uvas para vinho fino em ciclos de verão e inverno. *Rev. Bras. Frutic.* **2010**, *32*, 1127–1137. [CrossRef]
16. Regina, M.d.A.; Carmo, E.L.; Fonseca, A.R.; Purgatto, E.; Shiga, T.M.; Lajolo, F.M.; Ribeiro, A.P.; Mota, R.V. Influência da altitude na qualidade das uvas "Chardonnay" e "Pinot Noir" em Minas Gerais. *Rev. Bras. Frutic.* **2010**, *32*, 143–150. [CrossRef]

17. Favero, A.C.; De Amorim, D.A.; da Mota, R.V.; Soares, A.M.; de Souza, C.R.; de Albuquerque Regina, M. Double-pruning of 'Syrah' grapevines: A management strategy to harvest wine grapes during the winter in the Brazilian Southeast. *VITIS-J. Grapevine Res.* **2011**, *50*, 151–158.
18. Leão, P.C.S.; Ribeiro, T.P.; Nascimento, J.H.B. *Tropical Viticulture: Vineyard Management for Wine and Juice Production in Brazil*; Embrapa: Brasília, Brazil, 2020.
19. Tecchio, M.A.; Teixeira, L.A.J.; Terra, M.M.; Paioli-Pires, E.J.; Hernandes, J.L. *Uvas Finas para Mesa e Vinho (Vitis labrusca, Vitis vinífera, híbridos)—No Boletim 100: Recomendações de Adubação e Calagem Para o Estado de São Paulo*, 2nd ed.; Cantarella, H., Quaggio, J.A., Mattos, D., Jr., Boaretto, R.M., Van Raij, B., Eds.; Instituto Agronômico: Campinas, Brazil, 2022; pp. 308–313.
20. Eichhorn, K.W.; Lorenz, D.H. Phaenologische Entwicklungsstadien der Rebe. *Nachr. Dtsch. Pflanzenschutzd.* **1984**, *14*, 295–298.
21. Winkler, A.J. *Viticultura*; Companhia Editorial Continental: Cidade do México, México, 1965.
22. Singleton, V.L.; Rossi, J.A. Colorimetry of total phenolics with phosphomolybdic-phosphotungstic acid reagents. *Am. J. Enol. Vitic.* **1965**, *16*, 144–158. [CrossRef]
23. Popova, M.; Bankova, V.; Butovska, D.; Petkov, V.; Nikolova-Damyanova, B.; Sabatini, A.G.; Bogdanov, S. Validated methods for the quantification of biologically active constituents of poplar-type propolis. *Phytochem. Anal. An. Int. J. Plant Chem. Biochem. Tech.* **2004**, *15*, 235–240. [CrossRef]
24. Giusti, M.M.; Wrolstad, R.E. Anthocyanins. Characterization and measurement with UV-visible spectroscopy. *Curr. Protoc. Food Anal. Chem.* **2001**, *1*, 1–13. [CrossRef]
25. Brand-Williams, W.; Cuvelier, M.E.; Berset, C.L.W.T. Use of a free radical method to evaluate antioxidant activity. *LWT-Food Sci. Technol.* **1995**, *28*, 25–30. [CrossRef]
26. Ferreira, D.F. Sisvar: Um sistema computacional de análise estatística. *Ciênc. Agrotecnol.* **2011**, *35*, 1039–1042. [CrossRef]
27. Sánchez, C.A.P.C.; Tecchio, M.A.; Callili, D.; Souza, R.T.D.; Smarsi, R.C.; Leonel, S. Phenology, thermal requirement, and ripening of the 'BRS Isis' grape grafted on different rootstocks in a subtropical condition. *Bragantia* **2024**, *83*, e20230273. [CrossRef]
28. Monteiro, H.S.A.; Tecchio, M.A.; Brito, S.d.N.S.; Neto, F.J.D.; Sánchez, C.A.P.C.; Alonso, J.C.; Feliciano, D.E.F.; Maniero, C.R.; Cunha, P.H.H.; Silva, M.d.S. Phenological Development, Thermal Requirement, and Quality of 'BRS Núbia' (*Vitis vinifera* L. × *Vitis labrusca* L.) Grapes on Different Rootstocks. *Horticulturae* **2025**, *11*, 466. [CrossRef]
29. Koyama, R.; Borges, W.F.S.; Colombo, R.C.; Hussain, I.; Souza, R.T.D.; Roberto, S.R. Phenology and yield of the hybrid seedless grape 'BRS Melodia'grown in an annual double cropping system in a subtropical area. *Horticulturae* **2020**, *6*, 3. [CrossRef]
30. Mandelli, F.; Miele, A.; Tonietto, J. Uva em clima temperado. In *Agrometeorologia dos Cultivos: O Fator Meteorológico na Produção agrícola*; Monteiro, J.E.B.A., Ed.; Inmet: Brasília, Brazil, 2009; Volume 31, pp. 504–515.
31. Rouxinol, M.I.; Martins, M.R.; Barroso, J.M.; Rato, A.E. Wine grapes ripening: A review on climate effect and analytical approach to increase wine quality. *Appl. Biosci.* **2023**, *2*, 347–372. [CrossRef]
32. Steduto, P.; Hsiao, T.C.; Raes, D.; Fereres, E. AquaCrop—The FAO Crop Model to Simulate Yield Response to Water: I. Concepts and Underlying Principles. In *Crop Yield Prediction Under Water Deficit Conditions*; FAO: Rome, Italy, 2009; Volume 33, pp. 285–292.
33. Júnior, M.J.P.; Hernandes, J.L. Uva para vinho 'Syrah' cultivada durante a safra de inverno: Fenologia e evolução da maturação. *Rev. Bras. Frutic.* **2019**, *37*, 1–10.
34. Rogiers, S.Y.; Greer, D.H.; Liu, Y.; Baby, T.; Xiao, Z. Impact of climate change on grape berry ripening: An assessment of adaptation strategies for the Australian vineyard. *Front. Plant Sci.* **2022**, *13*, 1094633. [CrossRef]
35. Keller, M. *The Science of Grapevines: Anatomy and Physiology*; Academic Press: Burlington, ON, Canada, 2010.
36. Borghezan, M.; Gavioli, O.; Pit, F.A.; Silva, A.L. Comportamento vegetativo e produtivo da videira e composição da uva em São Joaquim, Santa Catarina. *Pesqui. Agropecu. Bras.* **2011**, *46*, 398–405. [CrossRef]
37. Radünz, A.L.; Schöffel, E.R.; Borges, C.T.; Malgarim, M.B.; Pötter, G.H. Necessidades térmicas de videiras na região da Campanha do Rio Grande do Sul- Brasil. *Ciênc. Rural.* **2015**, *45*, 626–632. [CrossRef]
38. Silva, A.R.; Pereira, G.E.; Lima, L.C.O.; Souza, C.R. Influence of harvest time and climate on disease control and phenolic composition in grapes under tropical conditions. *J. Appl. Vitic.* **2023**, *29*, 145–154.
39. Miranda, C.; Santesteban, L.G.; Royo, J.B. Evaluation and fitting of models for determining peach phenological stages at a regional scale. *Agric. For. Meteorol.* **2013**, *178–179*, 129–139. [CrossRef]
40. Leão, P.C.S.; Marques, A.T.B.; Silva, M.M.; Badji, C.A.; Barros, A.P.A.; Santos, J.P.O. *Cultivares de Videira para Elaboração de Vinhos Finos no Agreste Pernambucano*; Circular Técnica 136; EMBRAPA: Petrolina, PE, Brazil, 2024.
41. EPAGRI. *Sistema de Produção Para a Cultura da Videira em Santa Catarina*; Sistemas de Produção, 2; EPAGRI: Florianópolis, Brazil, 2005.
42. Hall, A.; Mathews, A.J.; Holzapfel, B.P. Potential effect of atmospheric warming on grapevine phenology and post-harvest heat accumulation across a range of climates. *Int. J. Biometeorol.* **2016**, *60*, 1405–1422. [CrossRef]
43. Conde, C.; Silva, P.; Fontes, N.; Dias, A.C.P.; Tavares, R.M.; Sousa, M.J.; Agasse, A.; Delrot, S.; Gerós, H. Biochemical changes throughout grape berry development and fruit and wine quality. *Food* **2007**, *1*, 1–22.

44. Tecchio, M.A.; Silva, M.J.R.D.; Sanchez, C.A.P.C.; Callili, D.; Vedoato, B.T.F.; Hernandes, J.L.; Moura, M.F. Yield performance and quality of wine grapes (*Vitis vinifera*) grafted onto different rootstocks under subtropical conditions. *Bragantia* **2022**, *81*, e1622. [CrossRef]
45. Poni, S.; Gatti, M.; Tombesi, S.; Squeri, C.; Sabbatini, P.; Lavado Rodas, N.; Frioni, T. Double Cropping in *Vitis vinifera* L. Pinot Noir: Myth or Reality? *Agronomy* **2020**, *10*, 799. [CrossRef]
46. Netzer, Y.; Suued, Y.; Harel, M.; Ferman-Mintz, D.; Drori, E.; Munitz, S.; Harari, G. Forever young? late shoot pruning affects phenological development, physiology, yield and wine quality of *Vitis vinifera* cv. *Malbec. Agric.* **2022**, *12*, 605. [CrossRef]
47. Rizzon, L.A.; Miele, A. Avaliação da cv. Merlot para elaboração de vinho tinto. *Food Sci. Technol.* **2003**, *23*, 156–161. [CrossRef]
48. Pedro Júnior, M.J.; Hernandes, J.L.; Blain, G.C.; Rolim, G.D. Microclima em vinhedos de 'Niagara rosada' em diferentes sistemas de condução durante safras de inverno e de verão. *Rev. Bras. Frutic.* **2013**, *35*, 151–158. [CrossRef]
49. Hidalgo, L. *Tratado de Viticultura General*; Mundi-Prensa: Madrid, Spain, 1999; p. 1172.
50. Barros, M.S. Irrigação, Fertirrigação, Tratos Culturais, Controle Fitossanitário e Pós-Colheita na Cultura da Uva na Fazenda Agrobras. Bachelor's Thesis, Universidade Federal Rural de Pernambuco (UFRPE), Recife, Brazil, 2019.
51. Leão, P.C.S.; Maia, J.D.G. Aspectos Culturais em Viticultura Tropical: Uvas de Mesa. 37 Informe Agropecuário. *Belo Horiz.* **1998**, *19*.
52. Falcão, J.V.; Lacerda, M.P.C.; Mendes, I.d.C.; Leão, T.P.; Carmo, F.F.D. Qualidade do solo cultivado com morangueiro sob manejo convencional e orgânico. *Pesqui. Agropecu. Trop.* **2013**, *43*, 450–459. [CrossRef]
53. Guerra, C.C. Maturação da Uva e Condução da Vinificação para a Elaboração de Vinhos Finos. In *Viticultura e Enologia: Atualizando Conceitos*; Regina, M.A., Ed.; EPAMIG/FECD: Caldas, Brazil, 2002; pp. 179–192.
54. Chitarra, M.I.F.; Chitarra, A.B. *Ós-Colheita de Frutas e Hortaliças: Fisiologia e Manuseio*, 2nd ed.; UFLA: Lavras, Brazil, 2005; p. 783.
55. Dai, Z.W.; Ollat, N.; Gomès, E.; Decroocq, S.; Tandonnet, J.-P.; Bordenave, L.; Pieri, P.; Hilbert, G.; Kappel, C.; van Leeuwen, C.; et al. Ecophysiological, genetic, and molecular causes of variation in grape berry weight and composition: A review. *Am. J. Enol. Vitic.* **2011**, *62*, 413–425. [CrossRef]
56. Kuhn, N.; Guan, L.; Dai, Z.W.; Wu, B.H.; Lauvergeat, V.; Gomès, E.; Li, S.H.; Godoy, F.; Arce-Johnson, P.; Delrot, S. Berry Ripening: Recent. Heard Through grapevine. *J. Exp. Bot.* **2013**, *65*, 4543–4559. [CrossRef]
57. Allamy, L.; van Leeuwen, C.; Pons, A. Impact of harvest date on aroma compound composition of Merlot and Cabernet-Sauvignon must and wine in a context of climate change: A focus on cooked fruit molecular markers. *OENO One* **2023**, *57*, 99–112. [CrossRef]
58. Ministério da Agricultura, Pecuária e Abastecimento (MAPA). *Instrução Normativa N° 14, de 8 de Fevereiro de 2018: Aplica-se ao Vinho e Derivados da Uva e do Vinho Comercializados em Todo o Território Nacional, Produzidos no Brasil e Importados*; Ministério da Agricultura, Pecuária e Abastecimento (MAPA): Brasília, Brazil, 2018.
59. Jubileu, B.S.; Sato, A.J.; Roberto, S.R. Caracterização fenológica e produtiva das videiras 'Cabernet Sauvignon' e' 'Alicante' (*Vitis vinifera* L.) produzidas fora de época no norte do Paraná. *Rev. Bras. Frutic.* **2010**, *32*, 451–462. [CrossRef]
60. Pedro Júnior, M.J.; Hernandes, J.L.; Blain, G.C.; Bardin-Camparotto, L. Curva de maturação e estimativa do teor de sólidos solúveis e acidez total em função de graus-dia: Uva IAC 138-22 'Máximo'. *Bragantia* **2014**, *73*, 81–85. [CrossRef]
61. Ojeda, H.; Andary, C.; Kraeva, E.; Carbonneau, A.; Deloire, A. Influence of pre-and postveraison water deficit on synthesis and concentration of skin phenolic compounds during berry growth of *Vitis vinifera* cv. Shiraz. *Am. J. Enol. Vitic.* **2002**, *53*, 261–267.
62. Brighenti, A.F.; Silva, A.L.; Brighenti, E.; Porro, D.; Stefanini, M. Desempenho vitícola de variedades autóctones italianas em condição de elevada altitude no Sul do Brasil. *Pesqui. Agropecu. Bras.* **2014**, *49*, 465–474. [CrossRef]
63. Ristic, R.; Downey, M.O.; Iland, P.G.; Bindon, K.; Francis, I.L.; Herderich, M.; Robinson, S.P. Exclusion of sunlight from Shiraz grapes alters wine colour, tannin and sensory properties. *Aust. J. Grape Wine Res.* **2007**, *13*, 53–65. [CrossRef]
64. Watson, B. Evaluation of winegrape maturity. In *Oregon Viticulture*; Hellman, E.W., Ed.; Oregon State University Press: Corvallis, OR, USA, 2003; pp. 235–245.
65. Yamamoto, L.Y.; Assis, A.M.; Roberto, S.R.; Bovolenta, Y.R.; Nixford, S.L.; Garcia-Romero, E.; Gomez-Alonso, S.; Hermanoin-Gutierrez, I. Application of abscisic acid (S-ABA to cv. Isabel grapes (*Vitis vinifera* × *Vitis labrusca*) for color improvement: Effects on color, phenolic composition and antioxidant capacity of their grape juice. *Food Res. Int.* **2015**, *77*, 572–583. [CrossRef]
66. Boulton, R. The general relationship between potassium, sodium and pH in grape juices and wines. *Am. J. Enol. Vitic.* **1980**, *31*, 182–186. [CrossRef]
67. Rojas-Lara, B.A.; Morrison, J.C. Differential effects of shading fruit or foliage on the development and composition of grape berries. *VITIS-J. Grapevine Res.* **1989**, *28*, 199–208.
68. Esteban, M.A.; Villanueva, M.J.; Lissarrague, J.R. Effect of irrigation on changes in the anthocyanin composition of the skin of Tempranillo (*Vitis vinifera* L.) grape berries during ripening. *J. Sci. Food Agric.* **2001**, *81*, 409–420. [CrossRef]
69. Winter, E.; Welter, L.J. *Perfil Químico e Atividade Antioxidante in Vitro de Mostos e Vinhos da Região Serrana Catarinense Elaborados com Vitis vinifera L. var—Sauvignon Blanc Cultivadas em Diferentes Altitudes*; Universidade Federal de Santa Catarina (UFSC): Florianópolis, Brazil, 2018.

70. Falcão, L.D.; Burin, V.M.; Chaves, E.S.; Vieira, H.J.; Brighenti, E.; Rosier, J.-P.; Bordignon-Luiz, M.T. Vineyard altitude and mesoclimate influences on the phenology and maturation of Cabernet-Sauvignon grapes from Santa Catarina State. *OENO One* **2010**, *44*, 135–150. [CrossRef]
71. Brighenti, A.F.; Brighenti, E.; Bonin, V.; Rufato, L. Caracterização fenológica e exigência térmica de diferentes variedades de uvas viníferas em São Joaquim, Santa Catarina—Brasil. *Ciênc. Rural.* **2013**, *43*, 1162–1167. [CrossRef]
72. Liu, H.; Wu, B.; Fan, P.; Li, S.; Li, L. Sugar and acid concentrations in 98 grape cultivars analyzed by principal component analysis. *J. Sci. Food Agric.* **2005**, *1536*, 1526–1536. [CrossRef]
73. Lee, J.; Steenwerth, K.L. Cabernet Sauvignon grape anthocyanin increased by soil conservation practices. *Sci. Hortic.* **2013**, *159*, 128–133. [CrossRef]
74. Neumann, P.A.; Matzarakis, A. Potential climate change impacts on winegrape must density and titratable acidity in southwest Germany. *Clim. Res.* **2014**, *59*, 161–172. [CrossRef]
75. Rizzon, L.A.; Miele, A. Acidez na vinificação em tinto das uvas isabel, cabernet sauvignon e cabernet franc the acidity in the vinification of isabella, cabernet sauvignon, and cabernet franc red grapes. *Ciênc. Rural.* **2002**, *32*, 2002. [CrossRef]
76. Ribéreau-Gayon, P.; Glories, Y.; Maujean, A.; Dubourdieu, D. *Handbook of Enology, the Chemistry of Wine: Stabilization and Treatments*, 2nd ed.; John Wiley & Sons: Hoboken, NJ, USA, 2006; Volume 2, pp. 1–441.
77. Cáceres-Mella, A.; Talaverano, M.I.; Villalobos-González, L.; Ribalta-Pizarro, C.; Pastenes, C. Controlled water deficit during ripening affects proanthocyanidin synthesis, concentration and composition in Cabernet Sauvignon grape skins. *Plant Physiol. Biochem.* **2017**, *117*, 34–41. [CrossRef]
78. Bravo, L. Chemistry, dietary sources, metabolism, and nutritional significance. *Nutr. Rev.* **1998**, *56*, 317–333. [CrossRef]
79. Mcrae, J.M.; Teng, B.; Bindon, K. Factors Influencing Red Wine Color from the Grape to the Glass. In *Encyclopedia of Food Chemistry*; Elsevier: Amsterdam, The Netherlands, 2019; pp. 97–106.
80. Rockenbach, I.I.; Rodrigues, E.; Gonzaga, L.V.; Caliari, V.; Genovese, M.I.; Gonçalves, A.E.S.S.; Fett, R. Phenolic compounds content and antioxidant activity in pomace from selected red grapes (*Vitis vinifera* L. and *Vitis labrusca* L.) widely produced in Brazil. *Food Chem.* **2011**, *127*, 174–179. [CrossRef]

MDPI AG
Grosspeteranlage 5
4052 Basel
Switzerland
Tel.: +41 61 683 77 34

Agriculture Editorial Office
E-mail: agriculture@mdpi.com
www.mdpi.com/journal/agriculture

www.ingramcontent.com/pod-product-compliance
Lightning Source LLC
Chambersburg PA
CBHW061505210926
51850CB00008B/1833
* 9 7 8 3 7 2 5 8 7 7 1 1 9 *